Chemical Bond and Intermolecular Interactions

Chemical Bond and Intermolecular Interactions

Editors

Qingzhong Li
Steve Scheiner
Zhiwu Yu

Basel • Beijing • Wuhan • Barcelona • Belgrade • Novi Sad • Cluj • Manchester

Editors

Qingzhong Li
School of Chemistry and
Chemical Engineering
Yantai University
Yantai
China

Steve Scheiner
Department of Chemistry
and Biochemistry
Utah State University
Logan
United States

Zhiwu Yu
Department of Chemistry
Tsinghua University
Beijing
China

Editorial Office
MDPI
St. Alban-Anlage 66
4052 Basel, Switzerland

This is a reprint of articles from the Special Issue published online in the open access journal *Molecules* (ISSN 1420-3049) (available at: www.mdpi.com/journal/molecules/special_issues/Chem_Bond).

For citation purposes, cite each article independently as indicated on the article page online and as indicated below:

Lastname, A.A.; Lastname, B.B. Article Title. *Journal Name* **Year**, *Volume Number*, Page Range.

ISBN 978-3-0365-9870-3 (Hbk)
ISBN 978-3-0365-9869-7 (PDF)
doi.org/10.3390/books978-3-0365-9869-7

Contents

About the Editors

Qingzhong Li

Prof. Dr. Qingzhong Li is from the School of Chemistry and Chemical Engineering, Yantai University, Yantai, China. His fields of interests are molecular spectroscopy, theoretical calculations, and intermolecular interactions.

Steve Scheiner

Prof. Dr. Steve Scheiner is from Department of Chemistry and Biochemistry, Utah State University, Logan, UT, USA. His fields of interests are hydrogen bonds, noncovalent bonds, halogen bonds, pnicogen bonds, and tetrel bonds.

Zhiwu Yu

Prof. Dr. Zhiwu Yu is from the Department of Chemistry, Tsinghua University, Beijing, China. His fields of interests are molecular spectroscopy, hydrogen and halogen bonds, phospholipids, and chemical thermodynamics.

Preface

This reprint focuses on the geometrical and spectroscopic features as well as the potential applications of halogen, chalcogen, pnictogen, tetrel, and triel bonds in chemical reactions, crystal engineering, molecular recognition, and biological systems. We aim to present cutting-edge studies concerning the bonds formed between some novel and important electron donors and acceptors, similarities and differences between these bonds, structures of the complexes composed of these bonds in the solution and gas phases, spectroscopic methods for measuring these bonds in solution and gas phases, and applications of these bonds in chemical reactions, crystal engineering, molecular recognition, and biological systems.

Qingzhong Li, Steve Scheiner, and Zhiwu Yu
Editors

Article

Triel Bond Formed by Malondialdehyde and Its Influence on the Intramolecular H-Bond and Proton Transfer

Qiaozhuo Wu, Shubin Yang and Qingzhong Li *

The Laboratory of Theoretical and Computational Chemistry, School of Chemistry and Chemical Engineering, Yantai University, Yantai 264005, China
* Correspondence: lqz@ytu.edu.cn

Abstract: Malondialdehyde (MDA) engages in a triel bond (TrB) with TrX_3 (Tr = B and Al; X = H, F, Cl, and Br) in three modes, in which the hydroxyl O, carbonyl O, and central carbon atoms of MDA act as the electron donors, respectively. A $H \cdots X$ secondary interaction coexists with the TrB in the former two types of complexes. The carbonyl O forms a stronger TrB than the hydroxyl O, and both of them are better electron donors than the central carbon atom. The TrB formed by the hydroxyl O enhances the intramolecular H-bond in MDA and thus promotes proton transfer in $MDA-BX_3$ (X = Cl and Br) and $MDA-AlX_3$ (X = halogen), while a weakening H-bond and the inhibition of proton transfer are caused by the TrB formed by the carbonyl O. The TrB formed by the central carbon atom imposes little influence on the H-bond. The BH_2 substitution on the central C-H bond can also realise the proton transfer in the triel-bonded complexes between the hydroxyl O and TrH_3 (Tr = B and Al).

Keywords: triel bond; hydrogen bonding; proton transfer; NBO

 check for
updates

Citation: Wu, Q.; Yang, S.; Li, Q. Triel Bond Formed by Malondialdehyde and Its Influence on the Intramolecular H-Bond and Proton Transfer. *Molecules* **2022**, *27*, 6091. https://doi.org/10.3390/ molecules27186091

Academic Editor: Nagatoshi Nishiwaki

Received: 28 August 2022
Accepted: 15 September 2022
Published: 18 September 2022

Publisher's Note: MDPI stays neutral with regard to jurisdictional claims in published maps and institutional affiliations.

1. Introduction

Malondialdehyde (MDA), a naturally occurring product in lipid peroxidation and prostaglandin biosynthesis [1], has been known as a biomarker of lipid oxidation induced by reactive oxygen species [2], a reliable biomarker for bipolar disorder [3], or an oxidative stress marker in oral squamous cell carcinoma [4]; thus, it has received much attention [5–13]. This molecule exhibits an intramolecular proton transfer with two equivalent forms separated by a barrier of medium height [5]. The intramolecular proton transfer in MDA involves a 2.2 kcal/mol lower barrier than that in its radical analogues [6]. When a NO_2 or BH_2 group is attached to the central carbon atom, the barrier is reduced to less than 1 kcal/mol [7]. Theoretical and experimental studies showed that 2-chloromalonaldehyde exhibits a weaker intramolecular H-bond than MDA [8,9]. However, the intramolecular H-bond in MDA is strengthened if strong electron donors and/or sterically hindered substituents are present in its two side carbon atoms [7]. The IR and UV spectra of MDA were also measured in the gas phase and water [10,11]. The MDA in water has a slightly red-shifted UV spectrum compared with that in the gas phase [10]. Tunnelling occurs in MDA, and its mechanisms can be understood through the isotope effect (IE), which is classified into primary IE and secondary IE [12]. The primary H/D kinetic IE on the intramolecular proton transfer in MDA is dominated by zero-point energy effects, and tunnelling plays a minor role at room temperature [13].

Intra- or intermolecular proton transfer reactions are considered to be one of the most fundamental and important processes in chemistry and biology, such as acid–base neutralisation reactions and enzymatic reactions [14]. The proton transfer in MDA can be regulated through cooperativity with other interactions. When a BeH_2 or BeF_2 group engages in a beryllium bond with the hydroxyl/carbonyl group of MDA, the intramolecular H-bond in MDA is strengthened or weakened [15], which is accompanied by the inhibition/promotion of proton transfer. This effect is also realised by adding F_2SiO to MDA, where a tetrel bond is formed [16]. In general, the stronger the interaction imposed, the more prominent the effect. The stronger interaction makes the binding distance of the weaker one undergo a larger change.

A triel bond (TrB) is an attractive interaction that occurs between the triel atom such as B or Al and an electron donor [17]. Such interactions are usually so strong that they have many of the characteristics of covalent bonds and can even be classified as typical covalent bonds [18]. Triel atoms are usually sp^2-hybridised with a π–hole above and below the molecular plane. Interestingly, when the sp^2-hybridised triel atom binds to a strong Lewis base, it may become sp^3-hybridised [19]. Thus, the trivalent centres in those complexes with such strong interactions follow the octet rule and can be classified as tetravalent centres, which usually have a tetrahedral structure, indicating a large geometric deformation of the interacting species, a feature of TrB formation. For example, an isolated BH_3 monomer has a planar triangular structure, while the $BH_3\cdots NH_3$ complex is tetrahedral [20]. This bond plays an important role in energy materials, chemical reactions, and biological systems [21–24]. For instance, the strong organoborane Lewis acid $B(C_6F_5)_3$ catalyses the hydrosilation of aromatic and aliphatic carbonyl functions at convenient rates, with loadings of $1-4\%$ [22]. TrB also contributes to the molecular hydrogen release process, which has also been explored as a research topic related to hydrogen storage materials [23]. Thus, TrB has garnered more attention in recent years [25–35]. Although different types of electron donors are utilised in the TrBs [32–35], the most common ones are from molecules with lone pairs such as N and O. A comparison was made for the TrBs with different chalcogen electron donors (H_2O, H_2S, and H_2Se), and it was found that H_2O forms a stronger TrB [36]. This bond displays some different properties from H-bond. In $HCN\cdots HCN\text{-}BF_3$, where a TrB and H-bond coexist, the strong TrB suffers a larger shortening than the weaker H-bond [37]. Thus, it is interesting to study the influence of TrB on the intramolecular H-bond in MDA and its proton transfer.

In this article, the complexes of MDA and TrX_3 (Tr = B and Al; X = H, F, Cl, and Br) are studied. There are two types of oxygen atoms (hydroxyl and carbonyl O atoms) in MDA; thus, their ability to bind with a triel atom is compared. In addition, the central carbon atom has negative molecular electrostatic potentials (MEPs), as shown in Figure 1. Therefore, this negative MEP region also binds with TrX_3, resulting in a π–π TrB. On the other hand, we focus on the influence of TrB on the intramolecular H-bond in MDA and the corresponding proton transfer. When a BH_2 group is attached to the central carbon atom, the barrier in the proton transfer reduces [7]. Thus, a combination of substitution and cooperative effects is used to regulate the proton transfer.

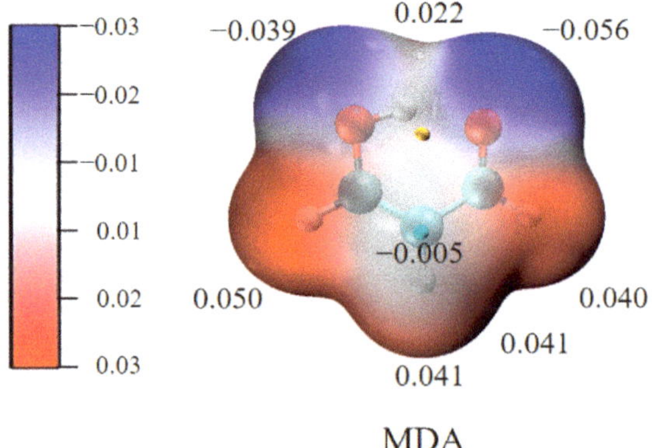

Figure 1. MEP maps of MDA. All are in a.u.

2. Results

2.1. Coplanar Triel-Bonded Complexes

Malondialdehyde (MDA) is often used as a model for studying intramolecular proton transfer, and TrX_3 can be added to MDA to form a TrB that in turn affects the proton transfer. As can be seen from the MEP diagram of MDA in Figure 1, there are both positive red and negative blue regions in MDA, indicating that it can act as both an electron donor and an acceptor. MDA contains two blue regions, which are located, respectively, at the hydroxyl oxygen and the carbonyl oxygen. It is clear that the latter has a more negative MEP value than the former. The three CH hydrogen atoms have red areas. Due to the delocalisation of the ring structure of MDA, the three carbon atoms have negative MEPs; thus, they can also bind with the π–hole of TrX_3, which has been studied in previous studies (Figure S1).

The π–hole on the T atom of TrX_3 may interact with the hydroxyl O(1) or carbonyl O(2) atom of MDA to form a TrB, respectively, designated by the "a" and "b" labels. Figures 2 and 3 show the structures of the a-type and b-type complexes, respectively. The $Tr\cdots O$ distance is shorter than 2 Å in most complexes, with an exception in $MDA\text{-}BF_3\text{-}a$. The shorter $Tr\cdots O$ distance means that the TrB formed between both molecules is very strong. The halogen substitution shortens the $Tr\cdots O$ distance relative to the TrH_3 complex in most cases. Only BF_3 elongates the $Tr\cdots O$ distance in spite of the largest π–hole on the B atom.

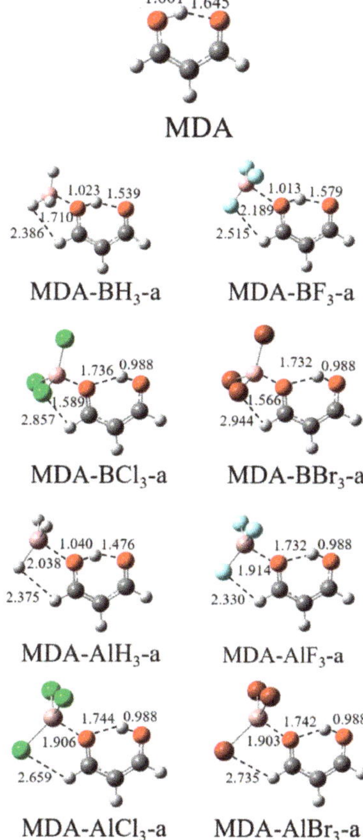

Figure 2. Structures of MDA and its coplanar complexes formed by the hydroxyl O with TrX_3. Distances are in Å.

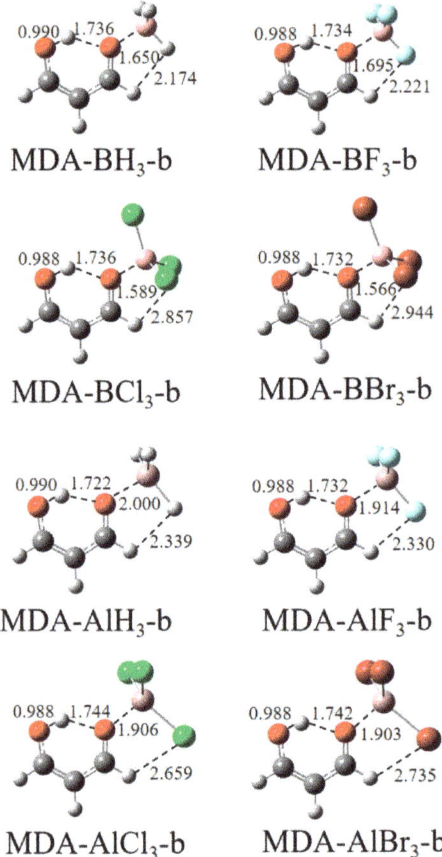

Figure 3. Structures of coplanar complexes of MDA formed by the carbonyl O with TrX$_3$. Distances are in Å.

Since Figure 1 confirms that the carbonyl O of MDA has a more negative MEP value than the hydroxyl O, it is not surprising that the former forms a stronger TrB with TrX$_3$ than the latter. This difference is reflected in the shorter Tr···O distance in the a-configuration, the more negative interaction energy (Table 1), and the greater electron density at the Tr···O BCP (Table 2). However, there are exceptions for the structures of MDA-BCl$_3$, MDA-BBr$_3$, MDA-AlF$_3$, MDA-AlCl$_3$, and MDA-AlBr$_3$ since both configurations have the same interaction energy, although they are formed in very different ways. As shown in Table 1, the interaction energy E_{int} in most of the MDA-AlX$_3$ complexes is larger than that of that in the MDA-BX$_3$ analogue, which may be due to the fact that the π–hole values of AlX$_3$ are larger than those of BX$_3$. When the H atoms of TrH$_3$ are replaced by halogen, E_{int} basically increases, except for MDA-BF$_3$-a, which also has the lowest E_{int} in all the complexes, although its π–hole is not the smallest. With the increase in halogen electronegativity, E_{int} increases for the MDA-AlX$_3$ complexes but decreases for the MDA-BX$_3$ complexes. This inconsistent change is mainly attributed to the distortion of TrX$_3$. BF$_3$ shows a smaller distortion relative to BCl$_3$ and BBr$_3$ in the complex; thus, an abnormal change occurs for the BF$_3$ complexes.

Table 1. Interaction energy (E_{int}), binding energy (E_b), and deformation energy (DE) of triel bond in the complexes, all in kcal/mol.

	E_{int}	E_b	DE
MDA-BH$_3$-a	−19.90	−10.83	9.07
MDA-BF$_3$-a	−7.49	−5.00	2.49
MDA-BCl$_3$-a	−37.36	−12.75	24.61
MDA-BBr$_3$-a	−38.10	−13.94	24.16
MDA-AlH$_3$-a	−20.98	−16.62	4.36
MDA-AlF$_3$-a	−41.59	−32.78	8.81
MDA-AlCl$_3$-a	−41.16	−31.83	9.33
MDA-AlBr$_3$-a	−39.86	−30.92	8.94
MDA-BH$_3$-b	−28.65	−17.60	11.05
MDA-BF$_3$-b	−29.27	−10.43	18.84
MDA-BCl$_3$-b	−37.36	−12.75	24.61
MDA-BBr$_3$-b	−38.10	−13.94	24.16
MDA-AlH$_3$-b	−26.97	−22.32	4.65
MDA-AlF$_3$-b	−41.59	−32.79	8.80
MDA-AlCl$_3$-b	−41.16	−31.83	9.33
MDA-AlBr$_3$-b	−39.86	−30.92	8.94
MDA-BH$_3$-c	−18.31	−9.68	8.63
MDA-BF$_3$-c	−2.79	−2.54	0.25
MDA-BCl$_3$-c	−4.12	−4.00	0.12
MDA-BBr$_3$-c	−4.40	−4.20	0.20
MDA-AlH$_3$-c	−11.97	−9.61	2.36
MDA-AlF$_3$-c	−21.26	−14.29	6.97
MDA-AlCl$_3$-c	−23.55	−14.73	8.82
MDA-AlBr$_3$-c	−23.12	−14.34	8.78

Table 2. Electron density (ρ), its Laplacian ($\nabla^2\rho$), energy density (H), kinetic energy density (D), and potential energy density (V) at the B···O/C BCPs in the complexes, all in a.u.

	ρ	$\nabla^2\rho$	H	G	V	|V|/G
MDA-BH$_3$-a	0.0638	0.4829	−0.0210	0.1037	−0.1359	1.3105
MDA-BF$_3$-a	0.0249	0.0732	−0.0028	0.0212	−0.0240	1.1321
MDA-BCl$_3$-a	0.1100	0.5153	−0.0652	0.1940	−0.2591	1.3356
MDA-BBr$_3$-a	0.1175	0.5575	−0.0716	0.2110	−0.2827	1.3398
MDA-AlH$_3$-a	0.0382	0.2914	0.0086	0.0643	−0.0558	0.8678
MDA-AlF$_3$-a	0.0575	0.4536	0.0072	0.1062	−0.0989	0.9313
MDA-AlCl$_3$-a	0.0593	0.4638	0.0064	0.1095	−0.1031	0.9416
MDA-AlBr$_3$-a	0.0600	0.4689	0.0062	0.1110	−0.1048	0.9441
MDA-BH$_3$-b	0.0788	0.5965	−0.0280	0.1771	−0.2051	1.1581
MDA-BF$_3$-b	0.0793	0.3505	−0.0416	0.1292	−0.1708	1.3220
MDA-BCl$_3$-b	0.1100	0.5153	−0.0652	0.1940	−0.2591	1.3356
MDA-BBr$_3$-b	0.1175	0.5575	−0.0716	0.2110	−0.2827	1.3398
MDA-AlH$_3$-b	0.0447	0.3366	0.0079	0.0762	−0.0683	0.8963
MDA-AlF$_3$-b	0.0575	0.4536	0.0072	0.1062	−0.0989	0.9313
MDA-AlCl$_3$-b	0.0593	0.4638	0.0064	0.1095	−0.1031	0.9416
MDA-AlBr$_3$-b	0.0600	0.4689	0.0062	0.1110	−0.1048	0.9441
MDA-BH$_3$-c	0.0600	0.0340	−0.0377	0.0472	−0.0851	1.8030
MDA-BF$_3$-c	0.0103	0.0278	0.0007	0.0064	−0.0058	0.9063
MDA-BCl$_3$-c	0.0085	0.0224	0.0008	0.0049	−0.0041	0.8367
MDA-BBr$_3$-c	0.0096	0.0242	0.0007	0.0054	−0.0047	0.8704
MDA-AlH$_3$-c	0.0254	0.0828	−0.0023	0.0225	−0.0250	1.1111
MDA-AlF$_3$-c	0.0374	0.1388	−0.0048	0.0385	−0.0436	1.1325
MDA-AlCl$_3$-c	0.0394	0.1335	−0.0064	0.0386	−0.0453	1.1736
MDA-AlBr$_3$-c	0.0404	0.1343	−0.0069	0.0402	−0.0471	1.1716

The binding energy E_b and deformation energy DE of these complexes are also given in the last two columns of Table 1. The binding energy is the difference between the energy of the complex relative to the sum of the energies of the isolated monomers (in their optimised geometry). In general, E_b has the same trend as E_{int}, but E_b is not as negative as E_{int}, and the difference between them is DE. The DE values in MDA-BCl$_3$ and MDA-BBr$_3$ are large enough (>24 kcal/mol), which is about double as much as E_b. This is also taken as a character of a TrB since the triel molecule is often distorted easily. This distortion easily occurs in the MDA-BX$_3$ (X = H, Cl, and Br) complexes, with a DE as much as double that in the MDA-AlX$_3$ analogues.

The formation of a TrB can be further confirmed by a complicated colour region in the NCI analysis (Figures S2 and S3). In most complexes, this colour region between the Tr and O atoms is overlapped with red and blue; thus, the TrB is very strong. In addition, a green region is found between one Tr-X bond of TrX$_3$ and the C-H bond adjoined with the O atom that binds with TrX$_3$, corresponding to a weak H-bond with a long H$\cdots$X distance. Both MDA and TrX$_3$ play a reverse role in both the TrB and H-bond; thus, both interactions display positive cooperativity with each other.

Table 2 collects the AIM data of the Tr$\cdots$O triel bond, including the important parameters of the electron density (ρ), its Laplacian ($\nabla^2\rho$), and the energy density (H) at the bond critical point (BCP) of the TrB. For the MDA-BX$_3$ complex, $\nabla^2\rho$ is positive but H is negative, indicating that the TrB is partially covalent interaction, which is also confirmed by the magnitude of $|V|/G$ (>1). However, for the MDA-AlX$_3$ complex, both $\nabla^2\rho$ and H are positive, and the magnitude of $|V|/G$ is smaller than 1, suggesting that this TrB is a completely closed shell interaction [38], which is inconsistent with its relatively strong interaction energy. Thus, the estimation of the nature of the TrB according to the topological parameters should be made with caution. In each type of complex, a good linear relationship is present between the electron density and the interaction energy for the B$\cdots$O TrB, while an opposite dependence is found for the Al$\cdots$O TrB. This again highlights the careful consideration necessary in studying a TrB according to the AIM parameters.

The charge transfer (CT) values for the different types of binary complexes are given in Table S1. In most complexes, CT is larger than 0.1 e, indicative of a stronger TrB. A good relationship is not found between CT and E_{int}, partly due to the coexistence of both the TrB and the H$\cdots$X H-bond with an opposite direction of CT.

To better understand the nature of the Tr$\cdots$O TrB, an energy decomposition analysis was carried out for these systems. As shown in Table S2, the interaction energy was decomposed into five terms, including the electrostatic energy (E^{ele}), the exchange energy (E^{ex}), the repulsion energy (E^{rep}), the polarisation energy (E^{pol}), and the dispersion energy (E^{disp}). In most cases, E^{ex} is the largest negative term, but it is usually cancelled with E^{rep}; thus, neither of these two terms is discussed. E^{ele} is larger than E^{pol} and E^{disp}, and E^{pol} is comparable with E^{ele} in most cases, which supports the conclusion that the TrB is very strong. E^{disp} is negative in the B$\cdots$O TrB but becomes positive in the Al analogue due to the very shorter Al$\cdots$O distance.

2.2. π–π Parallel Structures

A closer look at the MEP of the MDA molecule in Figure 1 shows that, in addition to the negatively charged blue region at the hydroxyl O-terminus and carbonyl O-terminus, there is also a relatively small negatively charged region above the C-atom at the centre of the MDA molecular plane, so when the π–hole on the Tr atom of TrX$_3$ approaches the central carbon atom of MDA from above, a face-to-face parallel π–π structure appears, as shown in Figure 4, like the π–π interactions in the aromatic systems.

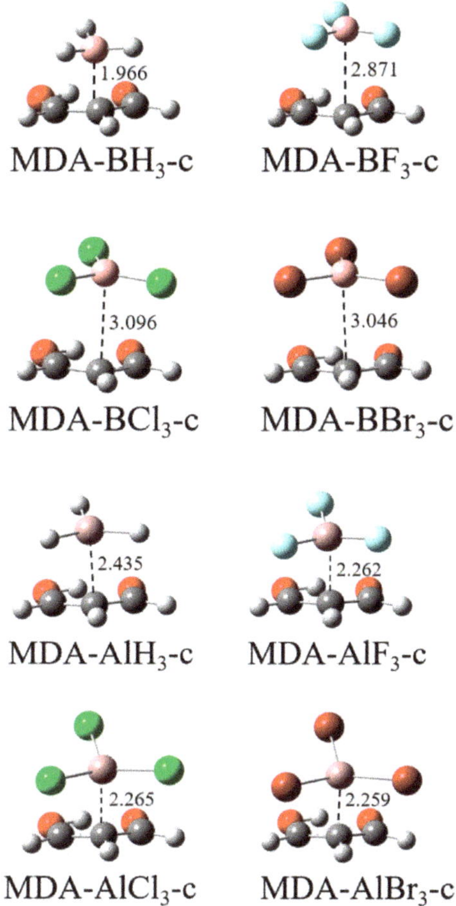

Figure 4. The π–π structures formed by the central carbon of MDA with TrX$_3$. Distances are in Å.

The π–π structure involving BH$_3$ has comparable stability with the a-type analogue but is much less stable for its halogen derivatives. AlX$_3$ also forms a weaker π–π structure with MDA again with the a-type analogue. The halogen substitution of AlX$_3$ has a similar enhancing effect on the stability of the π–π structure, while an opposite influence is found for the halogen substitution of BX$_3$. The effect of different halogen atoms has a small difference. Likely, AlX$_3$ engages in a stronger π–π TrB than BX$_3$ except for X = H. The above conclusions are obtained according to the interaction energy in Table 1 and the Tr···C distance in Figure 4.

The distribution of the NCI region between TrX$_3$ and MDA in the π–π structure (Figure S4) is like that in the π–π stacking of two benzene molecules [39]. The NCI region in the π–π structure of AlX$_3$ has a deeper and more complicated colour, consistent with a stronger TrB. The electron density at the Tr···C BCP does not reflect the change in the interaction energy in the π–π structure since this structure is not bounded only by a Tr···C TrB. This is also true for the Laplacian at the Tr···C BCP. The sign of the energy density is negative for the Tr···C BCP with the interaction energy larger than 10 kcal/mol.

Although MDA-BH$_3$-c has the smaller interaction energy, CT in MDA-BH$_3$-c is larger than that in MDA-BH$_3$-a and MDA-BH$_3$-b since the π electrons in the ring of MDA are easily lost.

If the interaction energy is smaller than 5 kcal/mol, the polarisation contribution is smallest, and the dispersion is even larger than the electrostatic energy in MDA-BCl$_3$-c and MDA-BBr$_3$-c due to the nature of π electron in the MDA ring and the longer distance (>3 Å). If the interaction energy is larger than 10 kcal/mol, the dispersion contribution is the smallest, and the polarisation energy is comparable with the electrostatic energy.

2.3. Proton Transfer

The formation of the TrB has an important influence on the intramolecular structures of MDA, particularly its intramolecular H-bond. When the Tr atom of TrX$_3$ participates in a TrB with the hydroxyl O of MDA, R_2(O-H) is stretched, and R_1(H$\cdots$O) is shortened. In MDA-AlF$_3$-a, MDA-BCl$_3$, and MDA-BBr$_3$, R_1(H$\cdots$O) is much shorter than R_2(O-H), which can be described as a partial proton transfer. However, when the Tr atom of TrX$_3$ engages in a TrB with the carbonyl O of MDA, R_2(O-H) is shortened, and the R_1(H$\cdots$O) is stretched, indicating that no proton transfer occurs. When the Tr atom of TrX$_3$ forms a π–π parallel structure with the central C atom of MDA, most of the structures have a small degree of R_2(O-H) elongation and a shortened R_1(H$\cdots$O), which can also be considered to have undergone proton transfer but to a much lesser extent than the a-configuration. It is also interesting to note that the R_2 (O-H) and R_1(H$\cdots$O) in both MDA-BH$_3$-c and MDA-BF$_3$-c are slightly elongated, a change that is negligible. The above bond lengths as well as their bond length variations are listed in Table 3.

Table 3. H$\cdots$O distance (R_1) and O-H bond length (R_2) in the complexes as well as their difference (ΔR) relative to the monomer, all in Å.

	R_1	ΔR_1	R_2	ΔR_2
MDA-BH$_3$-a	1.539	-0.106	1.023	0.022
MDA-BF$_3$-a	1.579	-0.066	1.013	0.012
MDA-BCl$_3$-a	0.988	-0.657	1.736	0.735
MDA-BBr$_3$-a	0.988	-0.657	1.732	0.731
MDA-AlH$_3$-a	1.476	-0.169	1.040	0.039
MDA-AlF$_3$-a	0.988	-0.657	1.732	0.731
MDA-AlCl$_3$-a	0.988	-0.657	1.744	0.743
MDA-AlBr$_3$-a	0.988	-0.657	1.742	0.741
MDA-BH$_3$-b	1.736	0.091	0.988	-0.013
MDA-BF$_3$-b	1.734	0.089	0.988	-0.013
MDA-BCl$_3$-b	1.736	0.091	0.988	-0.013
MDA-BBr$_3$-b	1.732	0.087	0.989	-0.012
MDA-AlH$_3$-b	1.722	0.077	0.990	-0.011
MDA-AlF$_3$-b	1.732	0.087	0.988	-0.013
MDA-AlCl$_3$-b	1.744	0.099	0.988	-0.013
MDA-AlBr$_3$-b	1.742	0.097	0.988	-0.013
MDA-BH$_3$-c	1.646	0.001	1.005	0.004
MDA-BF$_3$-c	1.646	0.001	1.002	0.001
MDA-BCl$_3$-c	1.642	-0.003	1.002	0.001
MDA-BBr$_3$-c	1.640	-0.005	1.003	0.002
MDA-AlH$_3$-c	1.639	-0.006	1.005	0.004
MDA-AlF$_3$-c	1.609	-0.036	1.012	0.011
MDA-AlCl$_3$-c	1.594	-0.051	1.015	0.014
MDA-AlBr$_3$-c	1.593	-0.052	1.016	0.015
BH$_2$-MDA-BH$_3$-a	0.992	-0.653	1.700	0.699
BH$_2$-MDA-BF$_3$-a	1.525	-0.120	1.027	0.026
BH$_2$-MDA-AlH$_3$-a	0.994	-0.651	1.692	0.691

Table 4 shows the AIM analysis for the intramolecular H$\cdots$O(2) and O(1)-H BCPs. In the MDA monomer, both $\nabla^2\rho$ and H at the O(1)-H BCP are negative, with a character of a covalent bond, while only H is negative at the H$\cdots$O(2) BCP, indicative of a partially covalent interaction. When the Tr atom of TrX$_3$ forms a TrB with the hydroxyl O of MDA, the ρ at the H$\cdots$O(2) BCP increases, while it is decreased for the O(1)-H BCP. Even the sign of $\nabla^2\rho$ at both types of BCPs is changed in MDA-BX$_3$-a (X = Cl and Br) and MDA-AlX$_3$-a (X = F, Cl, and Br). Specifically, $\nabla^2\rho$ becomes negative for the H$\cdots$O(2) BCP but positive for the O(1)-H BCP. In addition, H at the H$\cdots$O(2) BCP is more negative but less negative for the O(1)-H BCP. These changes demonstrate that the intramolecular H-bond is strengthened, and even a proton transfer occurs in the a-type complex with a very strong TrB. This enhancing effect is also found in the c-type complex except for MDA-BH$_3$-c and MDA-BF$_3$-c, but no proton charge occurs. When the Tr atom in TrX$_3$ forms a TrB with the carbonyl O of MDA, an opposite change is found for the ρ at the H$\cdots$O(2) and O(1)-H BCPs, and the signs of both $\nabla^2\rho$ and H are not changed. This means that the intramolecular H-bond is weakened.

Table 4. Electron density (ρ), Laplacians ($\nabla^2\rho$), and energy density (H) at the H$\cdots$O and O-H BCPs in the complexes, all in a.u.

	H$\cdots$O(2)			O(1)-H		
	ρ	$\nabla^2\rho$	H	ρ	$\nabla^2\rho$	H
MDA	0.0533	0.1352	−0.0131	0.3220	−2.5486	−0.6945
MDA-BH$_3$-a	0.0690	0.1369	−0.0241	0.2941	−2.2981	−0.6328
MDA-BF$_3$-a	0.0626	0.1380	−0.0193	0.3058	−2.4059	−0.6595
MDA-BCl$_3$-a	0.3364	−2.7220	−0.7339	0.0405	0.1366	−0.0044
MDA-BBr$_3$-a	0.3355	−2.7156	−0.7320	0.0409	0.1382	−0.0045
MDA-AlH$_3$-a	0.0810	0.1237	−0.0352	0.2766	−2.0763	−0.5819
MDA-AlF$_3$-a	0.3368	−2.7123	−0.7326	0.0416	0.1338	−0.0054
MDA-AlCl$_3$-a	0.3375	−2.7200	−0.7342	0.0404	0.1323	−0.0048
MDA-AlBr$_3$-a	0.3373	−2.7182	−0.7337	0.0406	0.1329	−0.0048
MDA-BH$_3$-b	0.0413	0.1325	−0.0053	0.3374	−2.7145	−0.7336
MDA-BF$_3$-b	0.0412	0.1339	−0.0051	0.3372	−2.7180	−0.7341
MDA-BCl$_3$-b	0.0405	0.1366	−0.0044	0.3364	−2.7220	−0.7339
MDA-BBr$_3$-b	0.0409	0.1382	−0.0045	0.3355	−2.7156	−0.7320
MDA-AlH$_3$-b	0.0429	0.1334	−0.0063	0.3353	−2.6940	−0.7286
MDA-AlF$_3$-b	0.0416	0.1338	−0.0054	0.3368	−2.7123	−0.7326
MDA-AlCl$_3$-b	0.0404	0.1323	−0.0048	0.3375	−2.7200	−0.7342
MDA-AlBr$_3$-b	0.0406	0.1329	−0.0048	0.3373	−2.7182	−0.7337
MDA-BH$_3$-c	0.0531	0.1325	−0.0131	0.3176	−2.5245	−0.6866
MDA-BF$_3$-c	0.0531	0.1350	−0.0129	0.3213	−2.5464	−0.6935
MDA-BCl$_3$-c	0.0537	0.1353	−0.0133	0.3210	−2.5393	−0.6921
MDA-BBr$_3$-c	0.0539	0.1353	−0.0135	0.3207	−2.5358	−0.6913
MDA-AlH$_3$-c	0.0540	0.1342	−0.0135	0.3170	−2.5186	−0.6854
MDA-AlF$_3$-c	0.0581	0.1350	−0.0163	0.3096	−2.4522	−0.6686
MDA-AlCl$_3$-c	0.0603	0.1351	−0.0179	0.3059	−2.4130	−0.6590
MDA-AlBr$_3$-c	0.0605	0.1349	−0.0181	0.3053	−2.4073	−0.6575
BH$_2$-MDA-BH$_3$-a	0.3319	−2.6740	−0.7228	0.0452	0.1362	−0.0752
BH$_2$-MDA-BF$_3$-a	0.0719	0.1342	−0.0267	0.2920	−2.2490	−0.6220
BH$_2$-MDA-AlH$_3$-a	0.3303	−2.6560	−0.7185	0.0463	0.1362	−0.0083

2.4. Substitution Effect

Besides external influences, the intramolecular H-bond of MDA can also be regulated by the presence of substituents. In order to enhance the H-bond within the MDA molecule, we selected the three structures of MDA-BH$_3$-a, MDA-BF$_3$-a, and MDA-AlH$_3$-a since no complete proton transfer occurs and replaces the H atom attached to the central C atom in these structures with a BH$_2$ group, as shown in Figure 5. It is clear from the figure that the structures of BH$_2$-MDA-BH$_3$-a and BH$_2$-MDA-AlH$_3$-a dramatically change, compared with the corresponding MDA-BH$_3$-a and MDA-AlH$_3$-a, while the structure of BH$_2$-MDA-BF$_3$-a has a slight change.

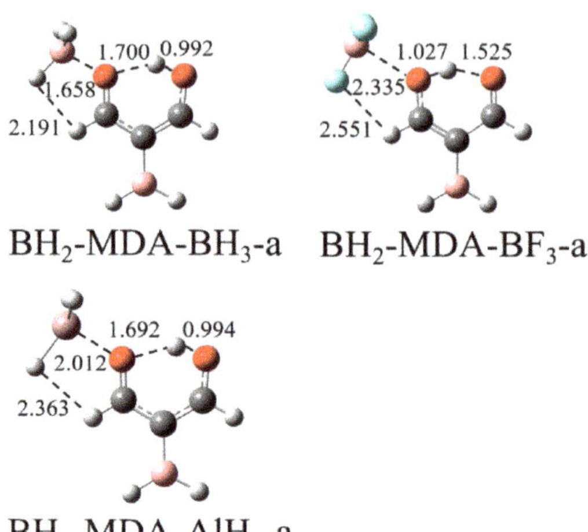

Figure 5. Structures of three a-type complexes substituted by a BH$_2$ group.

From Table 3, it can be seen that there is a large degree of contraction of R$_1$ and a significant degree of stretching of R$_2$ in both BH$_2$-MDA-BH$_3$-a and BH$_2$-MDA-AlH$_3$-a, compared with MDA-BH$_3$-a and MDA-AlH$_3$-a, which could indicate a significant enhancement in the H-bond and hence proton transfer. This conclusion is also deduced from the AIM analysis in Table 4, which shows that the ρ at the H$\cdots$O BCP in both the BH$_2$-MDA-BH$_3$-a and BH$_2$-MDA-AlH$_3$-a structures is considerably increased, and the sign of $\nabla^2\rho$ changes from positive in MDA-BH$_3$-a and MDA-AlH$_3$-a to negative, indicating that the H$\cdots$O H-bond in these structures has changed from the partially covalent interaction in MDA-BH$_3$-a and MDA-AlH$_3$-a to the present covalent bond. In contrast, the ρ at the O-H BCP in both BH$_2$-MDA-BH$_3$-a and BH$_2$-MDA-AlH$_3$-a decreases, and the sign of $\nabla^2\rho$ changes from negative to positive, indicating that the O-H changes in these structures from a covalent bond to a partially covalent interaction. However, when the H atom attached to the C atom in the centre in MDA-BF$_3$-a is replaced by a BH$_2$ group, there is only a small contraction of R$_1$ and a slight increase in ρ at the H$\cdots$O BCP, so we believe that the effect on the intramolecular H-bond is still very small in BH$_2$-MDA-BF$_3$-a, like MDA-BF$_3$-a.

The orbital interaction diagram also demonstrates that the addition of the BH$_2$ moiety enhances the H-bond and further promotes proton transfer. In MDA-BH$_3$-a (Figure 6a), the charge density moves from the lone pair orbital of the carbonyl O(2) atom into the antibonding orbital of the O(1)-H bond, while this orbital interaction becomes LP$_{O(1)}$$\rightarrow$$\sigma^*_{O(2)-H}$ in BH$_2$-MDA-BH$_3$-a (Figure 6c); thus, the direction of the orbital interaction is reversed. In addition, the overlapping between both orbitals is almost the same in both structures. This similar orbital interaction is also found in MDA-BCl$_3$-a (Figure 6b).

The addition of an electron-withdrawing group BH$_2$ to the central C atom of MDA also influences the strengths of the TrB and secondary H$\cdots$X interactions. The Tr$\cdots$O and H$\cdots$X distances are shortened in BH$_2$-MDA-BH$_3$-a and BH$_2$-MDA-AlH$_3$-a but elongated in BH$_2$-MDA-BF$_3$-a. Thus, both types of interactions are strengthened in BH$_2$-MDA-BH$_3$-a and BH$_2$-MDA-AlH$_3$-a but weakened in BH$_2$-MDA-BF$_3$-a. In turn, the change in TrB strength would impose an influence on the proton transfer. Therefore, the introduction of the BH$_2$ group can regulate the intramolecular proton transfer not only through a substitution effect but also through a cooperative effect.

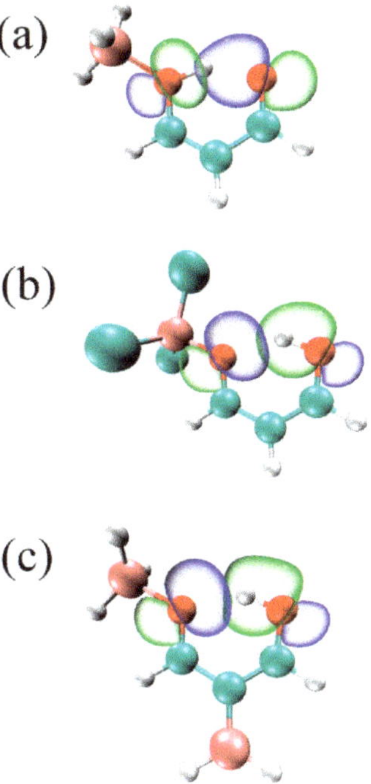

Figure 6. Diagrams of LpO→σ*H-O orbital interaction in (**a**) MDA-BH$_3$-a, (**b**) MDA-BCl$_3$-a, and (**c**) BH$_2$-MDA-BH$_3$-a.

3. Discussion

There is a hydroxyl O in MDA, which can form a TrB with TrX$_3$ with the interaction energy of 7–42 kcal/mol. The interaction energy exceeds 37 kcal/mol in MDA-BX$_3$-a (X = Cl and Br) and MDA-AlX$_3$-a (X = halogen) since the hydroxyl O becomes the carbonyl O upon the formation of a TrB. The carbonyl O has a larger negative MEP than the hydroxyl O; thus, the former participates in a stronger TrB. This supports the fact that the carbonyl O is a stronger electron donor in intermolecular interactions such as H-bond [40]. The intramolecular H-bond belongs to a resonance-assisted H-bond, and the charge density on the hydroxyl O is delocalised and thus reduced, which can be used to explain why the TrB formed by the hydroxyl O of MDA is weaker than that of H$_2$O (>20 kcal/mol) [36].

A comparison of the interaction energy between MDA-BH$_3$-a and MDA-BF$_3$-a shows that the F substituents on the B centre weaken the TrB, although the π–hole in the BF$_3$ molecule is larger than that in BH$_3$. This conclusion has been confirmed in previous studies [25,26,36]. This abnormality is primarily attributed to the back-bonding effect from the F atom into the outer p orbital of the boron atom, and this effect makes the BF$_3$ molecule not easily undergo distortion. DE amounts to 50% of the binding energy in MDA-BF$_3$-a and 84% in MDA-BH$_3$-a. Accordingly, the distortion has a prominent contribution to the interaction energy of TrB. When the carbonyl O binds with BX$_3$, the DE percentage is much larger due to the formation of a stronger TrB. MDA-BF$_3$-b has a larger DE contribution to the interaction energy than MDA-BH$_3$-b, and the variation in the interaction energy becomes normal in both complexes, consistent with the π–hole on the B atom. Whether the a-type or b-type complexes, the DE contribution to the binding energy in the AlX$_3$ complex is much smaller than that in the BX$_3$ analogue.

When the hydroxyl O of MDA engages in a very strong TrB in MDA-BX$_3$-a (X = Cl and Br) and MDA-AlX$_3$-a (X = halogen), the corresponding interaction energy is the same as that in the b-type analogue. This is also reflected in the related data of the geometries, AIM, and NBO. This shows that the a-type and the b-type structures of these complexes are the same. This transformation is described with MDA-BCl$_3$ as an example (Figure 7). Structure 1 is the initial complex before optimisation; thus, the geometrical parameters in 1 are the same as those in the isolated MDA. As the B⋯O distance is shorter in 2, the O(1)-H bond is elongated, and the H⋯O(2) distance is shortened. Structure 3 has an equivalent distance of the H atom with the two O atoms. Structure 4 is an enantiomer with 2. When the triel bond is strong enough in structure 5, the proton moves completely from the O(1) to O(2), and their roles are reversed.

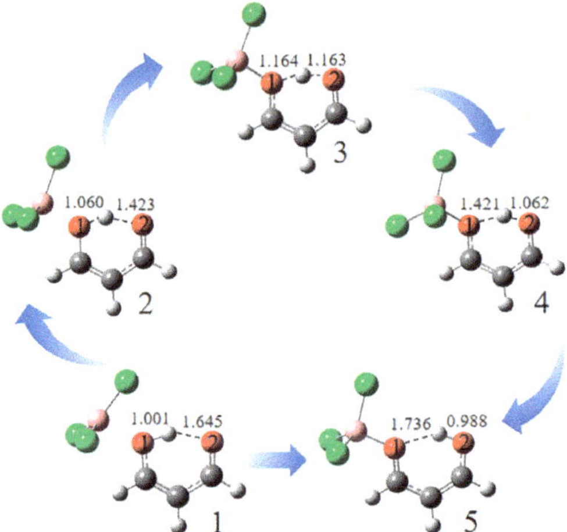

Figure 7. Different conformations in the formation process of MDA-BCl$_3$-a.

The intramolecular H-bond is strengthened if TrX$_3$ attacks the hydroxyl O of MDA, while an opposite effect is obtained if it is introduced into the carbonyl O. The former effect promotes the proton transfer, while the latter inhibits the proton transfer. Such an effect has also been imposed by introducing a beryllium bond [15] or a tetrel bond [16]. When a strong tetrel bond formed with F$_2$SiO (>42 kcal/mol) is introduced to the hydroxyl O of MDA, the ratio of R$_1$/R$_2$ is 0.57, which is almost equal to that caused by a stronger TrB (37–42 kcal/mol). This shows that the enhancement in the added interaction has a slight effect on the degree of proton transfer if its interaction energy exceeds a threshold value, and here, it may be 37 kcal/mol.

The π–π structure of the c-type complex between TrX$_3$ and MDA is very interesting. If the MDA is replaced by benzene, a similar π–π structure has been reported [41]. The strength of the corresponding TrB is also equal for the complexes of TrX$_3$ with both MDA and benzene except BH$_3$. The effect of halogen substitution on the triel donor is also the same. Specifically, the halogen substitution on the B centre weakens the π–π structure, while that on the Al centre has an opposite effect. If TrX$_3$ is changed into F$_2$TO (T = C and Si), the π–π structure obtained with MDA has an interaction energy value of 2.5 kcal/mol for F$_2$CO and 27 kcal/mol for F$_2$SiO [16]. The corresponding interaction energy is also in a similar range in the TrB formed by TrX$_3$.

The DE value in MDA-BBr$_3$-c is very small (0.2 kcal/mol); thus, the planar structures of both MDA and BBr$_3$ are held in the complex, resulting in a π–π structure. When the B⋯C distance in MDA-BBr$_3$-c is shortened to 1.8 Å (its structure is shown in Figure S5), the planar structures of both molecules are distorted with a high deformation energy value of 32 kcal/mol, and the corresponding π–π structure disappears. The interaction energy amounts to 33.64 kcal/mol for the distorted structure of MDA-BBr$_3$-c. Interestingly, its binding energy is very small (<2 kcal/mol), which is smaller than that in the corresponding π–π structure (~4 kcal/mol). We plotted the energy curve of MDA-BBr$_3$-c by changing the

B$\cdots$C distance from 1.5 to 3.5 Å (Figure S6). Two minima are found on the potential energy surface, corresponding to the structures in Figure 4 and Figure S5. The distorted structure is more stable than the π–π structure, and the barrier between both structures is 2 kcal/mol.

4. Conclusions

The π–hole above TrX_3 (Tr = B and Al; X = H, F, Cl, and Br) can form a TrB with the hydroxyl O, carbonyl O, and the central carbon atoms of MDA, marked with a, b, and c, respectively. Other than the TrB, there is also a H$\cdots$X secondary interaction in both the a-type complex and the b-type complex. As expected, the carbonyl O engages in a stronger TrB than the hydroxyl O, and both types of O atoms are better nucleophiles than the central carbon atom. When TrX_3 is introduced into the hydroxyl O in MDA-BX_3-a (X = Cl and Br) and MDA-AlX_3-a (X = halogen), the triel-bonded complexes formed have equal stability to the corresponding b-type complex, with a high degree of interaction energy (>37 kcal/mol). The halogen substitution in the triel donor has an enhancing effect on the strength of TrB with an exception in MDA-BF_3-a and MDA-BX_3-c. For each type of complex, AlX_3 shows a higher affinity to MDA than BX_3 except X = H.

The formation of TrB between TrX_3 and MDA has an effect on the strength of the intramolecular H-bond in MDA. When TrX_3 attacks the hydroxyl or carbonyl O atom of MDA, the former interaction strengthens the intramolecular H-bond, while the latter leads to a weakening H-bond. The π–TrB in the c-type complex also has an enhancing effect on the intramolecular H-bond except for MDA-BH_3-c and MDA-BF_3-c. Accompanied by the strengthening or weakening of the intramolecular H-bond, the proton transfer is promoted or inhibited. A complete proton transfer is seen in MDA-BX_3-a (X = Cl and Br) and MDA-AlX_3-a (X = halogen), and these complexes display an equal degree of proton transfer, independent of the strength of the TrB. An electron-withdrawing group BH_2 at the central carbon atom of MDA in BH_2-MDA-TrH_3-a (Tr = B and Al) can enhance the intramolecular H-bond and further cause a proton transfer. This substitution in BH_2-MDA-BF_3-a also strengthens the H-bond, but no proton transfer occurs.

5. Theoretical Methods

All the monomers and complexes were optimised using the MP2 method with an aug-cc-pVTZ basis set. Frequency calculations were performed at the same level to verify that the optimised structures were true minima on the potential energy surface, without imaginary frequencies. The interaction energies (E_{int}) were calculated using supramolecular methods involving monomers with their geometries adopted in the complexes. The binding energy (E_b) represents the difference between the energy of the complex and the sum of the monomer energies in the fully optimised structure. The difference between E_{int} and E_b is defined as the deformation energy (DE). These quantities were corrected for the basis set superposition error (BSSE) according to the equilibrium protocol proposed by Boys and Bernardi [42]. All the calculations were performed using the Gaussian 09 program [43].

The MEP maps of the monomers were plotted on 0.001 a.u. electron density isosurfaces at the MP2/aug-cc-pVTZ level using the Wave Function Analysis Surface Analysis Suite (WFA-SAS) software [44]. The topological parameters, including the electron density, its Laplacian, and the total energy density at the bond critical point (BCP), were calculated using the MultiWFN program [45]. Natural bond orbital (NBO) analysis was performed at the HF/aug-cc-pVTZ level to evaluate the charge transfer (CT) and interorbital interactions using the NBO 3.0 program [46]. Non-covalent interaction (NCI) [47] mapping was plotted using the Multiwfn [45] and VMD program [48]. The decomposition of the interaction energy comprised five physically significant components: the electrostatic energy (E^{ele}), the exchange energy (E^{ex}), the repulsive energy (E^{rep}), the polarisation energy (E^{pol}), and the dispersive energy (E^{disp}). These features were determined at the MP2/aug-cc-pVTZ level using the local molecular orbital energy decomposition analysis (LMO-EDA) method [49] in the GAMESS program [50].

Supplementary Materials: The following are available online at https://www.mdpi.com/article/10.3390/molecules27186091/s1, Figure S1: MEP maps of TrX$_3$. Color ranges are: red, greater than 0.02; yellow, between 0.02 and 0; green, between -0.02 and 0; blue, less than -0.02. All are in a.u., Figure S2: NCI diagram of binary complex formed by the hydroxyl O with TrX$_3$. Blue, green, and red areas represent strong attraction, weak attraction, and strong repulsion, respectively. Diagrams are drawn by the Multiwfn and the VMD programs, Figure S3: NCI diagram of binary complex of MDA formed by the carbonyl O with TrX$_3$. Blue, green, and red areas represent strong attraction, weak attraction, and strong repulsion, respectively, Figure S4: NCI diagram of π-π structures formed by the carbon center of MDA with TrX$_3$. Blue, green, and red areas represent strong attraction, weak attraction, and strong repulsion, respectively, Figure S5: Structure of MDA-BBr$_3$-c at a B$\cdots$C distance of 1.8 Å, Figure S6: The energy curve of MDA-BBr$_3$-c by changing the B$\cdots$C distance from 1.5 to 3.5 Å, Table S1: Charge transfer (CT, e) in the complexes, Table S2: Electrostatic (E^{ele}), exchange (E^{ex}), repulsion (E^{rep}), polarization (E^{pol}), and dispersion energies (E^{disp}) as well as the total interaction energy (ΔE^{total}) of triel bond in the binary complexes. All are in kcal/mol.

Author Contributions: Q.L. conceived of the idea for this project and wrote a first draft of the manuscript; Q.W. carried out the calculations and compiled the data; S.Y. supervised the calculations and helped with the final draft. All authors have read and agreed to the published version of the manuscript.

Funding: This research was funded by the Natural Science Foundation of Shandong Province (ZR2021MB123).

Institutional Review Board Statement: Not applicable.

Informed Consent Statement: Not applicable.

Data Availability Statement: Not applicable.

Conflicts of Interest: The author declares no conflict of interest.

Sample Availability: Samples of the compounds are not available from the authors.

References

1. Marnett, L.J. Lipid peroxidation—DNA damage by malondialdehyde. *Mutat. Res.-Fund. Mol. Mech.* **1999**, *424*, 83–95. [CrossRef]
2. Islayem, D.; FakihF, B.; Lee, S. Comparison of colorimetric methods to detect malondialdehyde, A biomarker of reactive oxygen species. *ChemistrySelect* **2022**, *7*, e202103627. [CrossRef]
3. Auxilia, A.M.; Caldiroli, A.; Capuzzi, E.; Clerici, M.; Ossola, P.; Buoli, M. Is malondialdehyde a reliable biomarker for bipolar disorder? *Eur. Neuropsychopharm.* **2021**, *53*, S297–S298. [CrossRef]
4. Mohideen, K.; Sudhakar, U.; Balakrishnan, T.; Almasri, M.A.; Al-Ahmari, M.M.; Dira, H.S.A.; Suhluli, M.; Dubey, A.; Mujoo, S.; Khurshid, Z.; et al. Malondialdehyde, an oxidative stress marker in oral squamous cell carcinoma—A systematic review and meta-analysis. *Curr. Issues Mol. Biol.* **2021**, *43*, 1019–1035. [CrossRef]
5. Pitsevich, G.A.; Malevich, A.E.; Kozlovskaya, E.N.; Yu, I.; Pogorelov, V.E.; Sablinskas, V.; Balevicius, V. Theoretical study of the C-H/O-H stretching vibrations in malonaldehyde. *Spectrochim. Acta A* **2015**, *145*, 384–393. [CrossRef]
6. Lin, C.; Kumar, M.; Finney, B.A.; Francisco, J.S. Intramolecular hydrogen bonding in malonaldehyde and its radical analogues. *J. Chem. Phys.* **2017**, *147*, 124309.
7. Hargis, J.C.; Evangelista, F.A.; Ingels, J.B.; Schaefer III, H.F. Short intramolecular hydrogen bonds: Derivatives of malonaldehyde with symmetric substituents. *J. Am. Chem. Soc.* **2008**, *130*, 17471–17478. [CrossRef]
8. Gutiérrez-Quintanilla, A.; Chevalier, M.; Platakyte, R.; Ceponkus, J.; Rojas-Lorenzo, G.; Crépin, C. 2-Chloromalonaldehyde, A model system of resonance-assisted hydrogen bonding: Vibrational investigation. *Phys. Chem. Chem. Phys.* **2018**, *20*, 12888–12897. [CrossRef]
9. Gutiérrez-Quintanilla, A.; Chevalier, M.; Platakyte, R.; Ceponkus, J.; Crépin, C. Selective photoisomerisation of 2-chloromalonaldehyde. *J. Chem. Phys.* **2019**, *150*, 034305. [CrossRef]
10. Xu, X.; Zheng, J.; Truhlar, D.G. Ultraviolet absorption spectrum of malonaldehyde in water is dominated by solvent-stabilized conformations. *J. Am. Chem. Soc.* **2015**, *137*, 8026–8029. [CrossRef]
11. Terranova, Z.L.; Corcelli, S.A. Monitoring intramolecular proton transfer with two-dimensional infrared spectroscopy: A computational prediction. *J. Phys. Chem. Lett.* **2012**, *3*, 1842–1846. [CrossRef] [PubMed]
12. Wu, F.; Ren, Y. Primary and secondary isotope effect on tunnelling in malonaldehyde using a quantum mechanical scheme. *Mol. Phys.* **2017**, *115*, 1700–1707. [CrossRef]
13. Huang, J.; Buchowiecki, M.; Nagy, T.; Vaníček, J.; Meuwly, M. Kinetic isotope effect in malonaldehyde determined from path integral Monte Carlo simulations. *Phys. Chem. Chem. Phys.* **2014**, *16*, 204–211. [CrossRef] [PubMed]

14. Douhal, A.; Lahmani, F.; Zewail, A.H. Proton-transfer reaction dynamics. *Chem. Phys.* **1996**, *207*, 477–498. [CrossRef]
15. Mó, O.; Yáne, M.; Alkorta, I.; Elguero, J. Modulating the strength of hydrogen bonds through beryllium bonds. *J. Chem. Theory Comput.* **2012**, *8*, 2293–2300. [CrossRef]
16. Wei, Y.X.; Li, Q.Z.; Scheiner, S. The π-tetrel bond and its influence on hydrogen bonding and proton transfer. *ChemPhysChem* **2018**, *19*, 736–743. [CrossRef]
17. Grabowski, S.J. Boron and other triel Lewis acid centers: From hypovalency to hypervalency. *ChemPhysChem* **2014**, *15*, 2985–2993. [CrossRef]
18. Buchberger, A.R.; Danforth, S.J.; Bloomgren, K.M.; Rohde, J.A.; Smith, E.L.; Gardener, C.C.; Phillips, J.A. Condensed-phase effects on the structural properties of FCH_2CN–BF_3 and $ClCH_2CN$–BF_3: A matrix-isolation and computational study. *J. Phys. Chem. B* **2013**, *117*, 11687–11696. [CrossRef]
19. Bhunya, S.; Malakar, T.; Ganguly, G.; Paul, A. Combining protons and hydrides by homogeneous catalysis for controlling the release of hydrogen from ammonia–borane: Present status and challenges. *ACS Catal.* **2016**, *6*, 7907–7934. [CrossRef]
20. Vyboishchikov, S.F.; Krapp, A.; Frenking, G. Two complementary molecular energy decomposition schemes: The Mayer and Ziegler–Rauk methods in comparison. *J. Chem. Phys.* **2008**, *129*, 144111. [CrossRef]
21. Ishita, B.; Sourav, B.; Ankan, P. Frustrated Lewis acid–base-pair-catalyzed amine-borane dehydrogenation. *Inorg. Chem.* **2020**, *59*, 1046–1056.
22. Parks, D.J.; Blackwell, J.M.; Piers, W.E. Studies on the mechanism of $B(C_6F_5)_3$-catalyzed hydrosilation of carbonyl functions. *J. Org. Chem.* **2000**, *65*, 3090–3098. [CrossRef]
23. Hamilton, C.W.; Baker, R.T.; Staubitz, A.; Manners, I. B–N compounds for chemical hydrogen storage. *Chem. Soc. Rev.* **2009**, *38*, 279–293. [CrossRef] [PubMed]
24. Diaz, D.B.; Yudin, A.K. The versatility of boron in biological target engagement. *Nat. Chem.* **2017**, *9*, 731–742. [CrossRef]
25. Grabowski, S.J. π-Hole bonds: Boron and aluminum Lewis acid centers. *ChemPhysChem* **2015**, *16*, 1470–1479. [CrossRef]
26. Bauzá, A.; Frontera, A. On the versatility of BH_2X (X = F, Cl, Br, and I) compounds as halogen-, hydrogen-, and triel-bond donors: An ab initio study. *ChemPhysChem* **2016**, *17*, 3181–3186. [CrossRef]
27. Mohajeri, A.; Eskandari, K.; Safaee, S.A. Endohedral pnicogen and triel bonds in doped C_{60} fullerenes. *New J. Chem.* **2017**, *41*, 10619–10626. [CrossRef]
28. Michalczyk, M.; Zierkiewicz, W.; Scheiner, S. Triel-bonded complexes between TrR_3 (Tr = B, Al, Ga; R = H, F, Cl, Br, CH_3) and Pyrazine. *ChemPhysChem* **2018**, *19*, 3122–3133. [CrossRef] [PubMed]
29. Grabowski, S.J. Bifurcated triel bonds—hydrides and halides of 1,2-bis(dichloroboryl) benzene and 1,8-bis(dichloroboryl) naphthalene. *Crystals* **2019**, *9*, 503. [CrossRef]
30. Grabowski, S.J. Triel bond and coordination of triel centres–Comparison with hydrogen bond interaction. *Chem. Rev.* **2020**, *407*, 213171. [CrossRef]
31. Yang, Q.; Zhou, B.; Li, Q.; Scheiner, S. Weak σ-hole triel bond between C_5H_5Tr (Tr = B, Al, Ga) and haloethyne: Substituent and cooperativity effects. *ChemPhysChem* **2021**, *22*, 481–487. [CrossRef] [PubMed]
32. Grabowski, S.J. Triel bonds, π-hole-π-electrons interactions in complexes of boron and aluminium trihalides and trihydrides with acetylene and ethylene. *Molecules* **2015**, *20*, 11297–11316. [CrossRef] [PubMed]
33. Esrafili, M.D.; Mohammadian-Sabet, F. Theoretical insights into nature of π-hole interactions between triel centers (B and Al) and radical methyl as a potential electron donor: Do single-electron triel bonds exist? *Struct. Chem.* **2016**, *27*, 1157–1164. [CrossRef]
34. Zhang, J.R.; Wei, Y.X.; Li, W.Z.; Cheng, J.B.; Li, Q.Z. Triel–hydride triel bond between ZX_3 (Z = B and Al; X = H and Me) and $THMe_3$ (T = Si, Ge and Sn). *Appl. Organomet. Chem.* **2018**, *32*, e4367. [CrossRef]
35. Chi, Z.Q.; Dong, W.B.; Li, Q.Z.; Yang, X.; Scheiner, S.; Liu, S.F. Carbene triel bonds between TrR_3 (Tr = B, Al) and N-heterocyclic carbenes. *Int. J. Quantum Chem.* **2019**, *119*, e25867. [CrossRef]
36. Chi, Z.Q.; Li, Q.Z.; Li, H.B. Comparison of triel bonds with different chalcogen electron donors: Its dependence on triel donor and methyl substitution. *Int. J. Quantum Chem.* **2020**, *120*, e26046. [CrossRef]
37. Fiacco, D.L.; Leopold, K.R. Partially bound systems as sensitive probes of microsolvation: A microwave and ab initio study of HCN$\cdots$HCN– BF_3. *J. Phys. Chem. A* **2003**, *107*, 2808–2814. [CrossRef]
38. Lipkowski, P.; Grabowski, S.J.; Robinson, T.L.; Leszczynski, J. Properties of the C−H$\cdots$H dihydrogen bond: An ab initio and topological analysis. *J. Phys. Chem. A* **2004**, *108*, 10865–10872. [CrossRef]
39. Silva, N.J.; Machado, F.B.C.; Lischka, H.; Aquino, A.J.A. π–π stacking between polyaromatic hydrocarbon sheets beyond dispersion interactions. *Phys. Chem. Chem. Phys.* **2016**, *18*, 22300–22310. [CrossRef]
40. Lommerse, J.P.M.; Price, S.L.; Taylor, R. Hydrogen bonding of carbonyl, ether, and ester oxygen atoms with alkanol hydroxyl groups. *J. Comput. Chem.* **1997**, *18*, 757–774. [CrossRef]
41. Grabowski, S.J. Triel bonds-complexes of boron and aluminum trihalides and trihydrides with benzene. *Struct. Chem.* **2017**, *28*, 1163–1171. [CrossRef]
42. Boys, S.F.; Bernardi, F. The calculation of small molecular interactions by the differences of separate total energies. Some procedures with reduced errors. *Mol. Phys.* **1970**, *19*, 553–566. [CrossRef]
43. Frisch, M.J.; Trucks, G.W.; Schlegel, H.B.; Scuseria, G.E.; Robb, M.A.; Cheeseman, J.R.; Scalmani, G.; Barone, V.; Mennucci, B.; Petersson, G.A.; et al. *Gaussian 09, RevisionD.01*; Gaussian Inc.: Wallingford, CT, USA, 2009.

44. Bulat, F.A.; Toro-Labbé, A.; Brinck, T.; Murray, J.S.; Politzer, P. Quantitative analysis of molecular surfaces: Areas, volumes, electrostatic potentials and average local ionization energies. *J. Mol. Model.* **2010**, *16*, 1679–1691. [CrossRef] [PubMed]
45. Lu, T.; Chen, F. Multiwfn: A multifunctional wavefunction analyzer. *J. Comput. Chem.* **2012**, *33*, 580–592. [CrossRef]
46. Reed, A.E.; Curtiss, L.A.; Weinhold, F. Intermolecular interactions from a natural bond orbital, donor-acceptor viewpoint. *Chem. Rev.* **1988**, *88*, 899–926. [CrossRef]
47. Johnson, E.R.; Keinan, S.; Mori-Sánchez, P.; Contreras-García, J.; Cohen, A.J.; Yang, W. Revealing noncovalent interactions. *J. Am. Chem. Soc.* **2010**, *132*, 6498–6506. [CrossRef]
48. Humphrey, W.; Dalke, A.; Schulten, K. VMD: Visual molecular dynamics. *J. Mol. Graph.* **1996**, *14*, 33–38. [CrossRef]
49. Schmidt, M.W.; Baldridge, K.K.; Boatz, J.A.; Elbert, S.T.; Gordon, M.S.; Jensen, J.H.; Koseki, S.; Matsunaga, N.; Nguyen, K.A.; Su, S. General atomic and molecular electronic structure system. *J. Comput. Chem.* **1993**, *14*, 1347–1363. [CrossRef]
50. Su, P.; Li, H. Energy decomposition analysis of covalent bonds and intermolecular interactions. *J. Chem. Phys.* **2009**, *131*, 014102. [CrossRef]

Article

Abnormalities of the Halogen Bonds in the Complexes between Y_2CTe (Y = H, F, CH_3) and XF (X = F, Cl, Br, I)

Ya-Qian Wang [1], Rui-Jing Wang [2], Qing-Zhong Li [2,*] and Zhi-Wu Yu [1,*]

1   MOE Key Laboratory of Bioorganic Phosphorous Chemistry and Chemical Biology, Department of Chemistry, Tsinghua University, Beijing 100084, China
2   The Laboratory of Theoretical and Computational Chemistry, School of Chemistry and Chemical Engineering, Yantai University, Yantai 264005, China
*   Correspondence: lqz@ytu.edu.cn (Q.-Z.L.); yuzhw@tsinghua.edu.cn (Z.-W.Y.)

Abstract: In this work, the hydrogen bonds and halogen bonds in the complexes between Y_2CTe (Y = H, F, CH_3) and XF (X = F, Cl, Br, I) have been studied by quantum chemical calculations. We found three interesting abnormalities regarding the interactions. Firstly, the strength of halogen bonds increases in the order of IF < BrF < ClF < F_2. Secondly, the halogen bonds formed by F_2 are very strong, with an interaction energy in the range between −199.8 and −233.1 kJ/mol. Thirdly, all the halogen bonds are stronger than the hydrogen bonds in the systems we examined. All these results are against the general understanding of halogen bonds. These apparent abnormal properties are reconciled with the high polarizability of the Te atom and the strong inducing effect of F on the Te atom of Y_2CTe. These findings provide a new perspective on halogen bonds. Additionally, we also proposed bonding distance-based methods to compare the strength of halogen/hydrogen bonds formed between different donor atoms and the same acceptor atom.

Keywords: halogen bond; hydrogen bond; abnormality; competition; AIM; NBO

Citation: Wang, Y.-Q.; Wang, R.-J.; Li, Q.-Z.; Yu, Z.-W. Abnormalities of the Halogen Bonds in the Complexes between Y_2CTe (Y = H, F, CH_3) and XF (X = F, Cl, Br, I). *Molecules* **2022**, *27*, 8523. https://doi.org/10.3390/molecules27238523

Academic Editor: Victor Mamane

Received: 1 November 2022
Accepted: 30 November 2022
Published: 3 December 2022

Publisher's Note: MDPI stays neutral with regard to jurisdictional claims in published maps and institutional affiliations.

1. Introduction

A halogen bond is formed between an electrophilic region of a halogen atom X (X = F, Cl, Br, and I) in a molecule R-X (R is an electron-withdrawing atom/group) and a nucleophilic region of a molecule Y-R′ [1], denoted as R-X···Y-R′. The electrophilic region, or the electron-deficient region of the X atom, is located along the R-X σ-bond, denoted as σ-hole, which is surrounded by a belt of negative electrostatic potential [2]. Nowadays, halogen bonds have received extensive attention due to their important roles in many fields such as supramolecular chemistry, organocatalysis, synthetic coordination chemistry, polymer chemistry, and drug discovery [3–13]. For example, halogen bonding has been a popular and much exploited supramolecular synthon in the crystal field [5,9]. The application of halogen atoms as pharmaceutically active ligand substituents has been widespread in recent medicinal chemistry [10,11].

The properties of halogen bonds are related to their strength, which is not only dependent on the halogen donor atom and the acceptor atom, but is also affected by substituents. Normally, the halogen bond becomes stronger with the halogen donor varying from F to I [1,12–16]. An electron-donating group in the halogen bond acceptor strengthens the halogen bond, while an electron-withdrawing group in the acceptor has a weakening effect [16]. The type of the halogen bond acceptor varies from anions and neutral molecules with lone pairs to π-electron molecules, radicals, metal hydrides, and carbenes [17,18]. Specially, the molecules containing N and O atoms are often taken as the halogen bond acceptor.

It is interesting to study the differences between hydrogen bonds and halogen bonds, since both types of interactions have comparable strength and may coexist in the same systems [19–27]. Usually, hydrogen bonds are stronger than halogen bonds, except for when an iodine atom acts as the halogen donor [24,28]. Thus, some studies have tried to make halogen bonds stronger than hydrogen bonds [28–30]. When the halogen bond acceptor H_2CO binds with the hydrogen/halogen donor HOBr, the interaction energy of the hydrogen bond is larger by 7 kJ/mol than that of the halogen bond [19]. Inversely, the interaction energy of the halogen bond is larger by 1 kJ/mol than that of the hydrogen bond if H_2CO is changed to H_2CS [28]. This difference is enlarged to 2 kJ/mol when one H atom of H_2CS is replaced by a Li atom [28]. These results indicate that the differences between hydrogen bonds and halogen bonds can be regulated by changing the halogen bond acceptor atoms and/or adding substituents. Nevertheless, these comparisons are not very convincing, because, for HOBr to participate in hydrogen bonding and halogen bonding interactions, the remaining moieties are -OBr and -OH, respectively, meaning that they are not identical. To overcome this difficulty, Li and coauthors designed a molecule called 6-OX-fulvene (X = H, Cl, Br, I), where the moiety of fulvene increases the acidity of the X atom. Then, they examined the interactions between this molecule and ZH_3/H_2Y (Z = N, P, As, and Sb; Y = O, S, Se, and Te) [31]. It was found that the hydrogen bond is weakened with the Lewis base atom growing in size; however, the effect of the same on halogen bonds is very limited [31]. If SbH_3 and H_2Te are selected as the acceptors, the halogen bonds are much stronger than the hydrogen bonds, and the largest difference in their interaction energies is 40 kJ/mol in the $SbH_3\cdots$6-OCl-fulvene complex [31].

H_2CTe is a homologue of H_2CO and H_2CS; thus, it can also work as an acceptor to form hydrogen bonds or halogen bonds. Considering that Te is a semimetal located on the dividing line between metals and non-metals, we expect that the halogen bond formed by it may have different patterns. In this study, we investigated the complexes between Y_2CTe (Y = H, F, and CH_3) and XF (X = H, F, Cl, Br, and I), wherein XF is a hydrogen/halogen bond donor and Y_2CTe is an acceptor. With the strong electronegativity of F, the designated molecules XF are expected to be prominent halogen bond donors. The following questions are addressed by the method of quantum chemical calculations: (1) Whether the halogen bond is stronger than the hydrogen bond. (2) Whether the strength of halogen bond follows the order of $F_2 <$ ClF < BrF < IF. (3) What is the nature of the hydrogen bond and halogen bond in these complexes?

2. Results

2.1. Molecular Electrostatic Potential (MEP) Analyses

It is well known that the MEP diagram of a molecule is helpful to effectively predict noncovalent interactions involving that molecule [32]. Figure 1 shows the MEP maps of two families of molecules: Y_2CTe (Y = F, H, and CH_3) and XF (X = H, F, Cl, Br, and I). The MEP distributions in both families are anisotropic. For Y_2CTe, we focus on the negative areas of the MEPs (blue colored areas). As expected, there are mainly two negative areas in each molecule, which correspond to the lone pairs of the Te atom. Compared with H_2CTe (−78.8 kJ/mol), the minimal MEP value of the Te atom decreases in F_2CTe (−52.5 kJ/mol) but increases in $(CH_3)_2CTe$ (−99.8 kJ/mol), which can be attributed to the electron-withdrawing nature of F atoms and the electron-donating ability of the methyl groups, respectively.

For XF, we focus on the positive areas of the MEPs (red areas). In the case of HF, the atom H exhibits positive electrostatic potential, while F is negative. For dihalogen molecules XF, there is a positive MEP region (σ-hole) at the X atom along the X-F bond. The magnitude of the σ-hole on the halogen atom increases with an increasing atomic mass of X. It is also found that the maximal MEP on the H atom is larger than that on the halogen atoms, including iodine.

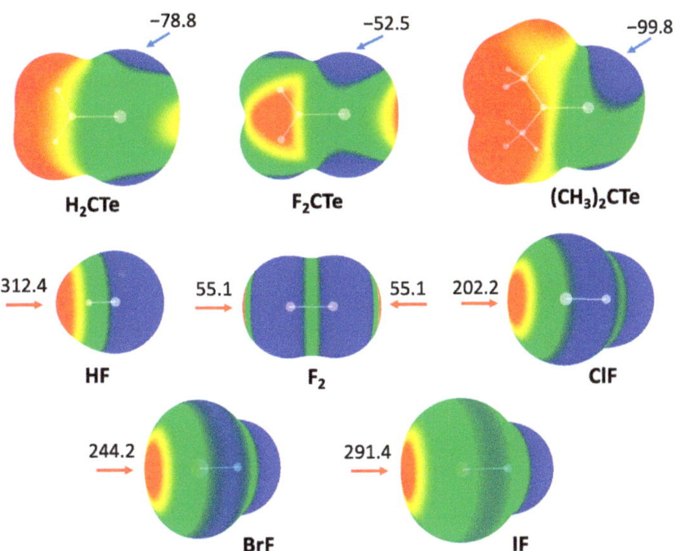

Figure 1. MEP diagrams of molecules studied in this work. Color ranges, in kJ/mol: red, greater than 52.5; yellow, between 52.5 and 0; green, between 0 and −52.5; blue, less than −52.5. Arrows refer to values of maxima and minima.

2.2. Geometries

For the hydrogen bonding or halogen bonding interactions ($Y_2CTe \cdots XF$), the general geometry of the complexes and the involved parameters are shown in Figure 2. We focus on the $Te \cdots X$ distance (R_1), the change in the X-F bond length (ΔR_2), and the $Te \cdots X$-F angle (α). The data of the optimized structures are listed in Table 1.

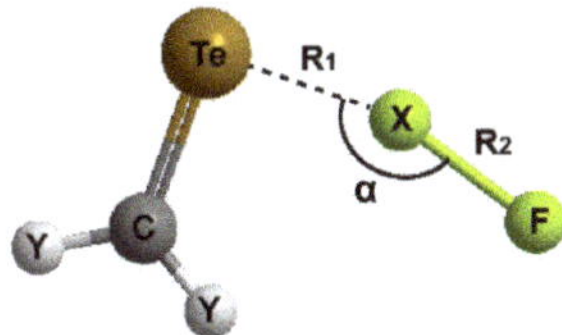

Figure 2. Illustration of the general structure of $Y_2CTe \cdots XF$ complex.

As can be seen in the table, all the values of R_1 are much shorter than the sum of the van der Waals radii of the respective atoms (3.3 Å for Te and H, 3.6 Å for Te and F, 4.0 Å for Te and Cl, 4.2 Å for Te and Br, and 4.4 Å for Te and I) [33,34]. This justifies the formation of hydrogen/halogen bonds. Further, the interactions between the electron-donor and acceptor molecules seem to be quite strong because stronger interaction is known to result in shorter bond length (R_1). To compare the relative strength between the halogen bonds between different interaction partners and the hydrogen bond, we define a quantity $\Delta R_1\%$ in the following equation:

$$\Delta R_1\% = \frac{R_c - R_1}{R_c} \times 100\% \tag{1}$$

where R_c is the sum of the van der Waals radii of the two atoms representing the critical distance to judge the presence of a hydrogen/halogen bond. After normalization with R_c, the shortening of the $Te \cdots X$ distance could be used to evaluate the strength of hydrogen/halogen bonds. Thus, for each of the three molecules (H_2CTe, F_2CTe, and $(CH_3)_2CTe$), the $\Delta R_1\%$ are all in the sequential order $F_2 > ClF > BrF > IF > HF$ when they form interaction pairs. This implies that all the halogen bonds are stronger than the hydrogen bonds. Most interestingly, the $\Delta R_1\%$ values suggest that the halogen bond strength decreases with an increasing size of the halogen atom in the donor molecule XF. This is different from the general understanding of halogen bonds.

The change in the X-F bond length R_2 of a donor could also reflect the interaction strength of the hydrogen/halogen bond. Here, we calculated the change of R_2 relative to the R_2 in the monomer, denoted as $\Delta R_2\%$, using the following formula:

$$\Delta R_2\% = \frac{\Delta R_2}{R_2} \times 100\% \tag{2}$$

The $\Delta R_2\%$ represents the elongation percentage of the X-F bond, and the larger $\Delta R_2\%$ implies more significant weakening of the bond and, thus, stronger interaction. As indicated by $\Delta R_2\%$, the value of X-F bond length is larger in the halogen-bonded complex than that in the hydrogen-bonded analogue. This relative elongation in the halogen-bonded complex decreases in the order of $F_2 >$ ClF $>$ BrF $>$ IF. These data are supportive of the conclusions from $\Delta R_1\%$.

The Te$\cdots$X-F angle (α) is in the range of 168–180°, confirming a good direction of the hydrogen/halogen bonds. The angles are less than 180° in the majority of the complexs due to the attraction between the Y atom/group in Y_2CTe and XF.

Table 1. Binding distance (R_1, Å), $\Delta R_1\%$; elongation of the X-F bond length (ΔR_2, Å), $\Delta R_2\%$; bond angle (α, deg) of the complexes.

	R_1	$\Delta R_1\%$	ΔR_2	$\Delta R_2\%$	α
H_2CTe$\cdots$HF	2.518	23.70%	0.015	1.63%	168.6
H_2CTe$\cdots$F$_2$	2.151	40.25%	0.359	25.62%	168.8
H_2CTe$\cdots$ClF	2.528	36.80%	0.201	12.26%	176.8
H_2CTe$\cdots$BrF	2.686	36.05%	0.140	7.96%	177.4
H_2CTe$\cdots$IF	2.904	34.00%	0.089	4.64%	177.9
F_2CTe$\cdots$HF	2.586	21.64%	0.010	1.08%	179.9
F_2CTe$\cdots$F$_2$	2.140	40.56%	0.371	26.48%	170.7
F_2CTe$\cdots$ClF	2.647	33.83%	0.137	8.36%	179.7
F_2CTe$\cdots$BrF	2.790	33.57%	0.098	5.57%	179.8
F_2CTe$\cdots$IF	2.996	31.91%	0.064	3.33%	179.2
$(CH_3)_2$CTe$\cdots$HF	2.483	24.76%	0.018	1.95%	170.7
$(CH_3)_2$CTe$\cdots$F$_2$	2.176	39.56%	0.343	24.48%	168.0
$(CH_3)_2$CTe$\cdots$ClF	2.518	37.05%	0.222	13.54%	178.1
$(CH_3)_2$CTe$\cdots$BrF	2.673	36.36%	0.158	8.99%	179.1
$(CH_3)_2$CTe$\cdots$IF	2.893	34.25%	0.101	5.26%	180.0

2.3. Energies

Here, we consider the interaction energy to be the most credible criteria to judge the strength of interactions. Therefore, we calculated the interaction energies (E_{int}) of the various complexes for comparing the hydrogen and the halogen bonds. We used the counterpoise correction method to eliminate the basis set superposition error (BSSE), and the corrected energy is denoted as $E_{int,BSSE}$. In addition, the more accurate energy $E_{int,CBS,BSSE}$ with complete basis set (CBS) was also calculated. The results with and without BSSE correction, as well as with CBS, are all listed in Table 2. The main concern of our study is that the changing trends of the interaction energy with the variation of X in XF are the same based on all the three methods. It is worth clarifying that the following discussions about energies in the full text are all according to their absolute values. As shown in Table 2, the interaction energies of hydrogen bonds in all of the three series of complexes are smaller than those of the halogen bonds, indicating that the hydrogen bonds are weaker than all of the halogen bonds. For the strength order of the halogen bonds, both $E_{int,BSSE}$ and $E_{int,CBS,BSSE}$ increased in the order of IF $<$ BrF $<$ ClF $<$ F$_2$ for the series of H_2CTe$\cdots$XF and

$(CH_3)_2CTe\cdots XF$ complexes. This result is abnormal compared to the common perception that the halogen bond becomes stronger with the halogen donor varying from F to I. For $F_2CTe\cdots XF$ complexes, the $E_{int,BSSE}/E_{int,CBS,BSSE}$ of ClF, BrF, and IF was close. For all of the three series of the complexes, F_2 molecules formed the strongest halogen bonds. The absolute values of the interaction energy were very large, up to 228.8 kJ/mol for $E_{int,BSSE}$ and 233.1 kJ/mol for $E_{int,CBS,BSSE}$. To compare the interaction energies of the halogen bonds formed by different acceptors (Y_2CTe), when Y is F, an electron-withdrawing atom, the $Y_2CTe\cdots XF$ interaction, was weakened and compared to that of H_2CTe. On the contrary, when Y was the electron-donating methyl group, the interaction was strengthened.

Table 2. Interaction energies (E_{int}) corrected with and without BSSE in the complexes at the MP2/aug-cc-pVTZ(PP) level, all in kJ/mol.

		HF	**F_2**	**ClF**	**BrF**	**IF**
H_2CTe	E_{int}	-27.8	-232.6	-107.0	-100.0	-94.8
	$E_{int,BSSE}$	-21.1	-220.7	-96.7	-87.0	-81.5
	$E_{int,CBS,BSSE}$	-22.3	-225.5	-106.5	-95.6	-89.9
F_2CTe	E_{int}	-21.1	-206.9	-61.2	-66.4	-68.8
	$E_{int,BSSE}$	-14.5	-194.4	-52.7	-54.8	-56.4
	$E_{int,CBS,BSSE}$	-15.6	-199.8	-60.5	-61.7	-63.6
$(CH_3)_2CTe$	E_{int}	-33.7	-242.2	-128.1	-120.0	-111.8
	$E_{int,BSSE}$	-26.3	-228.8	-116.4	-103.6	-94.7
	$E_{int,CBS,BSSE}$	-27.7	-233.1	-126.5	-112.5	-103.3

To understand the attribution of the interaction energy, we partitioned it into five terms: electrostatic energy (E^{es}), exchange energy (E^{ex}), repulsion energy (E^{rep}), polarization energy (E^{pol}), and dispersion energy (E^{disp}), and the data are listed in Table 3. Obviously, E^{ex} is the largest attractive term in each complex; thus, it plays the most important role in the stabilization of hydrogen/halogen bonds [35,36]. This term increases in the order of HF < IF < BrF < ClF < F_2, which is consistent with the results of orbital interaction discussed in the following section. The large E^{ex} of each complex suggests a strong orbital interaction between the two respective monomers. For the repulsive term, E^{rep} was very large and even exceeds 1000 kJ/mol in each F_2-related complex. This can be attributed to the much shorter $Te\cdots X$ distances. It is seen that E^{rep} was almost twice as much as E^{ex} and both terms have a good linear relationship (Figure S1), confirming their dependency each other.

Table 3. Electrostatic (E^{es}), exchange energy (E^{ex}), repulsion energy (E^{rep}), polarization (E^{pol}), and dispersion (E^{disp}) energies in the complexes, all in kJ/mol.

	E^{es}	E^{ex}	E^{rep}	E^{pol}	E^{disp}
$H_2CTe\cdots HF$	-33.0	-48.9	90.4	-21.3	-8.3
$H_2CTe\cdots F_2$	-246.0	-519.1	1097.3	-442.5	-110.3
$H_2CTe\cdots ClF$	-255.5	-474.9	985.5	-278.4	-73.4
$H_2CTe\cdots BrF$	-251.3	-417.7	865.0	-223.0	-60.1
$H_2CTe\cdots IF$	-183.2	-346.0	678.3	-182.8	-47.8
$F_2CTe\cdots HF$	-22.8	-38.4	70.8	-16.7	-7.4
$F_2CTe\cdots F_2$	-242.6	-524.2	1109.5	-423.0	-114.1
$F_2CTe\cdots ClF$	-170.5	-344.4	695.7	-169.2	-63.9
$F_2CTe\cdots BrF$	-171.7	-312.3	630.0	-148.2	-52.6
$F_2CTe\cdots IF$	-129.4	-265.6	510.3	-129.3	-42.3
$(CH_3)_2CTe\cdots HF$	-46.2	-64.5	118.7	-26.2	-8.1
$(CH_3)_2CTe\cdots F_2$	-253.4	-533.8	1116.9	-457.1	-101.5
$(CH_3)_2CTe\cdots ClF$	-287.7	-522.2	1080.9	-315.4	-72.1
$(CH_3)_2CTe\cdots BrF$	-287.6	-468.2	966.8	-253.4	-61.2
$(CH_3)_2CTe\cdots IF$	-211.6	-391.0	763.3	-205.4	-50.0

Now, we examine the three attractive terms (E^{es}, E^{pol}, and E^{disp}) in Table 3 in some detail (intuitively in Figure S2). For the hydrogen bond complex $Y_2CTe\cdots HF$, E^{es} was the largest attractive interaction among the three terms, followed by E^{pol}. For the interaction energies of halogen bonds formed by ClF, BrF, or IF with all the three acceptors Y_2CTe, the contributions of electrostatic and polarization interactions are comparable. However, for the interaction energies of the halogen bonds formed by F_2 with the Y_2CTe, the polarization interaction is the dominating contribution. This may be attributed to the special property of F, namely the largest electronegativity in the periodic table, thus possessing a very strong inducing ability.

2.4. Atoms in Molecules (AIM) Analyses

The hydrogen/halogen bonds can be characterized by the $Te\cdots X$ bond critical points (BCPs, Figure S3). The most important properties of each bond critical point are summarized in Table 4, where ρ refers to the electron density, $\nabla^2\rho$ its Laplacian, and H the energy density [37–39]. Generally, the larger electron density ρ reflects the stronger interaction. For all the investigated systems, ρ increases in the sequential order of HF < IF < BrF < ClF < F_2, in agreement with the order of interaction energies (Table 2). For the Laplacian, it was seen that $\nabla^2\rho > 0$ for all the complexes, demonstrating that the interactions studied were closed shell interaction. The energy density H is a more sensitive parameter than $\nabla^2\rho$. The negative values of H further demonstrate that the interactions are partially covalent in nature. In the complexes involving $(CH_3)_2CTe$, there are also BCPs between the methyl H and the halogen X in HX or XFs (Figure S3), indicating the presence of C-H$\cdots$F/X hydrogen bonds. The interaction energies of the C-H$\cdots$F/X hydrogen bonds were estimated by E = 0.5V(r) [40,41], where V(r) is the potential energy density at a BCP in each case. The corresponding data are -5.6, -17.8, -10.3, -9.9, and -8.1 kJ/mol for $(CH_3)_2CTe\cdots HF$, $(CH_3)_2CTe\cdots F_2$, $(CH_3)_2CTe\cdots ClF$, $(CH_3)_2CTe\cdots BrF$, and $(CH_3)_2CTe\cdots IF$, respectively. Clearly, these hydrogen bonds contributed to the stability of the complexes; however, their shares in the total interaction energies (Table 2) are small. Subtracting them from the total interaction energies, the residual results still have the same change trend with the total interaction energy. Thus, the presence of C-H$\cdots$F/X hydrogen bonds does not affect the abnormality of halogen bonds.

Table 4. Electron density (ρ), Laplacian ($\nabla^2\rho$), and total energy density (H) at the intermolecular BCP in the complexes (all values in a.u.).

	ρ	$\nabla^2\rho$	H
$H_2CTe\cdots HF$	0.022	0.032	-0.003
$H_2CTe\cdots F_2$	0.095	0.101	-0.036
$H_2CTe\cdots ClF$	0.079	0.024	-0.027
$H_2CTe\cdots BrF$	0.066	0.033	-0.020
$H_2CTe\cdots IF$	0.052	0.041	-0.013
$F_2CTe\cdots HF$	0.019	0.032	-0.001
$F_2CTe\cdots F_2$	0.096	0.112	-0.036
$F_2CTe\cdots ClF$	0.060	0.058	-0.015
$F_2CTe\cdots BrF$	0.052	0.053	-0.012
$F_2CTe\cdots IF$	0.042	0.049	-0.008
$(CH_3)_2CTe\cdots HF$	0.025	0.031	-0.004
$(CH_3)_2CTe\cdots F_2$	0.091	0.088	-0.033
$(CH_3)_2CTe\cdots ClF$	0.082	0.013	-0.029
$(CH_3)_2CTe\cdots BrF$	0.069	0.024	-0.022
$(CH_3)_2CTe\cdots IF$	0.054	0.036	-0.015

2.5. Natural Bond Orbital (NBO) Analyses

The charge transfers (CTs) from Y_2CTe to XF are listed in Table 5, which are calculated as the sum of the charge on all atoms in Y_2CTe in the complexes. The charge transfer reflects the electrons bias from electron donor (Y_2CTe) to electron acceptor (XF), providing information about the interaction strength in one aspect. As can be seen, the CTs are all larger than 0.2 e in the halogen bonds and even exceed 0.8 e in the F_2 complexes. On the contrary, the hydrogen-bonded complexes have much smaller CTs than the halogen-bonded analogues. Additionally, for all three of the series of complexes, the CT value decreases in the order F_2 > ClF > BrF > IF > HF, which is the same as the interaction strength order. Besides, when the Y is the electron-withdrawing atom F in Y_2CTe, the CT becomes smaller than that in H_2CTe. When the Y is the electron-donating methyl group, the CT increases.

There is an orbital interaction between the lone pair orbital on the Te atom of Y_2CTe and the anti-bonding orbital of the X-F bond ($Lp_{Te} \rightarrow \sigma^*_{X-F}$), and this orbital interaction can be measured with the second-order perturbation energy (E^2), which is also listed in Table 5. This orbital interaction is not detected in the F_2-containing complexes since the F-F bond is taken as two subunits in the NBO analysis. The E^2 has a consistent change order with the charge transfer. This orbital interaction is strong; therefore, it makes an important contribution to the formation of hydrogen/halogen bond. We also calculated the dipole moments of the complexes (Table 5). It was found that the order of the dipole moment is consistent with the interaction energy. Further, the relationship between the CTs and the population of the σ^*_{X-F} orbital was analyzed, and positive correlation was found (Figure S4). This suggests that the main destination/receiver of the CT is the σ^*_{X-F} orbital in each complex.

Table 5. Charge transfer (CT, e), second-order perturbation energy (E^2, kJ/mol), and dipole moment (μ, D) in the complexes.

	CT	E^2	μ
$H_2CTe\cdots HF$	0.043	70.0	3.17
$H_2CTe\cdots F_2$	0.893	-	10.52
$H_2CTe\cdots ClF$	0.511	781.6	8.23
$H_2CTe\cdots BrF$	0.414	761.4	7.87
$H_2CTe\cdots IF$	0.322	636.6	7.81
$F_2CTe\cdots HF$	0.031	51.0	2.88
$F_2CTe\cdots F_2$	0.843	-	9.53
$F_2CTe\cdots ClF$	0.334	462.3	5.62
$F_2CTe\cdots BrF$	0.293	417.7	5.93
$F_2CTe\cdots IF$	0.242	334.3	6.34
$(CH_3)_2CTe\cdots HF$	0.056	93.3	4.76
$(CH_3)_2CTe\cdots F_2$	0.943	-	11.73
$(CH_3)_2CTe\cdots ClF$	0.579	711.7	10.37
$(CH_3)_2CTe\cdots BrF$	0.462	696.9	9.96
$(CH_3)_2CTe\cdots IF$	0.350	613.1	9.84

3. Discussion

Hydrogen bonds and halogen bonds are two important noncovalent interactions, and they often coexist; thus, it is interesting to compare their strength. Generally speaking, hydrogen bonds are considered to be stronger than halogen bonds [28]. Interestingly, in the present study, using Y_2CTe (Y = H, F, and CH_3) as electron donors, we found that their halogen bonding interactions with dihalogen molecules XF (X = F, Cl, Br, and I) are stronger than their hydrogen bonding interactions with HF. This apparent abnormality was also seen in a previous study on the interactions between $PH_3/AsH_3/H_2Te$ and 6-OX-fulvene (X = H, Cl, Br, I) [31]. The second abnormality found in this work was that, when the X changes from F to I, the halogen bond becomes weaker, in contrast to the normal understanding that stronger halogen bonds accompany heavier halogen donors [16]. The abnormalities can be explained by the high polarizability of the Te atom, which is easily polarized when the electronegative XF approaches it. The greater the electronegativity of the approaching atom, the greater the polarization of the Te atom. Therefore, when the X atom of XF varies from I to F, the dipole moment of the complex increases, as seen in Table 5, and the polarization energy (the major contribution to the interaction energy) also increases. Based on the data in Tables 3 and 5, a near positive correlation between the polarization energy and the dipole moment of the complex is found (Figure S5).

Another surprising result was that F_2 participates in the strongest halogen bond with the interaction energy up to -233.1 kJ/mol in the $(CH_3)_2CTe\cdots F_2$ complex. Such large interaction energy is abnormal because it shows the least MEP at the end of the X atom among the four XF molecules. The apparent contradiction can be reconciled as follows. The F atom of F_2, due to it having the highest electronegativity among the halogens, would cause the largest polarization on the Te atom and, thus, the largest dipole moment of the $Y_2CTe\cdots XF$ complex. This is evidenced by the largest polarization energies being seen in the three $Y_2CTe\cdots F_2$ complexes. The polarization mechanism is also consistent with the charge transfer data, which are the biggest in the $Y_2CTe\cdots F_2$ complexes, even as big as 0.943e in the complex of $(CH_3)_2CTe\cdots F_2$. Such big charge transfers (>0.8e) mean that the molecule F_2 holds nearly an extra electron in each of the three complexes, similar to the process of becoming an anion.

In order to test if the above abnormal results are found only for Y_2CTe, the Te atom of H_2CTe was replaced by O, S, and Se. The corresponding interaction energies are listed in Table 6. For the lighter chalcogen atoms O and S, the halogen bond strength increases in the order of $F_2 <$ ClF $<$ BrF $<$ IF, which is the "normal" understanding of halogen bonds. For the heavier Se as the electron donor, the halogen bonds formed by ClF, BrF, and IF turned out to be comparable, while that formed by F_2 increased tremendously. This complicated situation is explained as follows. On one hand, a Se atom with a larger radius is more easily polarized than O and S. On the other hand, it is not as easily polarized as Te. Thus, only the most electronegative F_2 is able to assert marked influence on the electron distribution of H_2CSe, making it the strongest interaction in the $H_2CSe\cdots XF$ complexes.

Table 6. Interaction energies ($E_{int,BSSE}$, kJ/mol) of complexes between H_2CZ (Z = O, S, Se, and Te) and XF (X = H, F, Cl, Br, and I) at the MP2/aug-cc-pVTZ(PP) level.

	HF	F_2	ClF	BrF	IF
H_2CO	-34.4	-5.9	-25.2	-34.4	-42.3
H_2CS	-26.6	-7.3	-52.2	-60.6	-64.0
H_2CSe	-24.6	-171.7	-64.7	-68.2	-69.6
H_2CTe	-21.1	-220.7	-96.7	-87.0	-81.5

As discussed above, the size/polarizability of the chalcogen atoms in H_2CZ (Z = O, S, Se, and Te) plays a very important role in the strength of halogen bonds. The data in Table 6 demonstrate that, for a given XF (X = F, Cl, Br, and I), the strength of its halogen bond with H_2CZ increases monotonously when Z varies from O to Te. On the contrary, the strength of the hydrogen bond between HF and H_2CZ decreases monotonously. As a result, for H_2CO, only IF forms a stronger halogen bond than the hydrogen bond. For H_2CS, each dihalogen molecule, excluding F_2, participates in a stronger halogen bond than the hydrogen bond formed by HF. For H_2CSe and H_2CTe, all the dihalogen molecules form stronger halogen bonds than hydrogen bonds formed by HF.

The above discussions about the interaction strength of halogen/hydrogen bonds are based on their interaction energies, often regarded as golden criteria. Practically, other parameters such as electrostatic potential (MEP), halogen/hydrogen bond length (R_1), and donor bond length (R_2) are often used to compare the interaction strength. MEP sometimes provides correct indications. For example, the MEP order is Cl < Br < I at the end of X atom in XF, and the halogen bond strength order is ClF < BrF < IF in the $F_2CTe\cdots XF$ complexes. However, the F has the least MEP at the end of the X atom among the four XF molecules, but it forms the strongest interaction with F_2CTe. Clearly, MEP cannot always provide correct results because it only contains the information of isolated molecules. For R_1 and R_2, due to the different radii of the halogen atoms, we defined two quantities, $\Delta R_1\%$ and

$\Delta R_2\%$, to compare the halogen/hydrogen bond strength. The correlations between the two quantities and the respective interaction energies are plotted in Figure 3 (the trend comparisons are shown in Figure S6). Undoubtedly, they are all positively correlated, suggesting that both $\Delta R_1\%$ and $\Delta R_2\%$ can be used to compare the halogen/hydrogen bond strength qualitatively. Quantitatively, $\Delta R_2\%$ is a better choice.

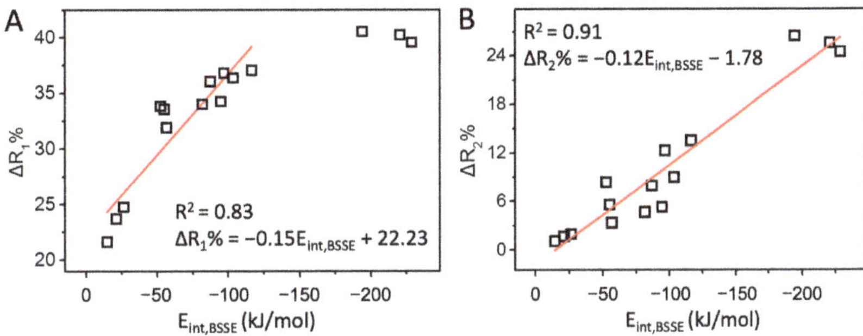

Figure 3. The correlation of $E_{int,BSSE}$ with $\Delta R_1\%$ (**A**) and $\Delta R_2\%$ (**B**).

4. Computational Methods

All calculations were carried out with the Gaussian 09 program [42]. The geometries of all the monomers and complexes were optimized at the MP2 level with the aug-cc-pVTZ basis set for all atoms except I and Te, where the aug-cc-pVTZ-PP basis set was adopted to account for relativistic corrections [43]. For all atoms in all the complexes, collectively, aug-cc-pVTZ(PP) represents the basis set used in this work. The extrema of molecular electrostatic potentials (MEPs) were calculated on the 0.001 a.u. isodensity surface at the MP2/aug-cc-pVTZ(PP) level using the WFA-SAS program [32]. The interaction energy (E_{int}) of a complex was computed as the difference between the energy of the complex and the sum of the energies of the monomers with their geometries frozen in the optimized complex. For this supermolecular method of calculating the MP2 interaction energy, the dispersion correction was taken into account, since MP2 contains certain correlation terms such as the uncoupled Hartree-Fock (UCHF) dispersion energy, the corresponding Hartree-Fock exchange-dispersion energy, and a deformation-correlation term [44]. Interaction energies were corrected for basis set superposition error (BSSE) using the counterpoise procedure (CP) proposed by Boys and Bernardi [45]. The two-point extrapolated energies with a complete basis set (CBS) proposed by Halkier et al. [46,47] were obtained with two basis sets of aug-cc-pVDZ(PP) and aug-cc-pVTZ(PP). The localized molecular orbital-energy decomposition analysis was used to decompose the interaction energy into five terms of electrostatic, exchange, repulsion, polarization, and dispersion at the MP2/aug-cc-pVTZ(PP) level with the GAMESS program [48]. The dispersion energy was obtained as a difference between the MP2 and CCSD(T) energy in the GAMESS program. The AIM2000 package [49] was used to assess the topological parameters at each bond critical point (BCP), including electron density, as well as its Laplacian, and energy density. Using the natural bond orbital (NBO) method [50] within the Gaussian 09 program, the charge transfer and second-order perturbation energy were obtained.

5. Conclusions

Quantum chemical calculations have been performed to study the interactions between Y_2CTe (Y = H, F, and CH_3) and XF (X = H, F, Cl, Br, and I). The results show that the electron-withdrawing groups F in F_2CTe weaken the interactions, while the electron-donating methyl groups in $(CH_3)_2CTe$ strengthen them. More importantly, we found three abnormalities regarding halogen bonds in this work. The first one is that the strength of halogen bond increases in the sequential order IF < BrF < ClF < F_2 in $H_2CTe\cdots XF$ and $(CH_3)_2CTe\cdots XF$ complexes. This is contrary to the normal understanding that the stronger halogen bonds accompany heavier halogen donors. The second one is that the halogen bonds formed by F_2 are very strong, even up to -233.1 kJ/mol with $(CH_3)_2CTe$. Contrary to this, the halogen bonds formed by F-R are normally considered to be very weak or even negligible. The last one is that all halogen bonds are stronger than the hydrogen bonds in the complexes, which is abnormal compared with the majority of previous studies. These abnormalities are discussed in the context of the high polarizability of the Te atom in the halogen acceptors. Because the Te atom is easily polarized when the electronegative XF approaches it, the greater the electronegativity of the X atom, the greater the polarization of the Te atom. Particularly, the F atom has the largest electronegativity in the periodic table and possesses a very strong inducing ability. Consequently, F-F forms tremendously strong interactions with Y_2CTe.

Supplementary Materials: The following supporting information can be downloaded at: https://www.mdpi.com/article/10.3390/molecules27238523/s1, Figure S1: The relationship between repulsion energy E^{rep} and exchange energy E^{ex} in the complexes between Y_2CTe (Y = H, F, and CH_3) and XF (X = H, F, Cl, Br, and I); Figure S2: Electrostatic (E^{es}), polarization (E^{pol}), and dispersion (E^{disp}) energies in complexes between Y_2CTe (Y = H, F, and CH_3) and XF (X = H, F, Cl, Br, and I); Figure S3: Molecular graphs of the complexes between Y_2CTe (Y = H, F, CH_3) and XF (X = H, F, Cl, Br, I). Small red balls indicate the Te$\cdots$X bond critical point; Figure S4: The relationship between the population of the $\sigma^*_{X\text{-}F}$ orbitals and the charge transfer in the complexes formed by Y_2CTe (Y = H, F, and CH_3) and XF (X = H, F, Cl, Br, and I); Figure S5: The relationship between the polarization energy E^{pol} and the dipole moment of the complexes formed by Y_2CTe (Y = H, F, and CH_3) and XF (X = F, Cl, Br, and I); Figure S6: Trend comparison of $E_{int,BSSE}$ with $\Delta R_1\%$ (A) and $\Delta R_2\%$ (B) in $H_2CTe\cdots XF$ systems; The coordinates of optimized monomer Y_2CTe (Y = H, F, CH_3), XF (X = H, F, Cl, Br, I), and their complexes.

Author Contributions: Data curation, Y.-Q.W. and R.-J.W.; Formal analysis, Y.-Q.W. and Q.-Z.L.; Supervision, Q.-Z.L. and Z.-W.Y.; Writing—original draft, Y.-Q.W. and Q.-Z.L.; Writing—review & editing, Z.-W.Y. All authors have read and agreed to the published version of the manuscript.

Funding: This work was supported by the National Natural Science Foundation of China (Nos. 21733011 and 22233006).

Institutional Review Board Statement: Not applicable.

Informed Consent Statement: Not applicable.

Data Availability Statement: Not applicable.

Conflicts of Interest: The author declares no conflict of interest.

Sample Availability: Samples of the compounds are not available from the authors.

References

1. Politzer, P.; Murray, J.S. Halogen Bonding: An Interim Discussion. *ChemPhysChem* **2013**, *14*, 278–294. [CrossRef] [PubMed]
2. Clark, T.; Hennemann, M.; Murray, J.S.; Politzer, P. Halogen Bonding: The Sigma-Hole. *J. Mol. Model.* **2007**, *13*, 291–296. [CrossRef] [PubMed]
3. Gilday, L.C.; Robinson, S.W.; Barendt, T.A.; Langton, M.J.; Mullaney, B.R.; Beer, P.D. Halogen Bonding in Supramolecular Chemistry. *Chem. Rev.* **2015**, *115*, 7118–7195. [CrossRef]
4. Bertani, R.; Metrangolo, P.; Moiana, A.; Perez, E.; Pilati, T.; Resnati, G.; Rico-Lattes, I.; Sassi, A. Supramolecular Route to Fluorinated Coatings: Self-Assembly Between Poly(4-vinylpyridines) and Haloperfluorocarbons. *Adv. Mater.* **2002**, *14*, 1197–1201. [CrossRef]
5. Fourmigue, M.; Batail, P. Activation of Hydrogen- and Halogen-Bonding Interactions in Tetrathiafulvalene-Based Crystalline Molecular Conductors. *Chem. Rev.* **2004**, *104*, 5379–5418. [CrossRef]
6. Jungbauer, S.H.; Walter, S.M.; Schindler, S.; Rout, L.; Kniep, F.; Huber, S.M. Activation of a Carbonyl Compound by Halogen Bonding. *Chem. Commun.* **2014**, *50*, 6281–6284. [CrossRef] [PubMed]
7. Libri, S.; Jasim, N.A.; Perutz, R.N.; Brammer, L. Metal Fluorides Form Strong Hydrogen Bonds and Halogen Bonds: Measuring Interaction Enthalpies and Entropies in Solution. *J. Am. Chem. Soc.* **2008**, *130*, 7842–7844. [CrossRef]
8. Mele, A.; Metrangolo, P.; Neukirch, H.; Pilati, T.; Resnati, G. A Halogen-Bonding-Based Heteroditopic Receptor for Alkali Metal Halides. *J. Am. Chem. Soc.* **2005**, *127*, 14972–14973. [CrossRef]

9. Adler, M.; Kochanny, M.J.; Ye, B.; Rumennik, G.; Light, D.R.; Biancalana, S.; Whitlow, M. Crystal Structures of Two Potent Nonamidine Inhibitors Bound to Factor Xa. *Biochemistry* **2002**, *41*, 15514–15523. [CrossRef]

10. Lu, Y.; Wang, Y.; Zhu, W. Nonbonding Interactions of Organic Halogens in Biological Systems: Implications for Drug Discovery and Biomolecular Design. *Phys. Chem. Chem. Phys.* **2010**, *12*, 4543–4551. [CrossRef]

11. Matter, H.; Nazaré, M.; Güssregen, S.; Will, D.W.; Schreuder, H.; Bauer, A.; Urmann, M.; Ritter, K.; Wagner, M.; Wehner, V. Evidence for C−Cl/C−Br···π Interactions as an Important Contribution to Protein–Ligand Binding Affinity. *Angew. Chem. Int. Ed.* **2009**, *48*, 2911–2916. [CrossRef] [PubMed]

12. Cavallo, G.; Metrangolo, P.; Milani, R.; Pilati, T.; Priimagi, A.; Resnati, G.; Terraneo, G. The Halogen Bond. *Chem. Rev.* **2016**, *116*, 2478–2601. [PubMed]

13. Erdelyi, M. Halogen Bonding in Solution. *Chem. Soc. Rev.* **2012**, *41*, 3547–3557. [CrossRef] [PubMed]

14. Dipaolo, T.; Sandorfy, C. On the Biological Importance of the Hydrogen Bond Breaking Potency of Fluorocarbons. *Chem. Phys. Lett.* **1974**, *26*, 466–473. [CrossRef]

15. Lu, Y.; Li, H.; Zhu, X.; Zhu, W.; Liu, H. How Does Halogen Bonding Behave in Solution? A Theoretical Study Using Implicit Solvation Model. *J. Phys. Chem. A* **2011**, *115*, 4467–4475. [CrossRef]

16. Politzer, P.; Murray, J.S.; Clark, T. Halogen bonding: An Electrostatically-Driven Highly Directional Noncovalent Interaction. *Phys. Chem. Chem. Phys.* **2010**, *12*, 7748–7757. [CrossRef]

17. Fourmigue, M. Halogen Bonding: Recent Advances. *Curr. Opin. Solid. St. M.* **2009**, *13*, 36–45. [CrossRef]

18. Tepper, R.; Schubert, U.S. Halogen Bonding in Solution: Anion Recognition, Templated Self-Assembly, and Organocatalysis. *Angew. Chem. Int. Ed.* **2018**, *57*, 6004–6016. [CrossRef]

19. Li, Q.Z.; Xu, X.S.; Liu, T.; Jing, B.; Li, W.Z.; Cheng, J.B.; Gong, B.A.; Sun, J.Z. Competition Between Hydrogen Bond and Halogen bond in Complexes of Formaldehyde with Hypohalous Acids. *Phys. Chem. Chem. Phys.* **2010**, *12*, 6837–6843. [CrossRef]

20. Politzer, P.; Murray, J.S.; Lane, P. σ-Hole Bonding and Hydrogen Bonding: Competitive Interactions. *Int. J. Quantum Chem.* **2007**, *107*, 3046–3052. [CrossRef]

21. Aakeroy, C.B.; Fasulo, M.; Schultheiss, N.; Desper, J.; Moore, C. Structural Competition between Hydrogen Bonds and Halogen Bonds. *J. Am. Chem. Soc.* **2007**, *129*, 13772–13773. [CrossRef] [PubMed]

22. Nagels, N.; Geboes, Y.; Pinter, B.; De Proft, F.; Herrebout, W.A. Tuning the Halogen/Hydrogen Bond Competition: A Spectroscopic and Conceptual DFT Study of Some Model Complexes Involving CHF$_2$I. *Chem. Eur. J.* **2014**, *20*, 8433–8443. [CrossRef] [PubMed]

23. An, X.L.; Yang, X.; Xiao, B.; Cheng, J.B.; Li, Q.Z. Comparison of Hydrogen and Halogen Bonds Between Dimethyl Sulfoxide and Hypohalous Acid: Competition and Cooperativity. *Mol. Phys.* **2017**, *115*, 1614–1623. [CrossRef]

24. An, X.L.; Zhuo, H.Y.; Wang, Y.Y.; Li, Q.Z. Competition Between Hydrogen Bonds and Halogen Bonds in Complexes of Formamidine and Hypohalous Acids. *J. Mol. Model.* **2013**, *19*, 4529–4535. [CrossRef] [PubMed]

25. Geboes, Y.; De Proft, F.; Herrebout, W.A. Towards a Better Understanding of the Parameters Determining the Competition Between Bromine Halogen Bonding and Hydrogen Bonding: An FTIR Spectroscopic Study of the Complexes Between Bromodifluoromethane and Trimethylamine. *J. Mol. Struct.* **2018**, *1165*, 349–355. [CrossRef]

26. Zheng, Y.Z.; Deng, G.; Zhou, Y.; Sun, H.Y.; Yu, Z.W. Comparative Study of Halogen- and Hydrogen-Bond Interactions between Benzene Derivatives and Dimethyl Sulfoxide. *ChemPhysChem* **2015**, *16*, 2594–2601. [CrossRef]

27. Zheng, Y.Z.; Wang, N.N.; Zhou, Y.; Yu, Z.W. Halogen-Bond and Hydrogen-Bond Interactions between Three Benzene Derivatives and Dimethyl Sulphoxide. *Phys. Chem. Chem. Phys.* **2014**, *16*, 6946–6956. [CrossRef]

28. Li, Q.Z.; Jing, B.; Li, R.; Liu, Z.B.; Li, W.Z.; Luan, F.; Cheng, J.B.; Gong, B.A.; Sun, J.Z. Some Measures for Making Halogen Bonds Stronger than Hydrogen Bonds in H$_2$CS-HOX (X = F, Cl, and Br) Complexes. *Phys. Chem. Chem. Phys.* **2011**, *13*, 2266–2271. [CrossRef]

29. Lv, H.; Zhuo, H.Y.; Li, Q.Z.; Yang, X.; Li, W.Z.; Cheng, J.B. Halogen Bonds with N-heterocyclic Carbenes as Halogen Acceptors: A Partially Covalent Character. *Mol. Phys.* **2015**, *112*, 3024–3032. [CrossRef]

30. Zhuo, H.Y.; Yu, H.; Li, Q.Z.; Li, W.Z.; Cheng, J.B. Some Measures for Mediating the Strengths of Halogen Bonds with the B-B Bond in Diborane(4) as an Unconventional Halogen Acceptor. *Int. J. Quantum Chem.* **2014**, *114*, 128–137. [CrossRef]

31. Hou, M.C.; Li, Q.Z.; Scheiner, S. Comparison between Hydrogen and Halogen Bonds in Complexes of 6-OX-Fulvene with Pnicogen and Chalcogen Electron Donors. *ChemPhysChem* **2019**, *20*, 1978–1984. [CrossRef] [PubMed]

32. Bulat, F.A.; Toro-Labbe, A.; Brinck, T.; Murray, J.S.; Politzer, P. Quantitative Analysis of Molecular Surfaces: Areas, Volumes, Electrostatic Potentials and Average Local Ionization Energies. *J. Mol. Model.* **2010**, *16*, 1679–1691. [CrossRef]

33. Pauling, L. *The Nature of the Chemical Bond*; Cornell University Press: Ithaca, NY, USA, 1960.

34. Pauling, L.; Pauling, P. *Chemistry*; W. H. Freeman Company: San Francisco, CA, USA, 1975.

35. Miranda, M.O.; Duarte, D.J.R. Halogen Bonds Stabilised by an Electronic Exchange Channel. *ChemistrySelect* **2021**, *6*, 680–684. [CrossRef]

36. Duarte, D.J.R.; Buralli, G.J.; Peruchena, N.M. Is σ-Hole an Electronic Exchange Channel in YX···CO Interactions? *Chem. Phys. Lett.* **2018**, *710*, 113–117. [CrossRef]

37. Koch, U.; Popelier, P.L.A. Characterization of C-H-O Hydrogen Bonds on the Basis of the Charge Density. *J. Phys. Chem.* **1995**, *99*, 9747–9754. [CrossRef]

38. Popelier, P.L.A. Characterization of a Dihydrogen Bond on the Basis of the Electron Density. *J. Phys. Chem. A* **1998**, *102*, 1873–1878. [CrossRef]

39. Arnold, W.D.; Oldfield, E. The Chemical Nature of Hydrogen Bonding in Proteins via NMR: J-Couplings, Chemical Shifts, and AIM Theory. *J. Am. Chem. Soc.* **2000**, *122*, 12835–12841. [CrossRef]
40. Espinosa, E.; Molins, E.; Lecomte, C. Hydrogen Bond Strengths Revealed by Topological Analyses of Experimentally Observed Electron Densities. *Chem. Phys. Lett.* **1998**, *285*, 170–173. [CrossRef]
41. Liu, N.; Li, Q.Z.; Scheiner, S.; Xie, X.Y. Resonance-Assisted Intramolecular Triel Bonds. *Phys. Chem. Chem. Phys.* **2022**, *24*, 15015–15024. [CrossRef]
42. Frisch, M.J.; Trucks, G.W.; Schlegel, H.B.; Scuseria, G.E.; Robb, M.A.; Cheeseman, J.R.; Scalmani, G.; Barone, V.; Mennucci, B.; Petersson, G.A.; et al. *Gaussian 09, Revision A.02*; Gaussian, Inc.: Wallingford, CT, USA, 2009.
43. Dunning, T.H. Gaussian Basis Sets for Use in Correlated Molecular Calculations. I. The Atoms Boron Through Neon and Hydrogen. *J. Chem. Phys.* **1989**, *90*, 1007–1023. [CrossRef]
44. Pitonak, M.; Hesselmann, A. Accurate Intermolecular Interaction Energies from a Combination of MP2 and TDDFT Response Theory. *J. Chem. Theory Comput.* **2010**, *6*, 168–178. [CrossRef]
45. Boys, S.F.; Bernardi, F. The Calculation of Small Molecular Interactions by the Differences of Separate Total Energies. Some Procedures with Reduced Errors. *Mol. Phys.* **1970**, *19*, 553–556. [CrossRef]
46. Halkier, A.; Helgaker, T.; Jorgensen, P.; Klopper, W.; Olsen, J. Basis-Set Convergence of the Energy in Molecular Hartree-Fock Calculations. *Chem. Phys. Lett.* **1999**, *302*, 437–446. [CrossRef]
47. Halkier, A.; Klopper, W.; Helgaker, T.; Jorgensen, P.; Taylor, P.R. Basis Set Convergence of the Interaction Energy of Hydrogen-Bonded Complexes. *J. Chem. Phys.* **1999**, *111*, 9157–9167. [CrossRef]
48. Su, P.F.; Li, H. Energy Decomposition Analysis of Covalent Bonds and Intermolecular Interactions. *J. Chem. Phys.* **2009**, *131*, 014102. [CrossRef] [PubMed]
49. Bader, R.F.W. *AIM2000 Program, v. 2.0*; McMaster University: Hamilton, ON, Canada, 2000.
50. Reed, A.E.; Curtiss, L.A.; Weinhold, F. Intermolecular Interactions from a Natural Bond Orbital, Donor-Acceptor Viewpoint. *Chem. Rev.* **1988**, *88*, 899–926. [CrossRef]

Article

Three for the Price of One: Concomitant I⋯N, I⋯O, and I⋯π Halogen Bonds in the Same Crystal Structure

Steven van Terwingen [1], Ruimin Wang [1,2] and Ulli Englert [1,2,*]

1 Institute of Inorganic Chemistry, RWTH Aachen University, Landoltweg 1, 52074 Aachen, Germany
2 Key Laboratory of Chemical Biology and Molecular Engineering of the Education Ministry, Shanxi University, 92 Wucheng Road, Taiyuan 030006, China
* Correspondence: ullrich.englert@ac.rwth-aachen.de; Tel.: +49-241-80-90064

Abstract: The ditopic molecule 3-(1,3,5-trimethyl-1*H*-4-pyrazolyl)pentane-2,4-dione (HacacMePz) combines two different Lewis basic sites. It forms a crystalline adduct with the popular halogen bond (XB) donor 2,3,5,6-tetrafluoro-1,4-diiodobenzene (TFDIB) with a HacacMePz:TFDIB ratio of 2:3. In a simplified picture, the topology of the adduct corresponds to a **hcb** net. In addition to the expected acetylacetone keto O and pyrazole N acceptor sites, a third and less common short contact to a TFDIB iodine is observed: The acceptor site is again the most electron-rich site of the pyrazole π-system. This iminic N atom is thus engaged as the acceptor in two orthogonal halogen bonds. Evaluation of the geometric results and of a single-point calculation agree with respect to the strength of the intermolecular contacts: The conventional N⋯I XB is the shortest (2.909(4) Å) and associated with the highest electron density (0.150 eÅ^{-3}) in the bond critical point (BCP), followed by the O⋯I contact (2.929(3) Å, 0.109 eÅ^{-3}), and the π contact (3.2157(3) Å, 0.075 eÅ^{-3}). If one accepts the idea of deducing interaction energies from energy densities at the BCP, the short contacts also follow this sequence. Two more criteria identify the short N⋯I contact as the most relevant: The associated C–I bond is significantly longer than the database average, and it is the only intermolecular interaction with a negative total energy density in the BCP.

Keywords: halogen bond; QTAIM; energy density; pyrazole; electron density; hcb net

Citation: van Terwingen, S.; Wang, R.; Englert, U. Three for the Price of One: Concomitant I⋯N, I⋯O, and I⋯π Halogen Bonds in the Same Crystal Structure. *Molecules* **2022**, 27, 7550. https://doi.org/10.3390/molecules27217550

Academic Editors: Qingzhong Li, Steve Scheiner and Zhiwu Yu

Received: 30 September 2022
Accepted: 31 October 2022
Published: 3 November 2022

Publisher's Note: MDPI stays neutral with regard to jurisdictional claims in published maps and institutional affiliations.

1. Introduction

Halogen bonds (XBs) arise from a local electron deficiency of a (mostly heavy) halogen on the opposing site of its σ-bond [1]. This so-called *σ-hole* [2] can interact with Lewis bases to form XBs; suitable acceptors are N-heterocycles [3–6], halides [7–9], or even π-systems [3,10–12]. In their most common appearance with a single atom bonded to the heavy halogen, XBs exhibit approximately linear geometry; together with their strongly electrostatic nature, they are related to hydrogen bonds [13,14]. Although their first explicit observation occurred in 1954 by Hassel et al. [15], it was not until the turn of the millennium that halogen bonds attracted the broad attention of both the theoretical and experimental crystal engineering community. Several groups have made significant contributions to the theoretical description of XB interactions, but only a few can be mentioned here [16–18]. Since 2005, research on this topic has rapidly expanded; it has been reviewed several times [12,18–20] and has become omnipresent in crystal engineering.

We have recently shown that the anion X$^-$ from hydrohalides of substituted N-heterocycles may form both halogen and hydrogen bonds in the same solid [8]; only shortly after, the Činčić group published multiple halogen and hydrogen bonded adducts of halopyridinium salts [9]. Herein, we report that the same substituted heterocycle can act as a multi-halogen bond acceptor, albeit without being protonated. In this solid, three different halogen bonds, namely I⋯N, I⋯O, and I⋯π, occur and can be compared directly. We decided for TFDIB as the XB donor, a particularly popular partner for XB-driven concrystallization: The CSD [21] comprises roughly 500 error-free entries in which a TFDIB iodine approaches an acceptor (N, O, Cl) at a distance shorter than the sum of the van der Waals radii. A chemical diagram of the asymmetric unit of our target cocrystal 3-(1,3,5-trimethyl-1*H*-4-pyrazolyl)pentane-2,4-dione (HacacMePz) with 1.5 equivalents of 2,3,5,6-tetrafluoro-1,4-diiodobenzene (TFDIB) **1** is given in Figure 1. We propose that all interactions shown in the figure may be exploited for crystal engineering purposes; similar motifs, where one molecule accepts multiple halogen bonds, have also been reported recently [22]. In addition to geometry arguments, which are based on crystallographic results, we corroborate our findings by a single-point calculation and subsequent analysis of the associated electron density according to Bader's Quantum Theory of Atoms in Molecules (QTAIM) [23].

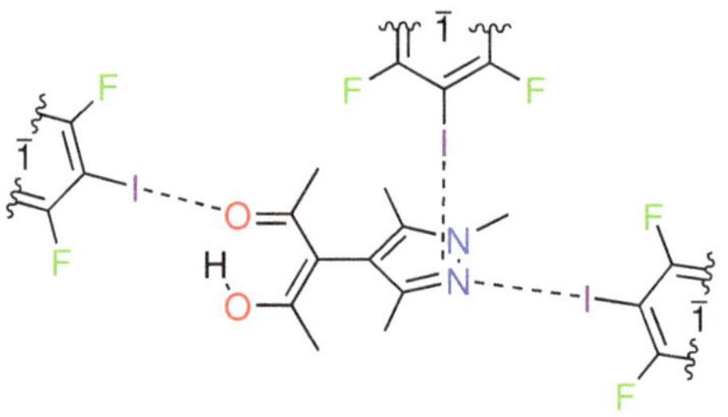

Figure 1. Chemical diagram of the asymmetric unit found in the crystal structure of HacacMePz · 1.5 TFDIB (**1**); inversion centers are marked with 1̄.

2. Materials and Methods

Searches in the Cambridge Structural Database [21] (CSD, version 5.43, including the update of September 2022) were restricted to perfluorinated iodobenzenes.

All chemicals were used as purchased without further purification. 3-(1,3,5-Trimethyl-1*H*-4-pyrazolyl)pentane-2,4-dione (HacacMePz) was synthesized as published before [24]. Single crystal X-ray intensity data was collected with a Bruker D8 goniometer equipped with an Incoatec microsource (Mo-K$_\alpha$ radiation, λ = 0.71073 Å, multilayer optics) and an APEX CCD area detector. A temperature of 100 K was maintained with an Oxford Cryostream 700 instrument, Oxfordshire, UK. Data was integrated with the Bruker SAINT program [25] and corrected for absorption by multiscan methods [26]. The structure was solved by intrinsic phasing [27] and refined with full matrix least squares procedures against F^2 [28]. Crystal data and refinement details are summarized in Table S1. The CIF for **1** has been deposited under CCDC No. 2209103. The powder X-ray diffraction pattern was recorded as a flat sample at room temperature with a STOE STADI-P diffractometer (Guinier geometry, Cu-K$_\alpha$ radiation, Johann Ge monochromator, STOE IP-PSD image plate detector, 0.005° 2θ step width). It shows that the bulk essentially corresponds to the phase established by single crystal diffraction (Supplementary Materials Figure S1). Thermogravimetric analysis (TGA) and differential scanning calorimetry (DSC) were performed using a Linseis STA PT 1600 instrument (Figure S2, Table S2). The sample was placed in a sealed Al_2O_3 crucible with a volume of 0.12 mL with a hole in the lid. Heating was applied at a rate of 5 K min^{-1} from room temperature to 500 °C under a constant flow of N_2 with a flow rate of 60 mL min^{-1}.

2.1. Synthesis and Crystallization

HacacMePz (10.4 mg, 0.05 mmol, 1.0 eq.) and TFDIB (30.1 mg, 0.075 mmol, 1.5 eq.) were dissolved in chloroform (2 mL). The solution was left unperturbed for slow evaporation at room temperature. After one week colorless, rod-shaped crystals formed. CHN: anal. calcd. for $C_{20}H_{16}F_6I_3N_2O_2$: C 29.6%, H 2.0%, N 3.5%; found: C 30.8%, H 2.2%, N 3.8%.

2.2. Computational Details

Before the single-point calculation was carried out, the C–H and O–H distances were idealized to values obtained from neutron diffraction experiments [29]. The theoretical electron density ρ was obtained from a single-point calculation of an expanded asymmetric unit (Figure S3) in the geometry established by X-ray diffraction; cartesian coordinates are available in the Supporting Information. The calculation was performed at the DFT level of theory with the M06-2X functional [30] and the MIDIX basis set [31] with the program Gaussian [32]. The derived electron density was analyzed with AIMAll [33] and Multiwfn [34] and interpreted with Bader's QTAIM [23]. Additionally, the kinetic energy density G and its ratio with the electron density G/ρ in the bond critical points (BCPs) were derived as suggested by Abramov [35]. Furthermore, the potential energy density V was calculated using the local virial theorem [36,37]. The interaction energies of the short contacts were estimated, as suggested by Espinosa et al. [36], by the equation $E_{XB} \approx 0.5 V_{BCP}$. E_{tot} was calculated with CrystalExplorer [38,39] with the "fast" setting (HF/3-21G level).

3. Results

3.1. Structural Features of 1

We first discuss the X-ray crystal structure of **1**. In order to achieve precise geometry data and account for the obviously large absorption in solid **1**, we collected data up to a high resolution of $\sin(\theta_{max})/\lambda \approx 1.0\ \text{Å}^{-1}$ with a redundancy of approximately 6.0, acceptably high for the triclinic symmetry. **1** crystallizes in space group $P\bar{1}$ with $Z = 2$; a displacement ellipsoid plot with important distances and angles is given in Figure 2.

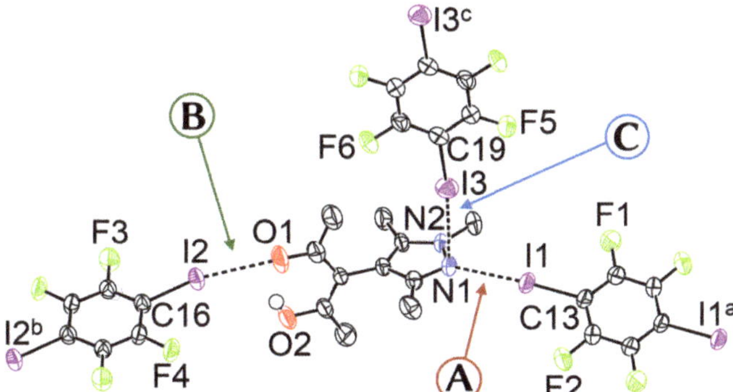

Figure 2. Displacement ellipsoid plot [40] of **1** with partial atom labeling (90% probability, carbon bonded hydrogen omitted). Selected distances and angles (Å, °): I1⋯N1 2.909(4), I2⋯O1 2.929(3), I3⋯Pz (Distance between I3 and the least squares plane of the pyrazole ring, consisting of the five atoms N1, N2, C7, C8, and C9) 3.2157(3), I1–C13 2.098(3), I2–C16 2.061(3), I3–C19 2.077(3), C13–I1⋯N1 172.05(12), C16–I2⋯O1 167.37(13), C19–I3⋯N1 173.35(12), ω 88.8(2). Symmetry operators: a = $-x, 1-y, -z$; b = $3-x, 1-y, 2-z$; c = $-x, -y, 1-z$.

The angle ω between the least squares planes of the pyrazole and the acetylacetone moiety is close to 90°; this is expected and within the energetically favored range of possible ω angles [24]. There are three independent TFDIB moieties, all located on different inversion centers. In the following, their shortest contacts to the substituted pyrazole molecule are referred to as capital letters **A**, **B**, and **C** (Figure 2).

A The pyrazole N⋯I halogen bond occurs between the TFDIB moiety located on Wyckoff position 1*c* and N1 at a N⋯I distance of 2.909(4) Å. For sufficiently precise data, anti-correlation between short I⋯donor XBs and long C–I bonds was reported [41]. Our data for **1** meets these requirements and allow us to discuss the competing XBs in the light of their associated C–I bonds. We found that C13–I1 is elongated and 0.02 Å is longer than the corresponding bond in pure TFDIB (CSD refcode ZZZAVM02 [42]). Only two contacts between a pyrazole and TFDIB were documented in the CSD; they amount to 2.860 Å in TOJBIE [43] and 2.934 Å in TIPKAH (In refcodes TIPKAH and AWUWOH, not *p*-TFDIB, but *o*-TFDIB was used) [44].

B   Another short contact exists between the acetylacetone keto O1 and I2 of the second TFDIB moiety, occupying the positions close to the inversion center with Wyckoff letter 1*e*; it amounts to 2.929(3) Å. As expected, the C–I bond in this moiety is less elongated than the C–I bond in the I⋯N halogen bond **A**. This is due to the weaker basicity of oxygen compared to the iminic nitrogen of the pyrazole moiety. For similar motifs, such as pyridyl substituted β-diketones, I⋯O$_{keto}$ amounts to about 3.05 Å (refcodes TAXYID [45], AWUWOH‡ [46]). In all cases of protonated β-diketones, the halogen bond acceptor is the keto oxygen, not the enol bond acceptor. Chemical intuition suggests that the keto oxygen is associated with the more negative charge. In several cocrystals of β-diketonato complexes with TFDIB, two oxygens of different β-diketonate ligands act as halogen bond acceptors; the XB is oriented more or less symmetrically bifurcated towards the midpoint between these two oxygen atoms [47,48].

C   Last but not least, I3 from the third symmetry independent TFDIB moiety, located around the inversion center with Wyckoff position 1*b*, acts as XB donor towards the pyrazole π-system with a distance of 3.2157(3) Å. As expected by the theoretical electrostatic potential for pyrazoles [49], the closest contact atom for I3 is the iminic N1 with a distance of 3.241(4) Å. Lewis basic π-systems as XB acceptors are known in literature, e.g., for cyclopentadienyl ligands [50], imidazoles [51], or carbazoles [52], and have been evaluated theoretically [53–55]; however, to the best of our knowledge, no pyrazole-π⋯I interactions with perfluoronated iodobenzenes have been reported to this date. This is also due to the competition with the more prominent I⋯N XB, as present in interaction **A**.

If the halogen bonds **A** and **B** are taken into account, an extended 1D structure can be derived. This chain expands along [3 0 2] in a "zig-zag" manner. Adding the third halogen bond **C** (C–I⋯π$_{Pz}$) to the contacts, a two-dimensional net can be perceived. It expands in the (−2 3 3) plane and no strict analogy can be found in the Reticular Chemistry Structure Resource (RCSR) [56]. If the N1 sites are perceived as triconnected vertices and the net is simplified by treating all edges as equivalent, its topology matches the honeycomb **hcb** net (Figure 3).

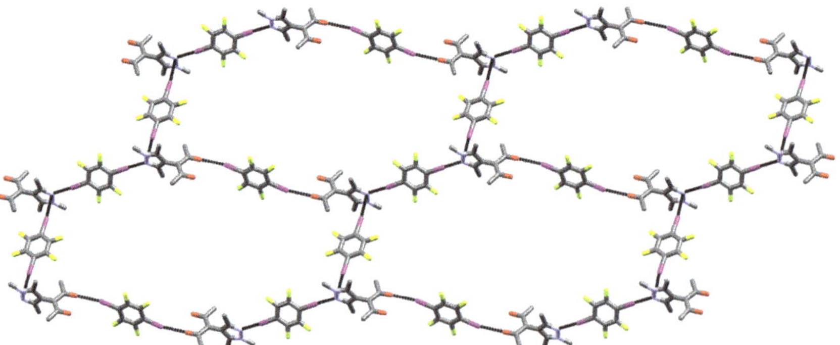

Figure 3. Excerpt of the two-dimensional net formed by the three halogen bonds towards one HacacMePz moiety in **1**, shown perpendicular to the (−2 3 3) plane (hydrogens omitted) [57].

Differential scanning calorimetry and thermogravimetry of **1** show that the melting point of the XB acceptor roughly corresponds to the decomposition point of the adduct (Figure S2). This behavior is commonly encountered for XB adducts [58]. Afterwards, at around 112 °C to 192 °C, a continuous weight loss of 64% is observed, which roughly corresponds to the loss of one HacacMePz moiety, together with two equivalents of TFDIB (calcd. 62.4%), leaving a stoichiometry of 1:1. Over the next approximate of 200 °C, further weight loss of 25% is observed, which corresponds to one TFDIB moiety (calcd. 24.8%).

3.2. Theoretical Electron Density Considerations

When d_{norm} is mapped on the Hirshfeld surface [59] about the HacacMePz moiety in **1**, the halogen bonds show up as close contacts (Figure 4). Additionally, a rather close contact between a methyl group to a fluorine (**D**) was highlighted.

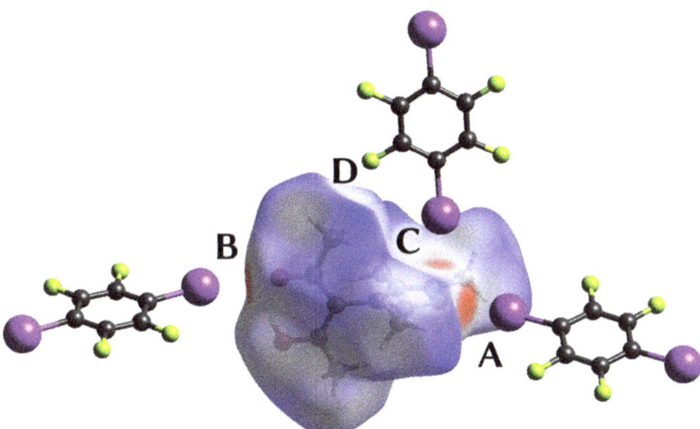

Figure 4. Depiction [38] of the Hirshfeld surface of the HacacMePz moiety mapped with d_{norm} (contacts **A** to **D** marked); regions marked in red represent close contacts.

In addition to the geometry arguments mentioned above, further insight about the coexisting XBs in **1** may come from the electron density and its derived properties, such as energy densities. For this purpose, a single-point calculation was performed, and the resulting electron density was analyzed by Bader's QTAIM [23]. Trajectory plots reveal that all contacts **A** to **D** are associated with essentially linear bond paths (Figures S4 and S5). In Table 1, characteristics of the aforementioned contacts in their BCPs are compiled.

Table 1. Short contacts in **1** with properties in their bond critical points (3, −1). BPL is the length of the bond path, ρ the electron density, $\nabla^2\rho$ the Laplacian of the electron density, G the kinetic, V the potential, and E the total energy density in the BCP.

Contact	BPL / Å	$\rho\,/\,e\,\text{Å}^{-3}$	$\nabla^2\rho\,/\,e\,\text{Å}^{-5}$	G / a.u.	G/ρ / a.u.	V / a.u.	E / a.u.
A	2.9120	0.1503	1.4847	0.01635	0.73	−0.01731	−0.00095
B	2.9352	0.1093	1.5402	0.01505	0.93	−0.01413	0.00092
C	3.2461	0.0746	0.9252	0.00846	0.77	−0.00733	0.00113
D	3.2720	0.0217	0.4890	0.00375	1.17	−0.00242	0.00132

We are not aware of charge density studies on halogen bonds to pyrazoles, but TFDIB represents a particular popular XB donor. The experimental electron density in its N⋯I contacts to other N heterocycles, such as pyridine [5,48,60] and O⋯I interactions to bipyridine oxide [61] and water [48], have been reported. Both N⋯I and O⋯I bonds involving TFDIB in the same crystal structure have been characterized by high resolution diffraction [48]; this study has found experimental electron densities, which closely match the outcome of the single point calculations for **A** and **B** reported here.

The electrostatic potential derived from the theoretical electron density offers an intuitive way to visualize XBs. Figure 5 shows the negative potential at the halogen acceptors **A** and **B** and the side-on interaction **C**. For each short contact the Laplacian of the electron density has been included.

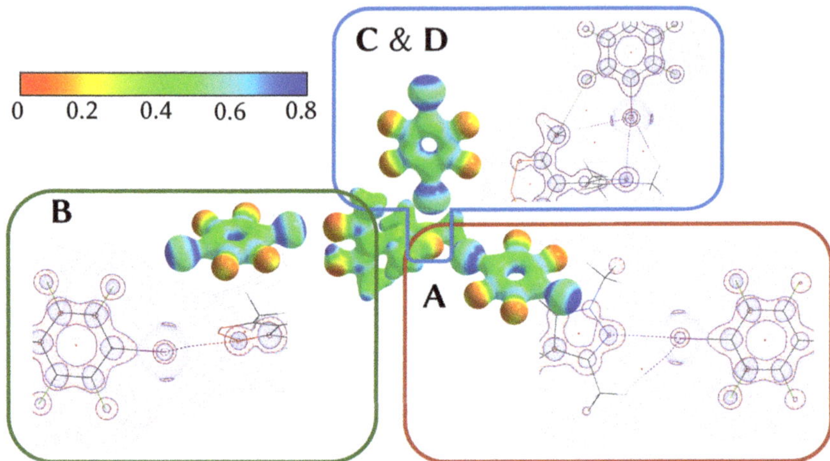

Figure 5. Center: Electrostatic potential of **1**, mapped onto an isosurface of $\rho = 0.07$ a.u. (scale given in the top left) [33]; excerpts of the Laplacians of contacts **A** to **D** are shown perpendicular to their respective TFDIB plane (contour lines drawn at $\pm 2^n \cdot 10^{-3}$ a.u. with $0 \leq n \leq 20$).

It is an attractive and somewhat controversial [62] idea that intermolecular interaction energies might be directly deduced from properties of the electron density ρ in the BCP between the contact atoms. If one accepts this concept, ρ in the BCPs represents the first criterion to gauge the strength of such interactions. From the more conventional and stronger halogen bonds **A** and **B** over the perpendicular π-type contact **C** to the presumably weak interaction **D** between a fluorine and a methyl group with their opposite charges, the electron density in the BCPs decreases (Table 1). Additional insight may come from energy density considerations: The (positive) kinetic energy density G and the ratio G/ρ_{bcp} have been suggested as qualifiers for chemical bonding. When expressed in atomic units, the ratio G/ρ_{bcp} typically assumes values around unity in closed-shell interactions [23], including hydrogen bonds [63], whereas much smaller G/ρ_{bcp} are associated with shared interactions, such as covalent bonds. For halogen bonds, the overall picture seems to be more complicated and intermediate values have been reported [64]. Espinosa et al. [65] have suggested to exploit the ratio $|V|/G$ between the (positive) kinetic energy density and (negative) potential energy density V to distinguish between pure closed-shell and incipient shared-shell interactions. With respect to this criterion, all interactions **A** to **D** are associated with values rather close to unity. Only **A**, apparently the strongest XB, can be assigned a significantly negative total energy density E.

We are well aware of the fact that much less data for halogen bonds than for hydrogen bonds are available, but attempts have been made to correlate electron density properties in the BCP and interaction energies for XBs [66,67]. Espinosa et al. have established relationships between the potential energy density V in the BCP and the interaction energy for hydrogen bonds [36,65]. Strictly speaking, this approach requires careful parametrization for each specific type of contact but it has also been applied to entirely different interactions, e.g., between neighboring azide groups [68]. When we applied the equation originally derived by Espinosa et al. for hydrogen bonds to halogen bonds and tentatively expressed the interaction energy as $E_{\mathrm{XB}} \approx 0.5 V_{\mathrm{BCP}}$, the potential energy densities in **1** afforded the interaction energies compiled in Table 2. CrystalExplorer [38] offers an alternative to estimate interaction energies according to benchmarked energy models [39]. In contrast to the approaches above, these interaction energies are not derived from electronic properties at the BCP of the contact atoms only.

Table 2. Interaction energies in the non-covalent contacts. E_{XB} was derived as suggested by Espinosa et al. [36,65] and E_{tot} was calculated with CrystalExplorer [38,39].

Contact	E_{XB} / kJ mol^{-1}	E_{tot} / kJ mol^{-1}
A	-22.7	-24.1
B	-18.5	-11.5
C + D	-12.8	-24.9

Individual components for electrostatic, polarization, dispersion, and repulsion energies thus obtained are compiled in the Supporting Information, and the total interaction energies E_{tot} for the "fast" energy model are included in Table 2 for comparison with the QTAIM-based approach. In either case, the interactions **C** and **D** occur between the same pair of molecules and have therefore been treated together in Table 2. The most obvious difference between both estimates is encountered for the less conventional π-type interaction, and there might be a good reason: CrystalExplorer, taking all energy contributions between neighboring molecules into account, assigns dispersion energy as dominant for **C + D**. In contrast, the approach via V_{bcp} focuses on specific short contacts and may be better suited for strongly directional interactions limited to just two or perhaps a few contact atoms. Correlation of V_{bcp} and interaction energies for halogen bonds seems an attractive task for the future.

To the best of our knowledge, no interaction energies have been determined for pyrazole N$\cdots$I halogen bonds. The closest match is the theoretical interaction energy between C_6F_5–I$\cdots$pyridine, which amounts to approx. $-23.4\,kJ\,mol^{-1}$ [69]; this value fits well with our estimates for contact **A**. For comparison with **B**, we found the interaction energy of C_6F_5–I$\cdots$O=CH$_2$ with a value of approximately $-19.6\,kJ\,mol^{-1}$ [70]. In this case, the literature value closely corresponds to the approximation for **B** established by the potential energy density in the BCP. There are not many interaction energies for $\pi\cdots$I contacts; the closest analogue we found was N$\equiv$C–I$\cdots$C$_6$H$_6$ with about $-20.6\,kJ\,mol^{-1}$ or C_6H_5–I$\cdots$C$_6$H$_6$ with about $-12.4\,kJ\,mol^{-1}$ [54]. No final conclusion can yet be drawn from these values in comparison with contacts **C + D**, for which our two estimates also differ, possibly for the reason given above. We want to recall that all these comparisons have to be taken with a grain of salt, mainly for two reasons: (a) the compared data comes from geometrically optimized molecules while we used crystallographic coordinates of **1** for the single-point calculation of ρ and (b) the compared data does not completely match the motif in **1**.

4. Conclusions

2,3,5,6-Tetrafluoro-1,4-diiodobenzene is intuitively perceived as a potential bridge between two halogen bond acceptors, and we indeed found this behavior for the shortest and strongest contacts. Competition between two perpendicular TFDIB bonds to the same N acceptor site was much less expected but also encountered in the crystal structure of **1**. This most unusual aspect will most likely also represent the major challenge for future work: How can crystal engineering enhance the frequency of structures in which orthogonal halogen bonds compete for the same acceptor, in our case the iminic N of the pyrazole heterocycle? Once this challenge has been met, fine tuning may target the sequence of the interaction energies and possibly invert the scenario, with stronger I$\cdots\pi$ and weaker I$\cdots$N contacts. The concomitant action of two different XB donor species may provide an additional synthetic degree of freedom for this purpose. We thank one of our reviewers for the following thought-provoking question: Should the three concomitant XBs in **1** be addressed as *competing* or rather as *cooperative* [71]? A competent answer to this question will require more structural input and therefore has to be postponed.

Supplementary Materials: The following supporting information can be downloaded at: https://www.mdpi.com/article/10.3390/molecules27217550/s1, Figure S1: Simulated (red) and experimental (black) powder patterns of **1**; Figure S2: Top: Thermogravimetric analysis curve with relative mass loss added. Bottom: Differential scanning calorimetry curve with three integrals added in different colors; Figure S3: Expanded asymmetric unit used for the single-point calculation to derive the electron density ρ in **1**; Figure S4: Trajectory plots in **1** to highlight contacts **A** and **B**; short contacts (hydrogen and halogen bonds) are shown as dashed lines, BCPs (3, −1) as pink spheres, ring critical points (3, 1) as light blue spheres; Figure S5: Trajectory plots in **1** to highlight contacts **C** and **D**; short contacts (hydrogen and halogen bonds) are shown as dashed lines, BCPs (3, −1) as pink spheres, ring critical points (3, 1) as light blue spheres; Table S1: Crystal data for compound **1** at 100(2) K; Table S2: Key data for the differential scanning calorimetry; Table S3: Calculated interaction energies E_{tot} and their components in kJ mol^{-1} in the non-covalent contacts **A** to **D** calculated with CrystalExplorer; Table S4: Cartesian coordinates of the expanded asymmetric unit (cf. Figure S3) used as Gaussian input for the single-point calculation.

Author Contributions: Conceptualization, S.v.T. and U.E.; methodology, U.E.; software, S.v.T. and R.W.; validation, S.v.T. and U.E.; formal analysis, S.v.T., R.W. and U.E.; investigation, S.v.T.; resources, U.E.; data curation, S.v.T. and U.E.; writing—original draft preparation, S.v.T., R.W. and U.E.; writing—review and editing, S.v.T., R.W. and U.E.; visualization, S.v.T. and R.W.; supervision, S.v.T. and U.E.; project administration, U.E.; funding acquisition, U.E. All authors have read and agreed to the published version of the manuscript.

Funding: This work was funded by a scholarship for doctoral students of the RWTH Graduierten-förderung to S.v.T.

Institutional Review Board Statement: Not applicable.

Informed Consent Statement: Not applicable.

Data Availability Statement: CCDC No. 2209103 contains the supplementary crystallographic data for this paper. These data can be obtained free of charge from The Cambridge Crystallographic Data Centre via www.ccdc.cam.ac.uk/data_request/cif (accessed on 25 May 2022).

Acknowledgments: The authors thank Simon Ernst for contributing to the experimental work for this submission and Anne Frommelius for conducting the DSC/TGA experiment.

Conflicts of Interest: The authors declare no conflict of interest. The funders had no role in the design of the study; in the collection, analyses, or interpretation of data; in the writing of the manuscript, or in the decision to publish the results.

Abbreviations

The following abbreviations are used in this manuscript:

BCP	bond critical point
BPL	bond path length
HacacMePz	3-(1,3,5-trimethyl-1*H*-4-pyrazolyl)pentane-2,4-dione
Pz	pyrazole
QTAIM	Quantum Theory of Atoms in Molecules
SCXRD	single-crystal X-ray diffraction
TFDIB	2,3,5,6-tetrafluoro-1,4-diiodobenzene
XB	halogen bond

References

1. Politzer, P.; Lane, P.; Concha, M.C.; Ma, Y.; Murray, J.S. An overview of halogen bonding. *J. Mol. Model.* **2007**, *13*, 305–311. [CrossRef] [PubMed]
2. Clark, T.; Hennemann, M.; Murray, J.S.; Politzer, P. Halogen bonding: The σ-hole. Proceedings of "Modeling interactions in biomolecules II", Prague, September 5th–9th, 2005. *J. Mol. Model.* **2007**, *13*, 291–296. [CrossRef] [PubMed]
3. Aakeröy, C.B.; Panikkattu, S.; Chopade, P.D.; Desper, J. Competing hydrogen-bond and halogen-bond donors in crystal engineering. *CrystEngComm* **2013**, *15*, 3125–3136. [CrossRef]
4. Wang, H.; Hu, R.X.; Pang, X.; Gao, H.Y.; Jin, W.J. The phosphorescent co-crystals of 1,4-diiodotetrafluorobenzene and bent 3-ring-N-heterocyclic hydrocarbons by C–I⋯N and C–I⋯π halogen bonds. *CrystEngComm* **2014**, *16*, 7942–7948. [CrossRef]

5. Wang, R.; Hartnick, D.; Englert, U. Short is strong: Experimental electron density in a very short N$\cdots$I halogen bond. *Z. Kristallogr.—Cryst. Mater.* **2018**, *233*, 733–744. [CrossRef]
6. Otte, F.; Kleinheider, J.; Hiller, W.; Wang, R.; Englert, U.; Strohmann, C. Weak yet Decisive: Molecular Halogen Bond and Competing Weak Interactions of Iodobenzene and Quinuclidine. *J. Am. Chem. Soc.* **2021**, *143*, 4133–4137. [CrossRef]
7. Sarwar, M.G.; Dragisic, B.; Sagoo, S.; Taylor, M.S. A Tridentate Halogen-Bonding Receptor for Tight Binding of Halide Anions. *Angew. Chem. Int. Ed.* **2010**, *122*, 1718–1721. [CrossRef]
8. van Terwingen, S.; Nachtigall, N.; Ebel, B.; Englert, U. N-Donor-Functionalized Acetylacetones for Heterobimetallic Coordination Polymers, the Next Episode: Trimethylpyrazoles. *Cryst. Growth Des.* **2021**, *21*, 2962–2969. [CrossRef]
9. Posavec, L.; Nemec, V.; Stilinović, V.; Cinčić, D. Halogen and Hydrogen Bond Motifs in Ionic Cocrystals Derived from 3-Halopyridinium Halogenides and Perfluorinated Iodobenzenes. *Cryst. Growth Des.* **2021**, *21*, 6044–6050. [CrossRef]
10. Ormond-Prout, J.E.; Smart, P.; Brammer, L. Cyanometallates as Halogen Bond Acceptors. *Cryst. Growth Des.* **2012**, *12*, 205–216. [CrossRef]
11. Shen, Q.J.; Pang, X.; Zhao, X.R.; Gao, H.Y.; Sun, H.L.; Jin, W.J. Phosphorescent cocrystals constructed by 1,4-diiodotetrafluorobenzene and polyaromatic hydrocarbons based on C–I$\cdots\pi$ halogen bonding and other assisting weak interactions. *CrystEngComm* **2012**, *14*, 5027. [CrossRef]
12. Wang, C.; Danovich, D.; Mo, Y.; Shaik, S. On The Nature of the Halogen Bond. *J. Chem. Theory Comput.* **2014**, *10*, 3726–3737. [CrossRef] [PubMed]
13. Legon, A.C. The halogen bond: An interim perspective. *Phys. Chem. Chem. Phys.* **2010**, *12*, 7736–7747. [CrossRef] [PubMed]
14. Aakeröy, C.B.; Welideniya, D.; Desper, J. Ethynyl hydrogen bonds and iodoethynyl halogen bonds: A case of synthon mimicry. *CrystEngComm* **2017**, *19*, 11–13. [CrossRef]
15. Hassel, O.; Hvoslef, J. The Structure of Bromine 1,4-Dioxonate. *Acta Chim. Scand.* **1954**, *8*, 873. [CrossRef]
16. Politzer, P.; Murray, J.S.; Clark, T. Halogen bonding: An electrostatically-driven highly directional noncovalent interaction. *Phys. Chem. Chem. Phys.* **2010**, *12*, 7748–7757. [CrossRef]
17. Politzer, P.; Murray, J.S. Halogen bonding: An interim discussion. *ChemPhysChem* **2013**, *14*, 278–294. [CrossRef]
18. Wolters, L.P.; Schyman, P.; Pavan, M.J.; Jorgensen, W.L.; Bickelhaupt, F.M.; Kozuch, S. The many faces of halogen bonding: A review of theoretical models and methods. *Comput. Mol. Sci.* **2014**, *4*, 523–540. [CrossRef]
19. Aakeröy, C.B.; Champness, N.R.; Janiak, C. Recent advances in crystal engineering. *CrystEngComm* **2010**, *12*, 22–43. [CrossRef]
20. Priimagi, A.; Cavallo, G.; Metrangolo, P.; Resnati, G. The halogen bond in the design of functional supramolecular materials: Recent advances. *Acc. Chem. Res.* **2013**, *46*, 2686–2695. [CrossRef]
21. Groom, C.R.; Bruno, I.J.; Lightfoot, M.P.; Ward, S.C. The Cambridge Structural Database. *Acta Crystallogr. B* **2016**, *72*, 171–179. [CrossRef] [PubMed]
22. Sušanj, R.; Nemec, V.; Bedeković, N.; Cinčić, D. Halogen Bond Motifs in Cocrystals of N,N,O and N,O,O Acceptors Derived from Diketones and Containing a Morpholine or Piperazine Moiety. *Cryst. Growth Des.* **2022**, *22*, 5135–5142. [CrossRef] [PubMed]
23. Bader, R.F.W. *Atoms in Molecules: A Quantum Theory*; Clarendon Press: Oxford, UK, 1990.
24. van Terwingen, S.; Brüx, D.; Wang, R.; Englert, U. Hydrogen-Bonded and Halogen-Bonded: Orthogonal Interactions for the Chloride Anion of a Pyrazolium Salt. *Molecules* **2021**, *26*, 3982. [CrossRef]
25. Bruker. *SAINT+: Program for Reduction of Data Collected on Bruker CCD Area Detector Diffractometer*; Bruker AXS Inc.: Billerica, MA, USA, 2009.
26. Bruker. *SADABS: Program for Empirical Absorption Correction of Area Detector Data*; Bruker AXS Inc.: Billerica, MA, USA, 2008.
27. Sheldrick, G.M. SHELXT—Integrated space-group and crystal-structure determination. *Acta Crystallogr. A* **2015**, *71*, 3–8. [CrossRef] [PubMed]
28. Sheldrick, G.M. Crystal structure refinement with SHELXL. *Acta Crystallogr. C* **2015**, *71*, 3–8. [CrossRef] [PubMed]
29. Allen, F.H.; Bruno, I.J. Bond lengths in organic and metal-organic compounds revisited: X-H bond lengths from neutron diffraction data. *Acta Crystallogr. B* **2010**, *66*, 380–386. [CrossRef]
30. Zhao, Y.; Truhlar, D.G. The M06 suite of density functionals for main group thermochemistry, thermochemical kinetics, noncovalent interactions, excited states, and transition elements: Two new functionals and systematic testing of four M06-class functionals and 12 other functionals. *Theor. Chem. Account* **2008**, *120*, 215–241. [CrossRef]
31. Easton, R.E.; Giesen, D.J.; Welch, A.; Cramer, C.J.; Truhlar, D.G. The MIDI! basis set for quantum mechanical calculations of molecular geometries and partial charges. *Theor. Chim. Acta* **1996**, *93*, 281–301. [CrossRef]
32. Frisch, M.J.; Trucks, G.W.; Schlegel, H.B.; Scuseria, G.E.; Robb, M.A.; Cheeseman, J.R.; Scalmani, G.; Barone, V.; Petersson, G.A.; Nakatsuji, H.; et al. *GAUSSIAN 16, Revision C.01*; Gaussian Inc.: Wallingford, CT, USA, 2016.
33. Keith, T.A. *AIMAll: Version 17.01.25*; TK Gristmill Software: Overland Park, KS, USA, 2017.
34. Lu, T.; Chen, F. Multiwfn: A multifunctional wavefunction analyzer. *J. Comput. Chem.* **2012**, *33*, 580–592. [CrossRef]
35. Abramov, Y.A. On the Possibility of Kinetic Energy Density Evaluation from the Experimental Electron-Density Distribution. *Acta Crystallogr. A* **1997**, *53*, 264–272. [CrossRef]
36. Espinosa, E.; Molins, E.; Lecomte, C. Hydrogen bond strengths revealed by topological analyses of experimentally observed electron densities. *Chem. Phys. Lett.* **1998**, *285*, 170–173. [CrossRef]
37. Espinosa, E.; Lecomte, C.; Molins, E. Experimental electron density overlapping in hydrogen bonds: Topology vs. energetics. *Chem. Phys. Lett.* **1999**, *300*, 745–748. [CrossRef]

38. Spackman, P.R.; Turner, M.J.; McKinnon, J.J.; Wolff, S.K.; Grimwood, D.J.; Jayatilaka, D.; Spackman, M.A. CrystalExplorer: A program for Hirshfeld surface analysis, visualization and quantitative analysis of molecular crystals. *J. Appl. Crystallogr.* **2021**, *54*, 1006–1011. [CrossRef] [PubMed]
39. Turner, M.J.; Grabowsky, S.; Jayatilaka, D.; Spackman, M.A. Accurate and Efficient Model Energies for Exploring Intermolecular Interactions in Molecular Crystals. *J. Phys. Chem. Lett.* **2014**, *5*, 4249–4255. [CrossRef]
40. Spek, A.L. Structure validation in chemical crystallography. *Acta Crystallogr. D* **2009**, *65*, 148–155. [CrossRef]
41. Wang, R.; George, J.; Potts, S.K.; Kremer, M.; Dronskowski, R.; Englert, U. The many flavours of halogen bonds—Message from experimental electron density and Raman spectroscopy. *Acta Crystallogr. C* **2019**, *75*, 1190–1201. [CrossRef]
42. Oh, S.Y.; Nickels, C.W.; Garcia, F.; Jones, W.; Friščić, T. Switching between halogen- and hydrogen-bonding in stoichiometric variations of a cocrystal of a phosphine oxide. *CrystEngComm* **2012**, *14*, 6110. [CrossRef]
43. Aakeröy, C.B.; Desper, J.; Fasulo, M.; Hussain, I.; Levin, B.; Schultheiss, N. Ten years of co-crystal synthesis; the good, the bad, and the ugly. *CrystEngComm* **2008**, *10*, 1816. [CrossRef]
44. Andree, S.N.L.; Sinha, A.S.; Aakeröy, C.B. Structural Examination of Halogen-Bonded Co-Crystals of Tritopic Acceptors. *Molecules* **2018**, *23*, 163. [CrossRef]
45. Brown, J.J.; Brock, A.J.; Pfrunder, M.C.; Sarju, J.P.; Perry, A.Z.; Whitwood, A.C.; Bruce, D.W.; McMurtrie, J.C.; Clegg, J.K. Co-Crystallisation of 1,4-Diiodotetrafluorobenzene with Three Different Symmetric Dipyridylacetylacetone Isomers Produces Four Halogen-Bonded Architectures. *Aust. J. Chem.* **2017**, *70*, 594. [CrossRef]
46. Martinez, V.; Bedeković, N.; Stilinović, V.; Cinčić, D. Tautomeric Equilibrium of an Asymmetric beta-Diketone in Halogen-Bonded Cocrystals with Perfluorinated Iodobenzenes. *Crystals* **2021**, *11*, 699. [CrossRef]
47. Stilinović, V.; Grgurić, T.; Piteša, T.; Nemec, V.; Cinčić, D. Bifurcated and Monocentric Halogen Bonds in Cocrystals of Metal(II) Acetylacetonates with *p*-Dihalotetrafluorobenzenes. *Cryst. Growth Des.* **2019**, *19*, 1245–1256. [CrossRef]
48. Merkens, C.; Pan, F.; Englert, U. 3-(4-Pyridyl)-2,4-pentanedione—A bridge between coordinative, halogen, and hydrogen bonds. *CrystEngComm* **2013**, *15*, 8153. [CrossRef]
49. Politzer, P.; Murray, J.S. Computational analysis of polyazoles and their N-oxides. *Struct. Chem.* **2017**, *28*, 1045–1063. [CrossRef]
50. Torubaev, Y.V.; Skabitsky, I.V.; Raghuvanshi, A. The structural landscape of ferrocenyl polychalcogenides. *J. Organomet. Chem.* **2021**, *951*, 122006. [CrossRef]
51. Aakeröy, C.B.; Wijethunga, T.K.; Desper, J. Practical crystal engineering using halogen bonding: A hierarchy based on calculated molecular electrostatic potential surfaces. *J. Mol. Struct.* **2014**, *1072*, 20–27. [CrossRef]
52. Zhu, Q.; Gao, Y.J.; Gao, H.Y.; Jin, W.J. Effect of N-methyl and ethyl on phosphorescence of carbazole in cocrystals assembled by CI···π halogen bond, π-hole···π bond and other interactions using 1,4-diiodotetrafluorobenzene as donor. *J. Photochem. Photobiol. A Chem.* **2014**, *289*, 31–38. [CrossRef]
53. Riley, K.E.; Vazquez, M.; Umemura, C.; Miller, C.; Tran, K.A. Exploring the (Very Flat) Potential Energy Landscape of R-Br···π Interactions with Accurate CCSD(T) and SAPT Techniques. *Chem. Eur. J.* **2016**, *22*, 17690–17695. [CrossRef]
54. Forni, A.; Pieraccini, S.; Rendine, S.; Gabas, F.; Sironi, M. Halogen-bonding interactions with π systems: CCSD(T), MP2, and DFT calculations. *ChemPhysChem* **2012**, *13*, 4224–4234. [CrossRef]
55. Wu, M.; Li, M.; Yuan, L.; Pan, F. "Useless Channels" in a Molecular Crystal Formed via F···F and F···π Halogen Bonds. *Cryst. Growth Des.* **2022**, *22*, 971–975. [CrossRef]
56. O'Keeffe, M.; Peskov, M.A.; Ramsden, S.J.; Yaghi, O.M. The Reticular Chemistry Structure Resource (RCSR) database of, and symbols for, crystal nets. *Acc. Chem. Res.* **2008**, *41*, 1782–1789. [CrossRef] [PubMed]
57. Macrae, C.F.; Sovago, I.; Cottrell, S.J.; Galek, P.T.A.; McCabe, P.; Pidcock, E.; Platings, M.; Shields, G.P.; Stevens, J.S.; Towler, M.; et al. Mercury 4.0: From visualization to analysis, design and prediction. *J. Appl. Crystallogr.* **2020**, *53*, 226–235. [CrossRef] [PubMed]
58. Nemec, V.; Cinčić, D. The Halogen Bonding Proclivity of the sp3 Sulfur Atom as a Halogen Bond Acceptor in Cocrystals of Tetrahydro-4H-thiopyran-4-one and Its Derivatives. *Cryst. Growth Des.* **2022**, *22*, 5796–5801. [CrossRef]
59. Spackman, M.A.; Jayatilaka, D. Hirshfeld surface analysis. *CrystEngComm* **2009**, *11*, 19–32. [CrossRef]
60. Bianchi, R.; Forni, A.; Pilati, T. The experimental electron density distribution in the complex of (E)-1,2-bis(4-pyridyl)ethylene with 1,4-diiodotetrafluorobenzene at 90 K. *Chem. Eur. J.* **2003**, *9*, 1631–1638. [CrossRef]
61. Bianchi, R.; Forni, A.; Pilati, T. Experimental electron density study of the supramolecular aggregation between 4,4'-dipyridyl-N,N'-dioxide and 1,4-diiodotetrafluorobenzene at 90 K. *Acta Crystallogr. B* **2004**, *60*, 559–568. [CrossRef]
62. Kuznetsov, M.L. Can halogen bond energy be reliably estimated from electron density properties at bond critical point? The case of the $(A)_nZ–Y···X^-$ (X, Y = F, Cl, Br) interactions. *Int. J. Quantum Chem.* **2019**, *119*, e25869. [CrossRef]
63. Şerb, M.D.; Wang, R.; Meven, M.; Englert, U. The whole range of hydrogen bonds in one crystal structure: Neutron diffraction and charge-density studies of N,N-dimethylbiguanidinium bis(hydrogensquarate). *Acta Crystallogr. B* **2011**, *67*, 552–559. [CrossRef]
64. Wang, A.; Wang, R.; Kalf, I.; Dreier, A.; Lehmann, C.W.; Englert, U. Charge-Assisted Halogen Bonds in Halogen-Substituted Pyridinium Salts: Experimental Electron Density. *Cryst. Growth Des.* **2017**, *17*, 2357–2364. [CrossRef]
65. Espinosa, E.; Alkorta, I.; Elguero, J.; Molins, E. From weak to strong interactions: A comprehensive analysis of the topological and energetic properties of the electron density distribution involving X–H···F–Y systems. *J. Chem. Phys.* **2002**, *117*, 5529–5542. [CrossRef]

66. Zeng, Y.; Zhang, X.; Li, X.; Meng, L.; Zheng, S. The role of molecular electrostatic potentials in the formation of a halogen bond in furan···XY and thiophene···XY complexes. *ChemPhysChem* **2011**, *12*, 1080–1087. [CrossRef]
67. Chakalov, E.R.; Tupikina, E.Y.; Ivanov, D.M.; Bartashevich, E.V.; Tolstoy, P.M. The Distance between Minima of Electron Density and Electrostatic Potential as a Measure of Halogen Bond Strength. *Molecules* **2022**, *27*, 4848. [CrossRef] [PubMed]
68. Lyssenko, K.A.; Nelubina, Y.V.; Safronov, D.V.; Haustova, O.I.; Kostyanovsky, R.G.; Lenev, D.A.; Antipin, M.Y. Intermolecular N_3···N_3 interactions in the crystal of pentaerythrityl tetraazide. *Mendeleev Commun.* **2005**, *15*, 232–234. [CrossRef]
69. Tsuzuki, S.; Wakisaka, A.; Ono, T.; Sonoda, T. Magnitude and origin of the attraction and directionality of the halogen bonds of the complexes of C_6F_5X and C_6H_5X (X = I, Br, Cl and F) with pyridine. *Chem. Eur. J.* **2012**, *18*, 951–960. [CrossRef]
70. Forni, A.; Rendine, S.; Pieraccini, S.; Sironi, M. Solvent effect on halogen bonding: The case of the I···O interaction. *J. Mol. Graph. Model.* **2012**, *38*, 31–39. [CrossRef]
71. Bedeković, N.; Piteša, T.; Eraković, M.; Stilinović, V.; Cinčić, D. Anticooperativity of Multiple Halogen Bonds and Its Effect on Stoichiometry of Cocrystals of Perfluorinated Iodobenzenes. *Cryst. Growth Des.* **2022**, *22*, 2644–2653. [CrossRef]

molecules

MDPI

Article

Dissecting Bonding Interactions in Cysteine Dimers

Santiago Gómez [1], Sara Gómez [2], Jorge David [3], Doris Guerra [1], Chiara Cappelli [2,*] and Albeiro Restrepo [1,*]

[1] Instituto de Química, Universidad de Antioquia UdeA, Calle 70 No. 52-21, Medellín 050010, Colombia
[2] Scuola Normale Superiore, Classe di Scienze, Piazza dei Cavalieri 7, 56126 Pisa, Italy
[3] Escuela de Ciencias y Humanidades, Departamento de Ciencias Básicas, Universidad Eafit, AA 3300, Medellín 050022, Colombia
* Correspondence: chiara.cappelli@sns.it (C.C.); albeiro.restrepo@udea.edu.co (A.R.)

Abstract: Neutral (*n*) and zwitterionic (*z*) forms of cysteine monomers are combined in this work to extensively explore the potential energy surfaces for the formation of cysteine dimers in aqueous environments represented by a continuum. A simulated annealing search followed by optimization and characterization of the candidate structures afforded a total of 746 structurally different dimers held together via 80 different types of intermolecular contacts in 2894 individual non-covalent interactions as concluded from Natural Bond Orbitals (NBO), Quantum Theory of Atoms in Molecules (QTAIM) and Non-Covalent Interactions (NCI) analyses. This large pool of interaction possibilities includes the traditional primary hydrogen bonds and salt bridges which actually dictate the structures of the dimers, as well as the less common secondary hydrogen bonds, exotic $X \cdots Y$ (X = C, N, O, S) contacts, and $H \cdots H$ dihydrogen bonds. These interactions are not homogeneous but have rather complex distributions of strengths, interfragment distances and overall stabilities. Judging by their Gibbs bonding energies, most of the structures located here are suitable for experimental detection at room conditions.

Keywords: cysteine dimers; NBO; NCI; QTAIM; stochastic optimization; hydrogen bonding; salt bridges; non-covalent interactions

Citation: Gómez, S.; Gómez, S.; David, J.; Guerra, D.; Capelli, C.; Restrepo, A. Dissecting Bonding Interactions in the Cysteine Dimers. *Molecules* **2022**, *27*, 8665. https:// doi.org/10.3390/molecules27248665

Academic Editors: Qingzhong Li, Steve Scheiner and Zhiwu Yu

Received: 11 November 2022
Accepted: 2 December 2022
Published: 7 December 2022

Publisher's Note: MDPI stays neutral with regard to jurisdictional claims in published maps and institutional affiliations.

1. Introduction

Cysteine, $HOOCCH(NH_2)CH_2SH$ is the only amino acid among the unique list of 20 found in proteins that possesses a thiol functional group [1]. This thiol group, which is comparatively a weaker Brønsted–Lowry acid than O–H in carboxylic acids, is extremely important in biochemistry. Among a large number of known functionalities, the S–H group is responsible for nucleophilic additions to $\alpha, \beta-$unsaturated carbonyl compounds via Michael reactions [2–5], serves as a deprotonation agent [6], and its strong nucleophilicity renders cysteine a key component of the active sites of several protease enzymes [7,8]. In addition to S–H, depending on the conditions, sulfur atoms in cysteine engage in S–S disulfide bonds, which are a central element determining secondary and tertiary structure in proteins [9–11] and are relevant in physiological redox activity. According to the SwissProt databank [12], six percent of all proteins contain at least one disulfide bridge, and the median number of disulfide bonds is two.

Protein$\cdots$protein interaction is one of the central problems in molecular biology. Unfortunately, with present days computational methods, a thorough understanding from a molecular perspective is unattainable because the number of explicit contacts grows exponentially with the size of the protein. For example, insulin is one of the smallest biologically active proteins containing a primary sequence of just 51 amino acids (six cysteines among them), for the dimer of this protein, not counting salt bridges and other intermolecular interactions, there are at least 6×10^4 possibilities for hydrogen bonding in the classical $X^{\delta-} \cdots {}^{+\delta}H-Y^{\delta-}$ description [13]. Typical proteins and other biomolecules contain in excess of 1000 amino acids and it is not uncommon to find very large proteins such as titin, which depending on the splice isoform, contains between 27,000 to 35,000 amino acids [14]; thus, the number of specific amino acid$\cdots$amino acid contacts quickly becomes intractable.

In an effort to understand the intricacies of protein$\cdots$protein interactions, the astonishingly large number of specific contacts calls for the use of reduced molecular models, often as gas phase isolated dimers of individual amino acids [13]. In this context, we attempt a detailed study of the cysteine dimers. This is a very complicated issue in its own right: First, there are the two enantiomeric forms of the amino acid. Second, there is the possibility of neutral and zwitterionic forms. Third, there is a large number of conformers of the monomer in a small energy window amenable to form dimers, which, for just the neutral form, has been estimated via ab initio computations to be 42, 51 or 71 depending on the sophistication of the model chemistry [1,9,15]. Notice that six well defined conformers have been experimentally identified via IR and MW spectroscopies [16,17]. Fourth, cysteine contains seven hydrogen atoms and four electronegative atoms; thus, ignoring salt bridges, dispersive dihydrogen interactions and other exotic contacts [13], from the classical $X^{\delta-}\cdots^{+\delta}H\text{–}Y^{\delta-}$ perspective, a total of 28 individual primary plus secondary hydrogen bonds are possible for each dimer. The number of possibilities is reduced to 20, distributed as 12 primary and 8 secondary HBs if the two $H\text{–}C_\beta$ bonds are grouped into just one type and if the two N–H bonds are considered as another type. Fifth, as seen for example in alanine [13], several dimers are attached by more than one contact. Sixth, S–H leads to considerably weaker interactions than O–H, then, the potential energy surface (PES) for the dimers is expected to be considerably richer in weakly bound pairs and thus high levels of electron correlation are needed to correctly describe the intermolecular interactions. Seventh, the environment sensibly impacts the ability of biomolecules to interact in biological settings; thus, using gas phase dimers as a reduced model for protein$\cdots$protein interactions does not seem enough and at least solvent effects must be included.

Cysteine has been thoroughly studied through experiments and computations. Besides the above mentioned publications dealing with the conformations of the monomer [1,9,15–17], Kaminski et al. [18] and Sadlej et al. [19] undertook somewhat exhaustive explorations of the conformational PES for the monomers to rationalize Raman Optical Activity (ROA) and Vibrational Circular Dichroism (VCD) spectra. For the dimers, early studies focused on gas phase and implicitly solvated models with limited explorations of the PES using a few hand constructed configurations [20], later studies considered both explicit water molecules and the neutral and zwitterionic forms [21,22]. There are reports dealing with the formation of the dimers, their stability and bonding (via density differences) when adsorbed in gold surfaces [23,24]. Group IA cations bonded to cysteine dimers have also been studied [25].

In view of the expected complexity arising from the multiple classical donor and acceptor hydrogen bonding sites of cysteine, which as a reduced molecular model has profound implications in the protein interaction problem, the brief summary of the scientific literature dealing with cysteine dimers just exposed reveals an unsatisfactory level of understanding not only of the potential energy surface but of the nature of the intermolecular bonding interactions for cysteine$\cdots$cysteine. The present work is an attempt to remedy this situation. To that end, we undertake systematic explorations of the neutral (n) and zwitterionic (z) pairs in $n\cdots n$, $n\cdots z$ and $z\cdots z$ combinations of low lying energy monomers via stochastic samplings of the corresponding PES, and dissected the nature of the interactions using formal quantum descriptors of bonding as provided by the Quantum Theory of Atoms in Molecules [26–29] (QTAIM), the Natural Bond Orbitals [30–33] (NBO) and the Non Covalent Interactions [34,35] (NCI), as discussed in the Section 2.

2. Computer Methods

Sampling the potential energy surfaces for all possible neutral (72 computed, 6 experimentally found) and zwitterionic forms (12 computed) is not only impossible but unnecessary under the premise that a few representative pairs would capture the vast majority of the specific contacts and thus would provide a sound picture applicable to all cases. Accordingly, we took two of the experimentally detected neutral monomers [17] (n_1, n_2 in Table 1) and two of the computed lowest energy zwitterions [9] (z_1, z_2 in Table 1) and exhausted all $x(x-1)/2 + x = 10$ possible dimeric combinations for $x = 4$. Each pair was superimposed at the center of a cubic box of 512 Å^3 (8 Å side) and was allowed to evolve under simulated annealing conditions [36–38] as implemented in the ASCEC program [39]. Superimposing the interacting system at the center of the box gives the algorithm the worst possible starting point (we call this the big bang initial conditions) and guarantees that the located stationary points within the corresponding PES are free of any structural bias. ASCEC [40,41], after its Spanish acronym Annealing Simulado Con Energía Cuántica, randomly explores the quantum energy landscape for the dimeric interaction, subjects the generated structures to a modified Metropolis acceptance test, and delivers a set of candidate structures that undergo further optimization via gradient following techniques and characterization as true minima via harmonic vibrational analysis. Each one of the 10 possible dimeric combinations was treated to duplicate ASCEC runs. All ASCEC runs and geometry optimizations were carried out in an aqueous environment represented by a continuum according to the PCM (ASCEC) and IEFPCM (optimization) models. [42–44].

Table 1. Structures of the B3LYP–D3/6–311++G(d, p) monomers of neutral (n_1, n_2) and zwitterionic (z_1, z_2) L–cysteine. In each case, $\Delta\Delta G$ are the corresponding differences in Gibbs energies at room conditions with respect to n_1, z_1, the lowest energy monomers. Descriptors of intramolecular bonding derived from QTAIM and NBO are included along with the specific NBO orbitals responsible for the interactions. $\Delta\Delta G, E^{(2)}_{d \to a}$ in kcal mol^{-1}, all other descriptors in a.u. n_1 and n_2 have been experimentally detected [17].

Monomer $\to$	n_1	n_2	z_1	z_2
Properties $\downarrow$				
$\Delta\Delta G$	0.0	2.2	0.0	0.0	0.3
A$\cdots$B Distance (Å)	1.86		2.63	1.91	1.94
$10^2 \times \rho(\mathbf{r}_c)$	4.07		1.37	3.29	3.11
$10^2 \times \nabla^2\rho(\mathbf{r}_c)$	10.87		4.63	11.78	11.29
$\lvert\mathcal{V}(\mathbf{r}_c)\rvert/\mathcal{G}(\mathbf{r}_c)$	1.12		0.82	0.93	0.92
$10^2 \times \mathcal{H}(\mathbf{r}_c)/\rho(\mathbf{r}_c)$	−9.18		12.73	5.62	7.04
$\phi_d \to \phi_a$	$n_{\mathrm{N}} \to \sigma^*_{\mathrm{O-H}}$		$n_{\mathrm{S}} \to \sigma^*_{\mathrm{N-H}}$	$n_{\mathrm{O}} \to \sigma^*_{\mathrm{N-H}}$	$n_{\mathrm{O}} \to \sigma^*_{\mathrm{N-H}}$
$E^{(2)}_{d \to a}$	14.9		2.3	8.3	7.3

Final equilibrium geometries and Gibbs energies for every located dimer computed with the Gaussian09 suite of programs [45] are reported here using the dispersion corrected B3LYP–D3/6–311++G(d, p) model chemistry. Binding using the Gibbs energies at room conditions (1 atm, 298.16 K) are calculated as the negative difference between the energy of the cluster and the energy of the fragments, $BE = -(E_{cluster} - \sum E_{fragments})$, in this way, positive binding energies indicate strongly bonded clusters.

Once the molecular wavefunctions and electron densities for the optimized geometries are recovered by the procedure just stated, we use them to gain insight into the nature of intermolecular bonding interactions using the tools provided by QTAIM, NBO, and NCI following strategies described elsewhere [46–53]. At this point, we state that we use those methods as well established analysis tools, the interested reader is directed to the specialized literature for detailed discussions of their merits and shortcomings and for a description of how the calculated descriptors are related to bonding [6,33,49]. In short, use the Multiwfn suite [54] to find the bond critical points (BCPs, $\mathbf{r}_c$) corresponding to intermolecular interactions, analyze their properties, i.e., the electron density $\rho(\mathbf{r}_c)$, its Laplacian $\nabla^2\rho(\mathbf{r}_c)$, the total, kinetic, and potential energy densities $\mathcal{H}(\mathbf{r}_c) = \mathcal{G}(\mathbf{r}_c) + \mathcal{V}(\mathbf{r}_c)$, and the virial ratio $|\mathcal{V}(\mathbf{r}_c)|/\mathcal{G}(\mathbf{r}_c)$. With the same program, we calculate the Laplacian of the electron density, a scalar field that gives direct information about the most probable regions to find the electrons. Then, we use NBO6 [55] to pinpoint the specific orbitals involved in the intermolecular interactions associated with each BCP and estimate the strength of the interaction via second order perturbation energy for the interaction between the donor and acceptor orbitals, $E^{(2)}_{d \to a}$. The NCIPLOT program [56] was used to derive the non-covalent interaction surfaces. Jmol and VMD [57,58] were used to visualize the molecular structures, and their related surfaces and orbital interactions.

3. Results

Table 1 shows the structures of the n_1, n_2 neutral and z_1, z_2 zwitterionic forms chosen in this work to study the dimers of cysteine. The four non-covalent intramolecular interactions, as derived from QTAIM are displayed as dotted lines along with the involved NBOs. QTAIM and NBO descriptors are included as well. Only n_2 has a structure free from intramolecular hydrogen bonds while z_1 exhibits two intramolecular contacts. Except for n_1, all intramolecular interactions are characterized as weak, long range contacts because of the positive Laplacians, relatively small accumulation of electron densities at BCPs, virial ratios smaller than 1, and positive bond degree parameters. However, the $n_N \to \sigma^*_{O-H}$ interaction in n_1 is uncharacteristically strong, with values for the bonding descriptors that in every case surpass those of the archetypal hydrogen bond in the water dimer [30,59]. These intramolecular contacts are quite important because the formation of the dimers will usually involve investing energy to eliminate those interactions in favor of dimeric contacts. The electrostatic potentials in Figure S1 of the Supplementary Information show the blue and red regions which are more susceptible to the formation of intermolecular contacts according to the classic electrostatic view of hydrogen bonding.

3.1. Structures and Energies

Complex and rich potential energy surfaces are uncovered by our stochastic searches in every case. A total of 746 distinct well defined dimers were located in the 10 monomer + monomer possible combinations. Table 2 lists the number of structural isomers for each PES and also shows that the vicinities of the putative global minima are populated with other close energy dimers; thus, all structures accounting for populations larger than 1% are within 2.1 kcal/mol of the $n \cdots n$ lowest energy structure, and so on. This point is emphasized by the results shown in Tables S1–S3 in the Supplementary information and in Figure 1, which clearly show that there are no dominant isomers.

Table 2. Summary of structural and energetical properties of the cysteine dimers. ΔG range: Gibbs energy difference between the highest and lowest energy structure in kcal/mol. $\%x_i$: isomer populations listed in Tables S1–S3 in the Supplementary Material.

Composition	Structures	ΔG Range	
		for $\%x_i > 1$	for All Structures
$n \cdots n$	244	2.1	21.4
$n \cdots z$	316	1.1	25.5
$z \cdots z$	186	2.3	27.9

Although 746 is a very large number of structures and is considerably higher than the numbers reported in any of the previous studies, we recognize that given the complexity of our problem, no stochastic or analytic search algorithm is able to locate all possible geometries. A representative set including only those dimers with populations exceeding 5% within each PES is shown in Figure 1, along with the NBOs responsible for the intermolecular interactions. Cartesian coordinates for all 746 structures located in this work are provided in the Supplementary Material.

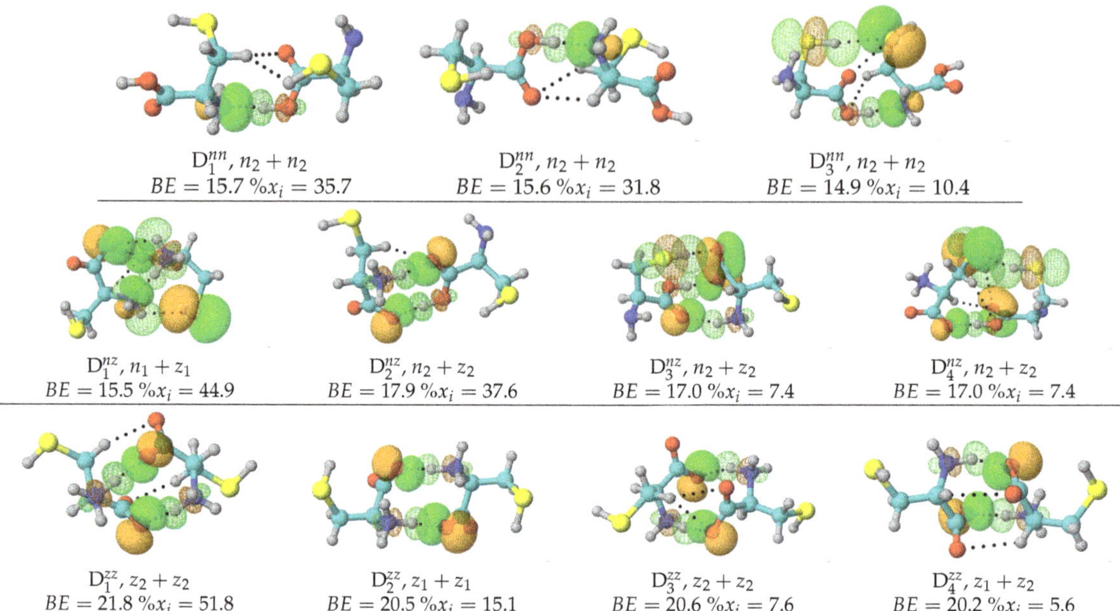

Figure 1. Lowest energy structures and the NBOs responsible for the strongest intermolecular interactions in the neutral $n \cdots n$ (top), neutral + zwitterionic $n \cdots z$ (middle) and zwitterionic $z \cdots z$ (bottom) B3LYP–D3/6–311++g(d, p) potential energy surfaces of the cysteine dimers under the continuum IEFPCM solvent model for water. Solid/meshed surfaces correspond to charge donor/acceptor orbitals, respectively. *BE*: binding energies in kcal/mol calculated using the Gibbs free energies at room conditions. See Table 1 for the structures of n_1, n_2, z_1, z_2, the isolated monomers. Only those structures with populations ($\%x_i$) higher than 5% within each PES are included. Energetics for the entire set of 746 dimers is provided in Tables S1–S3 of the supplementary material.

On the basis of purely ZPE–corrected electronic energies (Tables S1–S3), all cysteine dimers are stable towards fragmentation into the corresponding monomers, however, Figure 2 shows that consideration of temperature and entropy leads to 235 clusters (80 $n \cdots n$, 154 $n \cdots z$ and 1 $z \cdots z$) having negative binding energies as calculated from the Gibbs energies; thus, those particular structures correspond to unstable dimers and are not amenable to experimental detection at room conditions in aqueous environments, a fact that is emphasized by their $\%x_i \approx 0$ populations. Notice the contrast with the 416 $n \cdots n$ gas phase equilibrium structures reported for the Alanine dimers [13], which are all strongly bonded. Binding energies show a clear $BE_{n \cdots n} < BE_{n \cdots z} < BE_{z \cdots z}$ ordering; thus, there is a marked preference for charged cysteine dimers in aqueous environments.

Figure 2, showing distribution plots of the Gibbs binding energies leads to a few relevant observations: Dashed vertical lines indicate the expected values of the binding energies using the Boltzmann populations of the Gibbs energies within each PES as weighing factors. 14.3, 16.6 and 20.9 kcal/mol are obtained for $n \cdots n, n \cdots z, z \cdots z$, again showing a preference for charged dimers in aqueous environments. To put these binding energies in context, they are larger than the gas phase Gibbs binding energies of acetamide and acetic acid, which are 2.1 and 3.8 kcal/mol respectively, according to Copeland et al. [60] Notice that the same authors reported substantially higher binding energies when using only the ZPE-corrected electronic energies: 12.3 kcal/mol for acetamide and 14.7 kcal/mol for acetic acid. High ZPE-corrected binding energies have also been reported for the lowest energy structures in similar systems: 16.6 kcal/mol for the dimers of formic acid according to Kalescky et al. [61] and 12.7 according to Farfán et al. [62], 19.0 kcal/mol for the dimers of carbonic acid [63], and 20.9 kcal/mol for the alanine dimers [13]. Tables S1–S3 in the Supplementary material show exactly the same trend for all cysteine dimers calculated here, that is, comparatively much higher binding energies are obtained when only ZPE-corrected are considered with expected values of 25.9, 28.9, 33.7 kcal/mol for the $n \cdots n$, $n \cdots z$, and $z \cdots z$ cases, respectively. Notice that these numbers are up to over 6 times larger than the 5.0 kcal/mol binding energy arising because of a single hydrogen bond in the archetypal water dimer [59]. Finally, notice that those structures being unstable towards fragmentation ($BE < 0$) have minimal populations and thus do not contribute to the expected value of the binding energy. The role of dispersive interactions is clearly seen in the fact that when the D3 correction is removed from B3LYP, all strongly bound isomers become unstable towards fragmentation (values within parentheses in Tables S1–S3). For the cysteine dimers with positive binding energies, Tables S1–S3 show that the structures with the largest populations are strongly bonded.

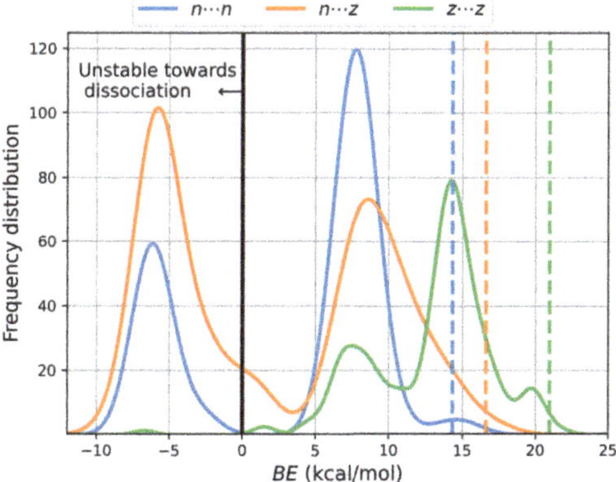

Figure 2. Distribution of binding energies of the cysteine dimers using the Gibbs energies. Dashed vertical lines mark the expected value for each potential energy surface using the Boltzmann populations at room conditions as weighing factors.

As general structural features of the cysteine dimers, we point out that in all cases where neutral monomers are involved, n_2 (no intramolecular HB, Table 1) leads to lower energy dimers. Additionally, except for D_1^{nz}, in all structures that contain n_1, the intramolecular HB in n_1 remains in the dimer. A surprising result is that contrary to the well known structures of the dimers of carboxylic acids, out of the 244 well characterized $n \cdots n$ local minima, only two (D_5^{nn}, $\%x_i = 1.7$ and D_7^{nn}, $\%x_i = 1.3$) exhibit the traditional eight atom, cyclic double C=O$\cdots$H–O stabilizing network. We attribute this to two factors: one, the influence of the solvent which favors other configurations, and two, the intramolecular hydrogen bond occupying the O–H bond in n_1 remains in all but one $n \cdots n$ dimer; thus, this bond is not available for intermolecular bonding (see the dissection of intermolecular bonding interactions below).

3.2. Bonding

The configurational space for the cysteine dimers is complex and rich. We located and characterized a total of 746 structures and there might as well be many more. This geometrical variety arises because of the large number of possible interactions discussed above. Our stochastic search and subsequent dissection of bonding interactions (see below) uncovers an astonishing total of 80 well characterized physically different types of direct intermolecular contacts listed in Tables 3 and 4. Gratifyingly, the found structures account for every single one of the 20 possible hydrogen bonds among the monomers as exposed in the Introduction and also reveal additional salt bridges, dihydrogen bonds, and a number of exotic $X \cdots Y$ (X, Y = O, S, N, C) and $C \cdots H–C$, $C \cdots H–S$ contacts. Notice that the $n \cdots n$ dimers exhibit a well balanced field of all non-charged interactions while the $z \cdots z$ dimers favor the salt bridges by a long shot (159 appearances) and, to a lesser extent, other interactions where only one of the fragments is charged. What should be clear is that the largest contributors to the stabilization of the dimers are $N \cdots H–O$ interactions in the $n \cdots n$ dimers, $C=O^{-} \cdots H–O$ in $n \cdots z$, and $C=O^{-} \cdots H–N^{+}$ salt bridges in $z \cdots z$. We think it is important to point out that, as a general rule, due to the comparatively larger interaction strength, it is the primary neutral, charged, or salt bridges forms of HBs that determine the molecular geometry of the dimers while secondary HBs and exotic contacts are a consequence of the structure (*vide infra*), however, the collective action of the multiple weak interactions on the stabilization energy of each cluster should not be ignored.

Our topological analysis of the electron densities of the 746 equilibrium structures located in this work affords a total of 2894 intermolecular contacts, which are collected into 80 different types in Tables 3 and 4. Without a single exception, positive Laplacians at bond critical points (see Figure S2 in the supplementary material) indicate that bonding in the $n \cdots n, n \cdots z$ and $z \cdots z$ cysteine dimers occurs via closed shell interactions, in the form of either ionic bonding or long range weak interactions. We dissect the nature of intermolecular interactions next.

Table 3. Properties of the 20 types of primary hydrogen bonds, 10 types of secondary hydrogen bonds, and 12 types of dihydrogen bonds stabilizing the cysteine dimers. $N_i^{A \cdots B}$ is the number of times that the interaction appears in the corresponding type of dimers. ϕ_d, ϕ_a are the charge donor and acceptor orbitals as identified from NBO. An example of a dimer containing each particular interaction is given in the rightmost column.

Label	Type	$N_i^{A \cdots B}$			$\phi_d \to \phi_a$	Example
		$n \cdots n$	$z \cdots z$	$n \cdots z$		
Primary hydrogen bonds						
1	$C=O \cdots H–O$	22	-	-	$n_O \to \sigma^*_{H-O}$	D_5^{nn}
2	$C=O^- \cdots H–O$	-	-	52	$n_O \to \sigma^*_{H-O}$	D_{38}^{nz}
3	$C=O \cdots H–N$	46	-	-	$n_O \to \sigma^*_{H-N}$	D_{119}^{nn}
4	$C=O \cdots H–N^+$	-	-	51	$n_O \to \sigma^*_{H-N}$	D_2^{nz}
5	$C=O^- \cdots H–N$	-	-	89	$n_O \to \sigma^*_{H-N}$	D_{55}^{nz}
6	$C=O^- \cdots H–N^+$	-	159	-	$n_O \to \sigma^*_{H-N}$	D_{16}^{zz}
7	$H–O \cdots H–O$	18	-	-	$n_O \to \sigma^*_{H-O}$	D_{172}^{nn}
8	$H–O \cdots H–N$	51	-	-	$n_O \to \sigma^*_{H-N}$	D_{137}^{nn}
9	$H–O \cdots H–N^+$	-	-	32	$n_O \to \sigma^*_{H-N}$	D_{207}^{nz}
10	$N \cdots H–O$	7	-	-	$n_N \to \sigma^*_{H-O}$	D_1^{nn}
11	$N \cdots H–N$	19	-	-	$n_N \to \sigma^*_{H-N}$	D_{174}^{nn}
12	$N \cdots H–N^+$	-	-	19	$n_N \to \sigma^*_{H-N}$	D_1^{nz}
13	$S \cdots H–S$	29	30	54	$n_S \to \sigma^*_{H-S}$	D_{61}^{nn}
14	$S \cdots H–N$	55	-	19	$n_S \to \sigma^*_{H-N}$	D_{51}^{nn}
15	$S \cdots H–N^+$	-	41	52	$n_S \to \sigma^*_{H-N}$	D_{175}^{zz}
16	$N \cdots H–S$	19	-	14	$n_N \to \sigma^*_{H-S}$	D_{108}^{nn}
17	$S \cdots H–O$	16	-	12	$n_S \to \sigma^*_{H-O}$	D_{107}^{nn}
18	$H–O \cdots H–S$	34	-	29	$n_O \to \sigma^*_{H-S}$	D_8^{nn}
19	$C=O \cdots H–S$	49	-	27	$n_O \to \sigma^*_{H-S}$	D_{18}^{nn}
20	$C=O^- \cdots H–S$	-	85	68	$n_O \to \sigma^*_{H-S}$	D_{142}^{zz}
Secondary hydrogen bonds						
21	$C=O \cdots H–C_\alpha$	49	-	26	$n_O \to \sigma^*_{H-C}$	D_{46}^{nn}
22	$C=O \cdots H–C_\beta$	57	-	44	$n_O \to \sigma^*_{H-C}$	D_{37}^{nn}
23	$C=O^- \cdots H–C_\alpha$	-	68	40	$n_O \to \sigma^*_{H-C}$	D_{145}^{zz}
24	$C=O^- \cdots H–C_\beta$	-	108	71	$n_O \to \sigma^*_{H-C}$	D_{144}^{zz}
25	$H–O \cdots H–C_\alpha$	48	-	20	$n_O \to \sigma^*_{H-C}$	D_{144}^{nn}
26	$H–O \cdots H–C_\beta$	71	-	41	$n_O \to \sigma^*_{H-C}$	D_{156}^{nn}
27	$N \cdots H–C_\alpha$	16	-	11	$n_N \to \sigma^*_{H-C}$	D_{188}^{nn}

Table 3. *Cont.*

Label	Type	$N_i^{A\cdots B}$			$\phi_d \to \phi_a$	Example
		$n\cdots n$	$z\cdots z$	$n\cdots z$		
28	$N\cdots H–C_\beta$	34	-	19	$n_N \to \sigma^*_{H–C}$	D^{nn}_{186}
29	$S\cdots H–C_\alpha$	52	24	57	$n_S \to \sigma^*_{H–C}$	D^{nn}_{39}
30	$S\cdots H–C_\beta$	59	37	81	$n_S \to \sigma^*_{H–C}$	D^{nn}_{44}
Dihydrogen contacts						
31	$C_\alpha–H\cdots H–N$	7	-	3	$\sigma_{C–H} \to \sigma^*_{H–N}$	D^{nn}_{135}
32	$C_\beta–H\cdots H–N$	13	-	3	$\sigma_{C–H} \to \sigma^*_{H–N}$	D^{nn}_{86}
33	$C_\beta–H\cdots H–N^+$	-	-	1	$\sigma_{C–H} \to \sigma^*_{H–N}$	D^{nz}_{139}
34	$N–H\cdots H–N$	3	-	-	$\sigma_{N–H} \to \sigma^*_{H–N}$	D^{nn}_{99}
35	$C_\alpha–H\cdots H–C_\alpha$	4	3	-	$\sigma_{C–H} \to \sigma^*_{H–C}$	D^{nn}_{152}
36	$C_\beta–H\cdots H–C_\beta$	42	15	23	$\sigma_{C–H} \to \sigma^*_{H–C}$	D^{nn}_{78}
37	$C_\alpha–H\cdots H–C_\beta$	27	20	22	$\sigma_{C–H} \to \sigma^*_{H–C}$	D^{nn}_{64}
38	$S–H\cdots H–S$	5	3	3	$\sigma_{S–H} \to \sigma^*_{H–S}$	D^{nn}_{62}
39	$S–H\cdots H–N$	5	-	3	$\sigma_{S–H} \to \sigma^*_{H–N}$	D^{nn}_{61}
40	$S–H\cdots H–N^+$	-	1	-	$\sigma_{S–H} \to \sigma^*_{H–N}$	D^{zz}_{53}
41	$S–H\cdots H–C_\alpha$	15	8	12	$\sigma_{S–H} \to \sigma^*_{H–C}$	D^{nn}_{159}
42	$S–H\cdots H–C_\beta$	23	14	21	$\sigma_{S–H} \to \sigma^*_{H–C}$	D^{nn}_{157}

Table 4. Properties of the "exotic" intermolecular contacts found in the cysteine dimers. $N_i^{A\cdots B}$ is the number of times that the interaction appears in the corresponding type of dimers. ϕ_d, ϕ_a are the charge donor and acceptor orbitals as identified from NBO. An example of a dimer containing each particular interaction is given in the rightmost column.

Label	Type	$N_i^{A\cdots B}$			$\phi_d \to \phi_a$	Example
		$n\cdots n$	$z\cdots z$	$n\cdots z$		
O$\cdots$C contacts						
43	$C=O\cdots C=O$	6	-	-	$n_O \to \pi^*_{C=O}$	D^{nn}_{113}
44	$C=O\cdots C=O^-$	-	-	4	$n_O \to \pi^*_{C=O}$	D^{nz}_{6}
45	$C=O^-\cdots C=O$	-	-	5	$n_O \to \pi^*_{C=O}$	D^{nz}_{90}
46	$C=O^-\cdots C=O^-$	-	7	-	$n_O \to \pi^*_{C=O}$	D^{zz}_{14}
47	$C=O\cdots C_\alpha–C$	5	-	-	$\pi_{C=O} \to \sigma^*_{C–C}$	D^{nn}_{112}
48	$C=O^-\cdots C_\alpha–C$	-	-	1	$n_O \to \sigma^*_{C–C}$	D^{nz}_{125}
49	$C=O\cdots C_\beta–S$	2	-	2	$n_O \to \sigma^*_{C–S}$	D^{nn}_{25}
50	$C=O^-\cdots C_\beta–S$	-	2	2	$n_O \to \sigma^*_{C–S}$	D^{zz}_{132}
51	$H–O\cdots C=O$	4	-	-	$n_O \to \pi^*_{C=O}$	D^{nn}_{161}
52	$H–O\cdots C_\beta–S$	2	-	1	$n_O \to \sigma^*_{C–S}$	D^{nn}_{80}
O$\cdots$O contacts						
53	$C=O\cdots O=C$	14	-	-	$\pi_{C=O} \to \pi^*_{O=C}$	D^{nn}_{102}
54	$C=O^-\cdots {}^-O=C$	-	17	-	$n_O \to \pi^*_{O=C}$	D^{zz}_{10}
55	$C=O^-\cdots O=C$	-	-	14	$n_O \to \pi^*_{O=C}$	D^{nz}_{23}
56	$C=O\cdots O–H$	26	-	-	$n_O \to \sigma^*_{O–H}$	D^{nn}_{230}
57	$C=O^-\cdots O–H$	-	-	13	$n_O \to \sigma^*_{O–H}$	D^{nz}_{217}
58	$H–O\cdots O–H$	11	-	-	$n_O \to \sigma^*_{O–H}$	D^{nn}_{45}
O$\cdots$N contacts						
59	$N\cdots O=C$	2	-	-	$n_N \to \pi^*_{O=C}$	D^{nn}_{11}
60	$N\cdots O–H$	2	-	-	$n_N \to \sigma^*_{O–H}$	D^{nn}_{35}
61	$C=O^-\cdots {}^+N–H$	-	16	-	$n_O \to \sigma^*_{N–H}$	D^{zz}_{117}
62	$C=O\cdots {}^+N–H$	-	-	3	$n_O \to \sigma^*_{N–H}$	D^{nz}_{12}
63	$H\text{-}O\cdots {}^+N–H$	-	-	1	$n_O \to \sigma^*_{N–H}$	D^{nz}_{15}
N$\cdots$C contacts						
64	$N\cdots C=O$	3	-	-	$n_N \to \pi^*_{C=O}$	D^{nn}_{130}
N$\cdots$N contacts						
65	$N\cdots N–H$	2	-	-	$n_N \to \sigma^*_{N–H}$	D^{nn}_{173}
C$\cdots$H contacts						
66	$H–C_\beta\cdots H–C_\alpha$	1	2	2	$\sigma_{H–C} \to \sigma^*_{H–C}$	D^{nn}_{70}
67	$H–C_\beta\cdots H–C_\beta$	2	2	2	$\sigma_{H–C} \to \sigma^*_{H–C}$	D^{nn}_{205}
68	$H–C_\beta\cdots H–S$	2	1	3	$\sigma_{H–C} \to \sigma^*_{H–S}$	D^{nn}_{105}
69	$^-O=C\cdots H–C_\alpha$	-	1	-	$\pi_{O=C} \to \sigma^*_{H–C}$	D^{zz}_{32}
70	$^-O=C\cdots H–C_\beta$	-	6	1	$\pi_{O=C} \to \sigma^*_{H–C}$	D^{zz}_{20}
S$\cdots$S contacts						
71	$S\cdots S–H$	12	7	11	$n_S \to \sigma^*_{S–H}$	D^{nn}_{71}
S$\cdots$C contacts						
72	$S\cdots C=O$	7	-	2	$n_S \to \pi^*_{C=O}$	D^{nn}_{135}
73	$S\cdots C=O^-$	-	12	2	$n_S \to \pi^*_{C=O}$	D^{zz}_{113}
74	$H–C_\alpha\cdots S–H$	-	1	-	$\sigma_{H–C} \to \sigma^*_{S–H}$	D^{zz}_{109}
75	$H–C_\beta\cdots S–H$	-	1	-	$\sigma_{H–C} \to \sigma^*_{S–H}$	D^{zz}_{110}

Table 4. *Cont.*

Label	Type	$N_i^{A\cdots B}$			$\phi_d \to \phi_a$	Example
		$n\cdots n$	$z\cdots z$	$n\cdots z$		
S$\cdots$O contacts						
76	C=O$\cdots$S–H	10	-	10	$n_O \to \sigma^*_{S-H}$	D^{nn}_{127}
77	C=O$^-\cdots$S–H	-	27	11	$\pi_{C=O} \to \sigma^*_{S-H}$	D^{zz}_{51}
78	S$\cdots$O–H	26	-	12	$n_S \to \sigma^*_{O-H}$	D^{nn}_{148}
S$\cdots$N contacts						
79	N$\cdots$S–H	9	-	3	$n_N \to \sigma^*_{S-H}$	D^{nn}_{173}
80	S$\cdots{}^+$N–H	-	3	3	$n_S \to \sigma^*_{N-H}$	D^{zz}_{178}

3.2.1. Interaction Distances

Figure 3 shows the distribution of the distances associated with individual intermolecular contacts separated by interaction type, that is, primary and secondary hydrogen bonds, dihydrogen bonds, and exotic contacts for all dimers. Remarkably, the spectrum of A$\cdots$B distances for direct intermolecular contacts covers a wide range, from the very short (1.50 Å for a C=O$^- \cdots$H–O in D^{nz}_{38}) to the very large (4.19 Å for the exotic S$\cdots$S in D^{nn}_{77}), which sensibly departs in both directions from the reference 1.98 Å in the isolated gas phase water dimer. Notice that regardless of the constituting monomers, only primary hydrogen bonds fall below 1.98 Å. In a classical sense, a zwitterion may be conveniently seen as two remote charge islands within the same molecule, in this view, the effect of the charges in the structural complexity of the cysteine dimers is clear: on one hand, intermolecular distances are reduced for the dimers with more charge islands, i.e., $r^{nn}_{AB} > r^{nz}_{AB} > r^{zz}_{AB}$, on the other, the structural complexity is also sensibly reduced for the more charge-separated species because the $n\cdots n$ distributions have more peaks than $n\cdots z$ which in turn have more peaks than $z\cdots z$. In addition, it may be argued that among all the types of interactions stabilizing the cysteine dimers, salt bridges should be the strongest and thus the most important structural determining factor whenever they occur. Indeed, the lowest energy $z\cdots z$ dimers with populations larger than 5% shown in Figure 1, are actually stabilized by two salt bridges. Notice that the center of the peak for the distribution of C=O$^- \cdots$H-N$^+$ distances (1.66 Å, Figure 3C) is actually larger than 1.57 Å, the center of the peak for the distribution of C=O$^- \cdots$H-O interactions (Figure 3B), which are *a priori* not as strong as the salt bridges but which dictate the structures of the $n\cdots z$ dimers, the reason for this apparent contradiction is that formation of the two salt bridges confers structural rigidity to the clusters.

When immersed in a continuum aqueous environment, there is partial dissociation of the O–H bonds upon the formation of the dimers. Figure 4 shows the changes in the corresponding distances and Wiberg Bond Indices (WBI) compared against the reference monomer. Evidence for partial dissociation is provided by the peak centered at $\approx$0.58 WBI, which actually corresponds to O–H groups of the low energy, high population dimers where neutral monomers are involved.

3.2.2. Electron Densities at Bond Critical Points, $\rho(\mathbf{r}_c)$

The relationship between electron density at bond critical points and the nature of the interaction is clear: large accumulations of electron densities at BCPs indicate that the electrons are shared between two fragments or atoms, otherwise known as covalent bonding. Conversely, small electron densities at BCPs indicate that the electrons are displaced towards the nuclei, thus signaling either ionic bonding or long range interactions. Figure 5 shows the values for the calculated electron densities at the 2894 bond critical points associated to intermolecular interactions in the 746 cysteine dimers. Electron densities at those points cover the $[9.1 \times 10^{-4}, 7.6 \times 10^{-2}]$ a.u. interval. These values are sensibly smaller than the 0.24 and 0.35 a.u obtained for the covalent C–C and O–H bonds in D^{nn}_1. The smaller electron densities correspond to secondary HBs and exotic contacts while among primary HBs, those with the smallest densities involve the S–H group. Only some primary HBs and salt bridges have larger densities than the reference water dimer. The distributions plotted in Figure 5 are wide; thus, there are many possibilities for the same type of interaction. Finally, as expected [49,64–66], there seems to be an inverse correlation between interaction distance and electron density at intermolecular BCPs.

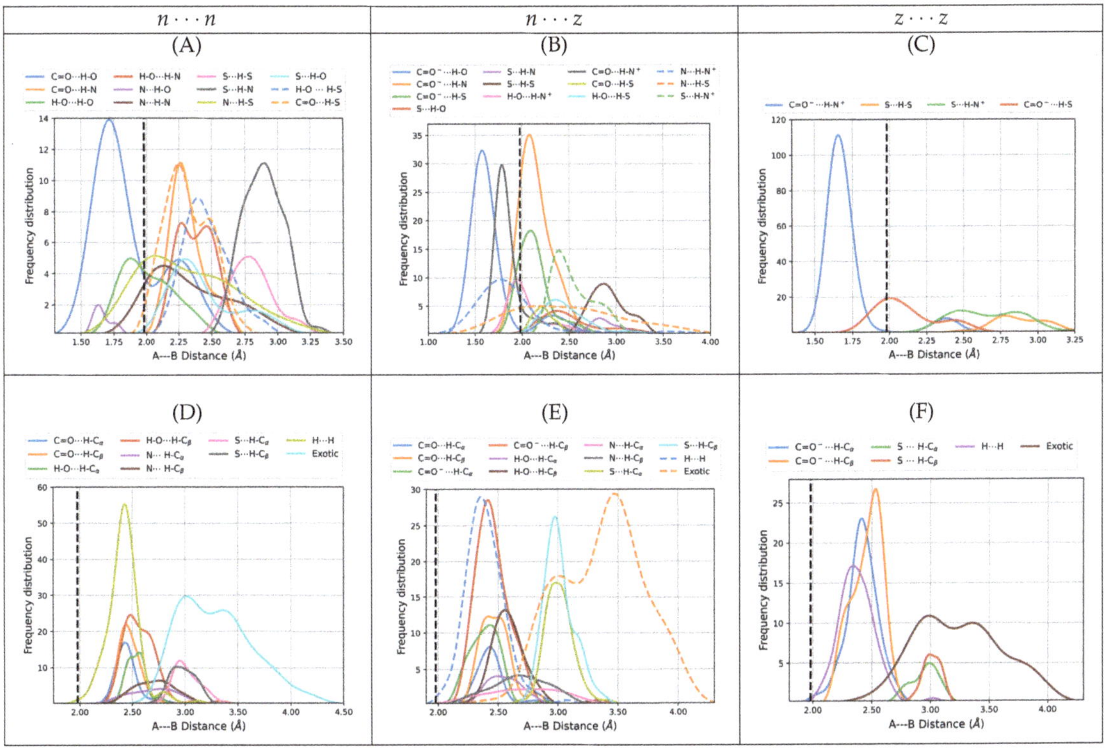

Figure 3. Distributions of the A$\cdots$B distances for intermolecular contacts for all dimers of cysteine found in this work. The distributions are fitted to the actual histograms, so the center of the peaks of the distributions are statistically relevant. The top row shows only primary hydrogen bonds including salt bridges. The bottom row shows secondary hydrogen bonds, dihydrogen bonds, and all exotic interactions. The left column is reserved for the $n\cdots n$ dimers (subfigures (**A,D**)), the middle column for $n\cdots z$ (subfigures (**B,E**)) and the right column for $z\cdots z$ (subfigures (**C,F**)). All distances taken from the B3LYP–D3/6–311++G(d,p) potential energy surfaces with water represented as a continuum solvent. The dashed vertical lines at 1.98 Å mark the reference H$\cdots$O distance in the gas phase water dimer.

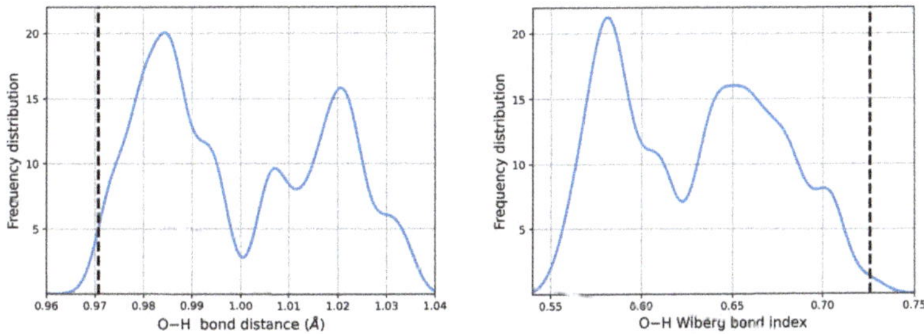

Figure 4. Variation of the O–H bond properties as a consequence of intermolecular interactions. The reference values for the n_2 cysteine monomer are included as vertical dashed lines.

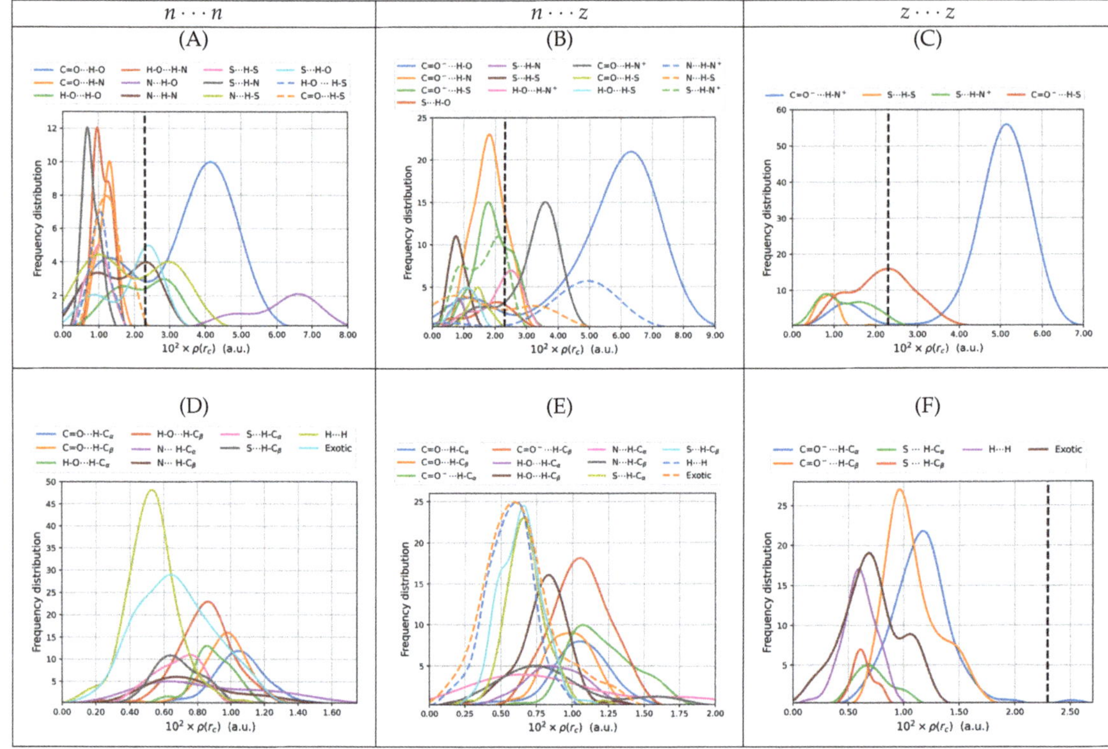

Figure 5. Electron densities at bond critical points for intermolecular contacts for all dimers of cysteine found in this work. The top row shows only primary hydrogen bonds including salt bridges. The bottom row shows secondary hydrogen bonds, dihydrogen bonds, and all exotic interactions. The left column is reserved for the $n \cdots n$ dimers (subfigures (**A**,**D**)), the middle column for $n \cdots z$ (subfigures (**B**,**E**)) and the right column for $z \cdots z$ (subfigures (**C**,**F**)). All values taken from the B3LYP–D3/6–311++G(d, p) potential energy surfaces with water represented as a continuum solvent. The dashed vertical lines mark the reference value for the gas phase water dimer.

3.2.3. Bond Degree Parameters $\mathcal{H}(\mathbf{r}_c)/\rho(\mathbf{r}_c)$

The bond degree parameter is related to chemical bonding as follows. Kinetic energy is everywhere positive and repulsive ($mv^2/2 = p^2/2m > 0$ in classical mechanics) while potential energy is everywhere negative and attractive. The total energy is the sum of kinetic and potential energies, $\mathcal{H} = \mathcal{G} + \mathcal{V}$; thus, its sign reveals the winner of the local kinetic vs potential energy tug of war and dictates the nature of the interaction. Indeed, positive total energies at BCPs are obtained when there is a local dominance of the repulsive kinetic energy, indicating local depletion of electrons in the internuclear region and displacement of the electron density associated with the particular bonding interactions towards the nuclei. Conversely, negative total energies are obtained when there is a local dominance of the attractive potential energy indicating that there is shared electron density concentrated in the internuclear region and signaling an increasingly covalent character of the interaction. An alternative rigorous physical meaning to energy densities is offered by a dimensional analysis: energy density has units of pressure ($E/V = F/A = P$); thus, local negative energy densities may be equated to negative quantum pressures which strongly attract electrons towards the BCP, indicating increasingly covalent interactions while local positive energy densities correspond to positive quantum pressures that push electrons away from the BCPs towards the nuclei, indicating anionic or long range interactions.

It is well known that the sign of $\nabla^2\rho(\mathbf{r}_c)$ is not a sufficient criterium to establish the nature of the interaction in every case [28,67–70], specifically, it is quite often the case that a particular interaction has both positive Laplacian and negative bond degree parameter at the same time. Thus, the bond degree parameter is used in conjunction with the Laplacian of the electron density at BCPs to remove any ambiguity according to Rozas et al. [68]: weak to medium strength hydrogen bonds have both $\nabla^2\rho(\mathbf{r}_c), \mathcal{H}(\mathbf{r}_c)/\rho(\mathbf{r}_c) > 0$, strong hydrogen bonds have $\nabla^2\rho(\mathbf{r}_c) > 0, \mathcal{H}(\mathbf{r}_c)/\rho(\mathbf{r}_c) < 0$ and very strong HBs have both $\nabla^2\rho(\mathbf{r}_c), \mathcal{H}(\mathbf{r}_c)/\rho(\mathbf{r}_c) < 0$. Figure 6 plots distributions of the bond degree parameters for all dimers found in this work. It is clear from the distributions of $\mathcal{H}(\mathbf{r}_c)/\rho(\mathbf{r}_c)$ that all intermolecular contacts found here cover a wide spectrum of possibilities with a substantial number of only primary hydrogen bonds or salt bridges having $\mathcal{H}(\mathbf{r}_c)/\rho(\mathbf{r}_c) < 0$ (Figure 6A–C), thus should be considered as strong contacts by the above criteria. The wide spectrum of bond degree parameters, the large number of structural possibilities and the strong character of the interactions have deep implications in the biological role of cysteine and of the aminoacids that make up proteins and biomolecules: similar results have been obtained for example in the interactions between the spike protein of SARS-COV-2 and the ACE2 receptors [48,71] and between the envelope protein of the Zika virus and the glycosaminoglycans that act as receptors [47]. In the case of SARS-COV-2, the formation of strong salt bridges and hydrogen bonds is one of the main factors of the pressure driving the evolution of the virus towards new variants. For the cysteine dimers, many of the primary hydrogen bonds with positive bond degree parameters are located to the left of the reference isolated gas phase water dimer, which confers them medium to strong character. All secondary and exotic contacts (Figure 6D–F) exhibit positive bond degree parameters and many areas actually to the right of the reference water dimer; thus, they are classified as weak. As a general rule, hydrogen bonds involving the carbonyl, carboxylate and amino groups as electron donors and the hydroxyl, amino and ammonium groups as electron acceptors, are the ones with highly negative $\mathcal{H}(\mathbf{r}_c)/\rho(\mathbf{r}_c)$ values. Some HBs involving the thiol group, either as donor or acceptor, have slightly negative bond degree parameters.

3.2.4. Virial Ratios, $|\mathcal{V}(\mathbf{r}_c)|/\mathcal{G}(\mathbf{r}_c)$

Analysis of the virial ratios at bond critical points serves as a more quantitative description of the nature of the interactions than the Laplacians of the electron density and the bond degree parameters. See the works of Grabowski [28] and of Rozas et al. [68] for a formal description of how the virial ratio is related to bonding. In short, local depletion of electron density (local dominance of the repulsive kinetic energy), which is indicative of ionic or long interactions have $0 < |\mathcal{V}(\mathbf{r}_c)|/\mathcal{G}(\mathbf{r}_c) < 1$, local concentration of electron density (local dominance of the attractive potential energy), indicative of covalent interactions have $|\mathcal{V}(\mathbf{r}_c)|/\mathcal{G}(\mathbf{r}_c) > 2$, and the $1 < |\mathcal{V}(\mathbf{r}_c)|/\mathcal{G}(\mathbf{r}_c) < 2$ interval describes interactions with mixed contributions.

Figure 7 shows the distribution of the virial ratios for all dimers found in this work, which cover the [0.61, 1.49] interval for primary hydrogen bonds and salt bridges (Figure 7A–C) and the [0.56, 0.92] interval for secondary HBs, exotic and dihydrogen contacts (Figure 7A–C). Notice that as in the analysis of the previous descriptors, the fitted distributions go a little beyond the actual limits. It is quite revealing that a large number of contacts, especially those involving the carbonyl group have virial ratios larger than 1, which confers them a high degree of covalency while not being formal bonds. Interestingly, these include the charged carboxylate which may naively be thought as being involved in highly ionic contacts. Virial ratios larger than 1 transcend the carbonyl group, which is indeed the case for the following HBs: $N\cdots H{-}O, C{=}O\cdots H{-}O, S\cdots H{-}O, N\cdots H{-}S$ for $n\cdots n$ dimers, $C{=}O^-\cdots H{-}O, N\cdots H{-}N^+, S\cdots H{-}N^+, C{=}O\cdots H{-}N^+$ for $n\cdots z$ and for all salt bridges in $z\cdots z$. This high covalency of the *a priori* ionic contacts has been reported for other cases, including for example the microsolvation of charged species [49,53,72]. Most secondary HBs, $H\cdots H$, and exotic contacts have virials smaller than the water dimer reference.

Surprisingly, the thiol group in a number of cases is involved in stronger interactions than the H–O$\cdots$H–O and N$\cdots$H–O contacts.

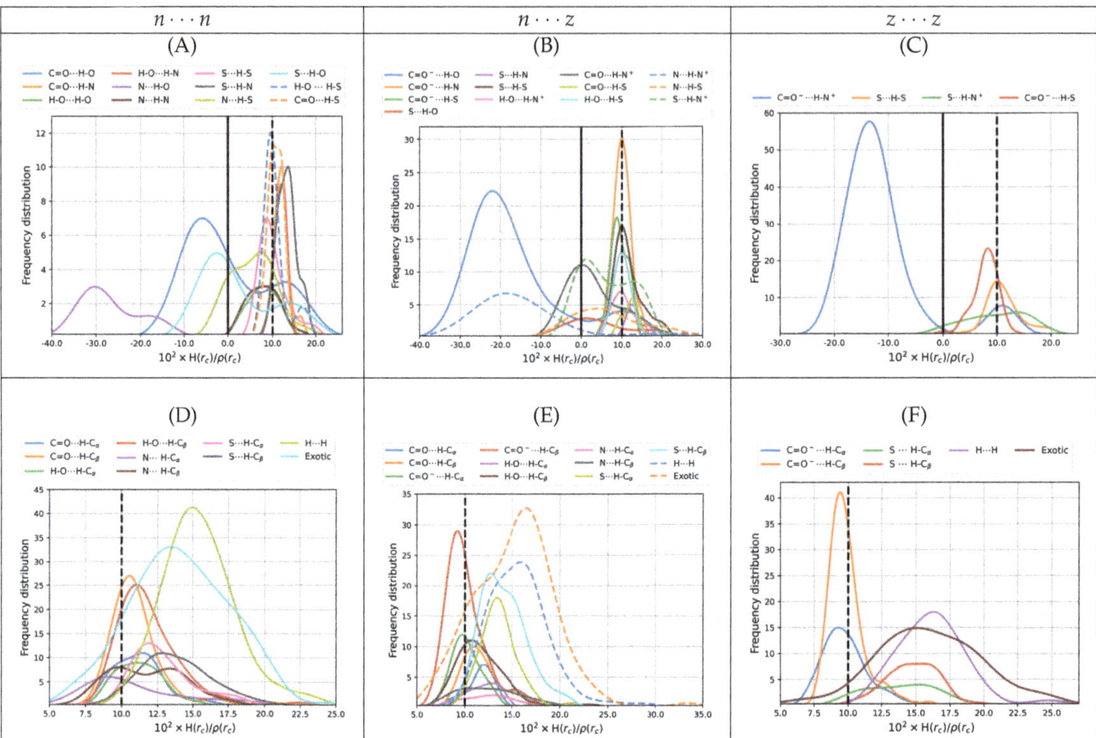

Figure 6. Bond degree parameters for intermolecular contacts for all dimers of cysteine found in this work. The top row shows only primary hydrogen bonds including salt bridges. The bottom row shows secondary hydrogen bonds, dihydrogen bonds, and all exotic interactions. The left column is reserved for the $n \cdots n$ dimers (subfigures (**A**,**D**)), the middle column for $n \cdots z$ (subfigures (**B**,**E**)) and the right column for $z \cdots z$ (subfigures (**C**,**F**)). All values taken from the B3LYP–D3/6–311++G(d, p) potential energy surfaces with water represented as a continuum solvent. Solid vertical lines mark the QTAIM boundaries separating locally stabilizing from the locally destabilizing interactions. Dashed vertical lines mark the reference value for the gas phase water dimer.

3.2.5. NBO and NCI Picture of Intermolecular Interactions

Intermolecular interactions have been successfully studied under the NBO formalism in a wide range of problems [30,33]. In the particular case of the cysteine dimers, we proceeded to identify the localized donor Lewis orbitals from which charge is transferred to acceptor orbitals according to the $\phi_d \rightarrow \phi_a$ scheme. Tables 3 and 4 list the involved orbitals for each one of the 80 types of interactions found in this work, Figure 1 provides the corresponding surfaces for those dimers with populations larger than 5%. Once identified, we quantified the strength of the orbital interaction by second order perturbation theory on the Fock matrix as given by $-E_{d \rightarrow a}^{(2)} = q_d |\langle \phi_d | \mathcal{F} | \phi_a \rangle|^2 / (E_a - E_d)$. With this procedure, the strength of the interaction is directly related to the magnitude of $E_{d \rightarrow a}^{(2)}$.

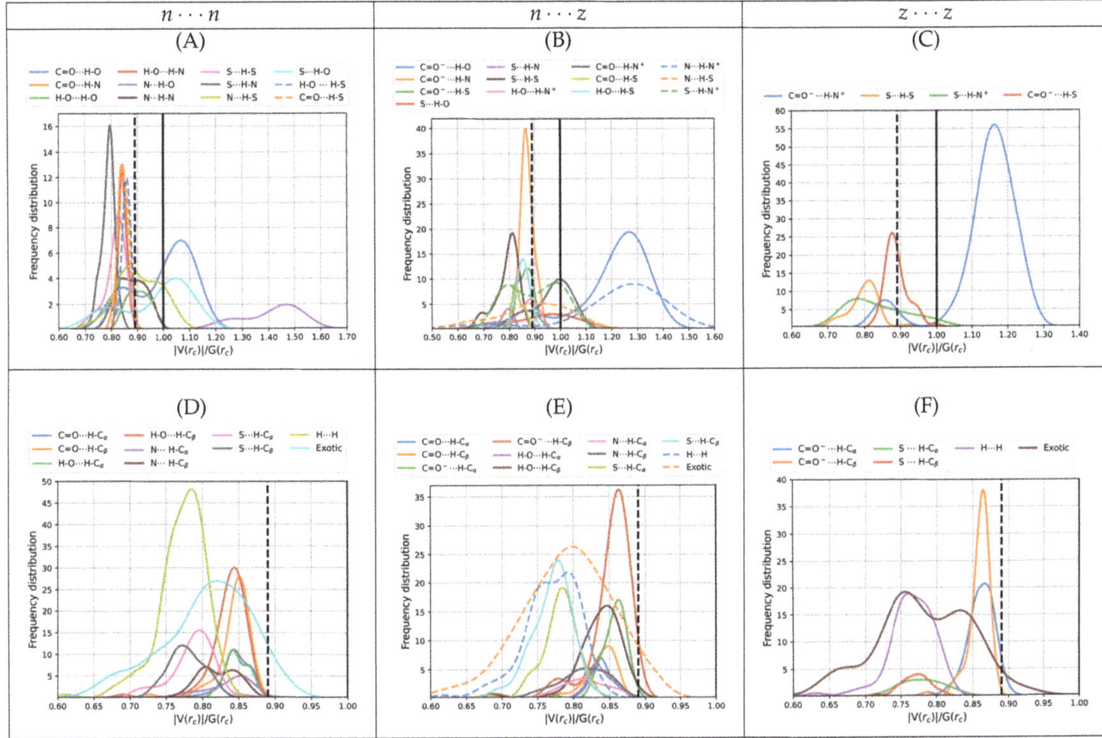

Figure 7. Virial ratios at bond critical points for intermolecular contacts for all dimers of cysteine found in this work. The top row shows only primary hydrogen bonds including salt bridges. The bottom row shows secondary hydrogen bonds, dihydrogen bonds, and all exotic interactions. The left column is reserved for the $n \cdots n$ dimers (subfigures (**A,D**)), the middle column for $n \cdots z$ (subfigures (**B,E**)) and the right column for $z \cdots z$ (subfigures (**C,F**)). All values taken from the B3LYP–D3/6–311++G(d, p) potential energy surfaces with water represented as a continuum solvent. Vertical solid lines mark the QTAIM boundaries separating long range from intermediate character interactions. Dashed vertical lines mark the reference value for the gas phase water dimer.

Donor → acceptor orbital interactions resulting in primary hydrogen bonds and salt bridges in $n \cdots n$, $n \cdots z$ and $z \cdots z$ cysteine dimers include everything from the very weak to the very strong, covering the wide [0.06, 50.30] kcal/mol interval (Figure 8A–C). Salt bridges exhibit an uncommonly complex distribution of energies with several shoulders. Interestingly, the strongest contacts are not salt bridges but rather $N \cdots H$–O primary hydrogen bonds, listed as interaction **10** in Table 3 and shown in Figure 8A. This interaction type, which is also the strongest intermolecular contact found in alanine dimers [13], arises from $n_N \rightarrow \sigma^*_{H-O}$ charge transfer in $n \cdots n$ dimers. Next in the strength hierarchy are the highly ionic $C=O^- \cdots H$–O and $N \cdots H$–N^+ contacts arising from $n_O \rightarrow \sigma^*_{H-O}$ and $n_N \rightarrow \sigma^*_{H-N}$ charge transfers in $n \cdots z$ dimers. These are listed as interactions **2, 12** with the corresponding distributions shown in Figure 8B. $C=O^- \cdots H$–N^+ salt bridges in $z \cdots z$ dimers come only third in the interaction energy scale. They arise from $n_O \rightarrow \sigma^*_{H-N}$ charge transfer, are listed as interaction **6** and the corresponding distributions are shown in Figure 8C. These sets of interactions are present on the structures with populations higher than 5%. In a manner consistent with the QTAIM descriptors analyzed above, a large number of primary HBs and salt bridges have interaction energies larger than 6.63 [30] kcal/mol, the orbital interaction energy for the reference water dimer, however, no secondary hydrogen bond, no exotic contact and no dihydrogen bond exceed the reference.

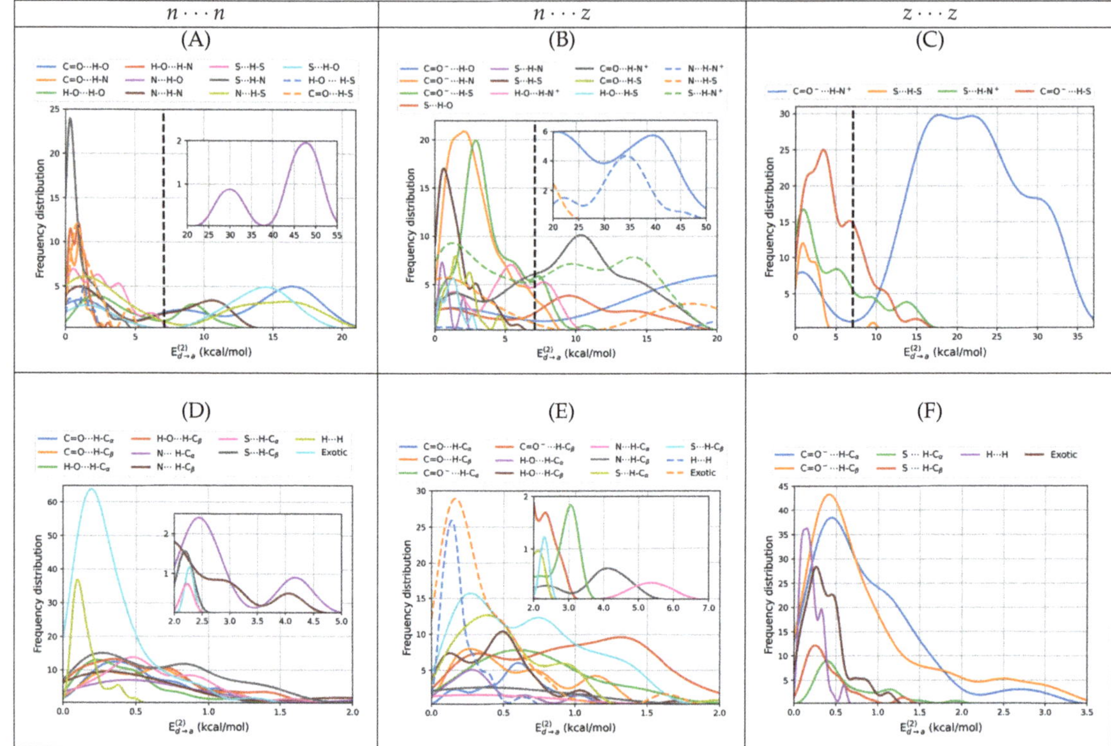

Figure 8. Donor $\cdots$ acceptor NBO energies for intermolecular contacts for all dimers of cysteine found in this work. The top row shows only primary hydrogen bonds including salt bridges. The bottom row shows secondary hydrogen bonds, dihydrogen bonds, and all exotic interactions. The left column is reserved for the $n \cdots n$ dimers (subfigures (**A**,**D**)), the middle column for $n \cdots z$ (subfigures (**B**,**E**)) and the right column for $z \cdots z$ (subfigures (**C**,**F**)). All values taken from the B3LYP–D3/6–311++G(d, p) potential energy surfaces with water represented as a continuum solvent. The dashed vertical lines mark the reference value for the gas phase water dimer.

As stated above, primary HBs and salt bridges determine the structure of the dimers. According to Table 3, they always arise from orbital interactions of the $n_X \rightarrow \sigma^*_{H-Y}$ type with X,Y = O, N, S. As also stated above, secondary HBs, dihydrogen bonds and exotic contacts are usually a consequence of the structure. Notwithstanding, the orbitals involved in the weaker interactions offer a quite interesting and rather uncommon picture. First, notice that all exotic O$\cdots$O contacts (**53–58** in Table 4) put the two negative ends of the fragments with various degrees of negative character in direct contact, with the most severe case being interaction **54** connecting two formal negative charges with no intermediaries. Second, notice that the 139 interactions grouped into **58, 61–63, 65, 71, 76, 78–80** may all be described by the general $n_X \rightarrow \sigma^*_{Y-H}$ charge transfer scheme with X, Y = O, N, S. These correspond to what David et al. [13] have called inverted hydrogen bonds because the lone pair on X donates electron charge to an antibonding σ^*_{Y-H} orbital which is inverted from the usual σ^*_{H-Y}, in other words, the charge donation occurs between two orbitals overlapping from one negative atom to another negative atom with no bridging proton. Anti electrostatic hydrogen bonds of the type described in this paragraph have been reported by Weinhold and Klein [73].

Figure 9 shows the obtained non covalent surfaces as well as the thoroughs of the reduced gradients for the largest binding energy dimers in the $n \cdots n, n \cdots z, z \cdots z$ potential energy surfaces, we include the corresponding interacting NBOs to help visualize the source of the NCI surfaces. See Figures S3–S5 in the supplementary material for the corresponding surfaces in all dimers with populations larger than 5%. The standard NCI color code [34,35] assigns green surfaces to weakly bonding contacts and blue surfaces to strong interactions, for the thoroughs, negative values of $\text{sign}(\lambda_2)\rho$ reveal bonding interactions [34] which are weaker when $\text{sign}(\lambda_2)\rho \approx 0$. NCI reveals that the two salt bridges in the $z \cdots z$ dimers are actually very similar (in fact, they cannot be told apart in the thoroughs) and that in most cases, large stabilizing surfaces arising from individual contacts transferring tiny amounts of charge to the interstitial region have significant contributions to the overall binding of the dimers. Unexpectedly, these charge transfer contributions are major contributors to the charged cases. Notice that charge transfer to the interstitial region appears to be the norm when several molecular units are stabilized via non covalent interactions: these fluxional surfaces of charge have been found to be a major player in the molecular interpretation of hydrophobicity [46], in the initial recognition and attachment of viruses to cell receptors [47,48,71], in the microsolvation and encapsulation of charged and neutral species, in the microscopic structure of ionic liquids [51], etc.

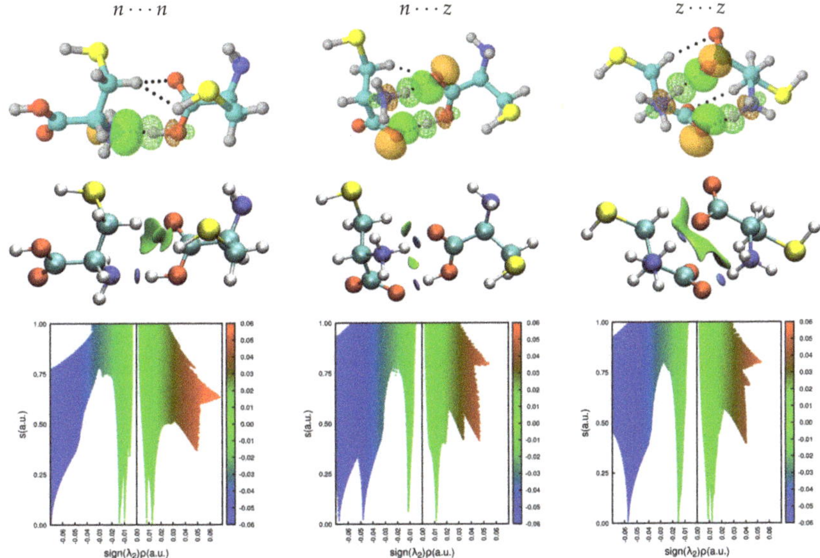

Figure 9. NCI surfaces and thoroughs describing the intermolecular contacts for the highest binding energy cysteine dimers. NBO pictures are also included to ease visualization of the interactions.

4. Discussion and Context

Accurate description and characterization of chemical bonding is a notoriously hard problem in chemistry, whose difficulty is magnified when dealing with weak intermolecular non covalent interactions. When studying molecules and their interactions, a large portion of the conceptual framework developed by experimentalists and theoreticians invokes a number of useful ideas that correspond to non observable quantities (partial atom charges, orbital interactions, virial ratios at BCPs, etc.); thus, there are no quantum mechanical operators whose expected values may be used to calculate them and therefore, approximate methods, however accurate, are used to determine these quantities. This approach has a fundamental problem: each quantity may be obtained by several methods and the results quite often vary among them. In this context, it is impressive and certainly reassuring that QTAIM, NBO and NCI, which are conceptually and methodologically substantially different, afford a consistent, complementary picture of intermolecular bonding in the dimers of cysteine.

The large number of interactions, isomers, and types of binary contacts are intimately connected to the problem of molecular evolution and to the complexity of life observable on this planet. By virtue of the large number of accessible states, molecular systems where specific cysteine to cysteine contacts are observed are thermodynamically favored because of the ever increasing entropy of the universe, in other words, this type of systems will evolve towards equilibrium states with large structural diversity. Since the interactions dissected here are responsible for the molecular interactions between all aminoacid pairs, this argument of entropy driving molecular evolution readily applies to large proteins and biomolecules.

5. Summary and Conclusions

An intensive exploration of the potential energy surfaces for the interaction of neutral and charged cysteine monomers to form dimers in an aqueous environment represented by a continuum afforded a large number of isomers, amounting to 746 well characterized local minima. The isomers with the largest population are distributed within small energy differences of the putative global minima. Ten potential energy surfaces were explored in total for the $n \cdots n$, $n \cdots z$, and $z \cdots z$ combinations with two neutral (n) and two zwitterionic (z) forms. A number of strongly bound dimers were found, with interaction energies exceeding 20 kcal/mol in several cases and with interaction distances covering the very small to the very large in the [1.50, 4.19] Å interval. The nature of intermolecular bonding interactions was dissected using QTAIM, NCI, and NBO, three conceptually different methods, which for the present case afford consistent, complementary pictures. A total of 80 types of different intermolecular contacts were found in this complex and large universe of dimers. As a general rule, primary hydrogen bonds and salt bridges are the strongest of the interactions and determine the molecular geometry, conversely, secondary hydrogen bonds, exotic $X \cdots Y$ (X = C, N, O, S) and $H \cdots H$ dihydrogen contacts are weaker and most often a consequence of the structure. All interactions, even the highly ionic, may be described by the $\phi_d \rightarrow \phi_a$ orbital charge transfer scheme, leading to accumulation or depletion of electron density at the bond critical points as revealed by topological analysis of the electron densities. The large binding energies mentioned above are the result of unusually strong charge assisted hydrogen bonds and salt bridges. We found that the highly ionic salt bridges have large degrees of covalency; thus, a simplistic electrostatic attraction between positively and negatively charged fragments does not suffice for a proper account of bonding interactions whenever the zwitterions are involved. The weaker secondary hydrogen bonds, exotic $X \cdots Y$ and dihydrogen contacts in no few cases are stronger than, for example, the archetypal hydrogen bond in the water dimer. Moreover, independent of how strong or weak individual interactions are, their collective action cannot be ignored because they lead to the formation of large attractive non-covalent surfaces in the interstitial region between the fragments. A few antielectrostatic contacts [73] as well as a few inverted hydrogen bonds [13] in which the charge transfer occurs between the two negative ends of the fragments, were found; thus, they appear to be of common occurrence in nature.

Supplementary Materials: The following supporting information can be downloaded at: https://www.mdpi.com/article/10.3390/molecules27248665/s1, Figure S1: Electrostatic potential surfaces for the cysteine monomers.; Figure S2: Distributions of the Laplacians of the electron densities at bond critical points for all dimers.; Figures S3–S5: Additional NBO and NCI plots of descriptors of bonding interactions for the dimers with populations larger than 5%; Table S1: Binding energies and energy differences for neutral dimers.; Table S2: Binding energies and energy differences for mixed dimers. Table S3: Binding energies and energy differences for zwitterionic dimers. Cartesian coordinates for the entire set of 746 dimers are also provided.

Author Contributions: Conceptualization, S.G. (Sara Gómez), C.C. and A.R.; Data curation, S.G. (Santiago Gómez), S.G. (Sara Gómez), J.D. and D.G.; Formal analysis, S.G. (Santiago Gómez), S.G. (Sara Gómez), J.D., D.G., C.C. and A.R.; Funding acquisition, C.C. and A.R.; Investigation, S.G. (Santiago Gómez), S.G. (Sara Gómez), J.D. and D.G.; Methodology, S.G. (Sara Gómez) and A.R.; Project administration, A.R.; Resources, C.C. and A.R.; Supervision, S.G. (Sara Gómez) and A.R.; Validation, S.G. (Santiago Gómez), J.D. and D.G.; Visualization, S.G. (Santiago Gómez); Writing—original draft, S.G. (Santiago Gómez), S.G. (Sara Gómez), J.D., D.G. and A.R.; Writing—review and editing, S.G. (Santiago Gómez), S.G. (Sara Gómez), J.D., D.G., C.C. and A.R. All authors have read and agreed to the published version of the manuscript.

Funding: This research was funded by Universidad de Antioquia, grant "Estrategia para la sostenibilidad".

Institutional Review Board Statement: Not applicable.

Informed Consent Statement: Not applicable.

Data Availability Statement: All data are available within the article or Supplementary Materials.

Acknowledgments: Internal support from Universidad de Antioquia via "Estrategia para la sostenibilidad" is acknowledged. We gratefully acknowledge the Center for High Performance Computing (CHPC) at SNS for providing the computational infrastructure.

Conflicts of Interest: The authors declare no conflict of interest.

References

1. Wilke, J.J.; Lind, M.C.; Schaefer, H.F., III; Császár, A.G.; Allen, W.D. Conformers of gaseous cysteine. *J. Chem. Theory Comput.* **2009**, *5*, 1511–1523. [CrossRef] [PubMed]
2. Giraldo, C.; Gómez, S.; Weinhold, F.; Restrepo, A. Insight into the Mechanism of the Michael Reaction. *ChemPhysChem* **2016**, *17*, 2022–2034. [CrossRef] [PubMed]
3. Cardona, W.; Guerra, D.; Restrepo, A. Reactivity of δ-substituted α, β-unsaturated cyclic lactones with antileishmanial activity. *Mol. Simul.* **2014**, *40*, 477–484. [CrossRef]
4. Clayden, J.; Greeves, N.; Warren, S. *Organic Chemistry*; Oxford University Press: Oxford, UK, 2012.
5. Santos, M.; Moreira, R. Michael acceptors as cysteine protease inhibitors. *Mini-Rev. Med. Chem.* **2007**, *7*, 1040–1050. [CrossRef]
6. Gómez, S.; Ramírez-Malule, H.; Cardona-G, W.; Osorio, E.; Restrepo, A. Double-ring epimerization in the biosynthesis of clavulanic acid. *J. Phys. Chem. A* **2020**, *124*, 9413–9426. [CrossRef] [PubMed]
7. Lu, X.; Galkin, A.; Herzberg, O.; Dunaway-Mariano, D. Arginine deiminase uses an active-site cysteine in nucleophilic catalysis of L-arginine hydrolysis. *J. Am. Chem. Soc.* **2004**, *126*, 5374–5375. [CrossRef] [PubMed]
8. Li, H.; Thomas, G.J., Jr. Studies of virus structure by Raman spectroscopy. Cysteine conformation and sulfhydryl interactions in proteins and viruses. 1. Correlation of the Raman sulfur-hydrogen band with hydrogen bonding and intramolecular geometry in model compounds. *J. Am. Chem. Soc.* **1991**, *113*, 456–462. [CrossRef]
9. Dobrowolski, J.C.; Rode, J.E.; Sadlej, J. Cysteine conformations revisited. *J. Mol. Struct. THEOCHEM* **2007**, *810*, 129–134. [CrossRef]
10. Wiedemann, C.; Kumar, A.; Lang, A.; Ohlenschläger, O. Cysteines and disulfide bonds as structure-forming units: insights from different domains of life and the potential for characterization by NMR. *Front. Chem.* **2020**, *8*, 280. [CrossRef]
11. Fraser, R.; MacRae, T.P.; Sparrow, L.G.; Parry, D. Disulphide bonding in α-keratin. *Int. J. Biol. Macromol.* **1988**, *10*, 106–112. [CrossRef]
12. Bairoch, A.; Apweiler, R. The SWISS-PROT protein sequence database and its supplement TrEMBL in 2000. *Nucleic Acids Res.* **2000**, *28*, 45–48. [CrossRef]
13. David, J.; Gómez, S.; Guerra, D.; Guerra, D.; Restrepo, A. A Comprehensive Picture of the Structures, Energies, and Bonding in the Alanine Dimers. *ChemPhysChem* **2021**, *22*, 2401–2412. [CrossRef]
14. Tonino, P.; Kiss, B.; Strom, J.; Methawasin, M.; Smith, J.E.; Kolb, J.; Labeit, S.; Granzier, H. The giant protein titin regulates the length of the striated muscle thick filament. *Nat. Commun.* **2017**, *8*, 1–11. [CrossRef]
15. Gronert, S.; O'Hair, R.A. Ab initio studies of amino acid conformations. 1. The conformers of alanine, serine, and cysteine. *J. Am. Chem. Soc.* **1995**, *117*, 2071–2081. [CrossRef]
16. Dobrowolski, J.C.; Jamróz, M.H.; Kołos, R.; Rode, J.E.; Sadlej, J. Theoretical Prediction and the First IR Matrix Observation of Several l-Cysteine Molecule Conformers. *ChemPhysChem* **2007**, *8*, 1085–1094. [CrossRef]
17. Sanz, M.E.; Blanco, S.; López, J.C.; Alonso, J.L. Rotational probes of six conformers of neutral cysteine. *Angew. Chem. Int. Ed.* **2008**, *47*, 6216–6220. [CrossRef] [PubMed]
18. Kaminski, M.; Kudelski, A.; Pecul, M. Vibrational optical activity of cysteine in aqueous solution: A comparison of theoretical and experimental spectra. *J. Phys. Chem. B* **2012**, *116*, 4976–4990. [CrossRef]

19. Sadlej, J.; Dobrowolski, J.C.; Rode, J.E.; Jamróz, M.H. Density functional theory study on vibrational circular dichroism as a tool for analysis of intermolecular systems:(1: 1) cysteine- water complex conformations. *J. Phys. Chem. A* **2007**, *111*, 10703–10711. [CrossRef]

20. Liu, S.Z.; Wang, H.Q.; Zhou, Z.Y.; Dong, X.L.; Gong, X.L. Theoretical study of helical structure caused by chirality of cysteine dimer. *Int. J. Quant. Chem.* **2005**, *105*, 66–73. [CrossRef]

21. Tiwari, S.; Mishra, P. Vibrational spectra of cysteine zwitterion and mechanism of its formation: Bulk and specific solvent effects and geometry optimization in aqueous media. *Spectrochim. Acta A Mol. Biomol. Spectrosc.* **2009**, *73*, 719–729. [CrossRef]

22. Bachrach, S.M.; Nguyen, T.T.; Demoin, D.W. Microsolvation of cysteine: A density functional theory study. *J. Phys. Chem. A* **2009**, *113*, 6172–6181. [CrossRef]

23. Chapman, C.R.; Ting, E.C.; Kereszti, A.; Paci, I. Self-assembly of cysteine dimers at the gold surface: A computational study of competing interactions. *J. Phys. Chem. C* **2013**, *117*, 19426–19435. [CrossRef]

24. Buimaga-Iarinca, L.; Calborean, A. Electronic structure of the ll-cysteine dimers adsorbed on Au (111): A density functional theory study. *Phys. Scr.* **2012**, *86*, 035707. [CrossRef]

25. Ieritano, C.; Carr, P.J.; Hasan, M.; Burt, M.; Marta, R.A.; Steinmetz, V.; Fillion, E.; McMahon, T.B.; Hopkins, W.S. The structures and properties of proton-and alkali-bound cysteine dimers. *Phys. Chem. Chem. Phys.* **2016**, *18*, 4704–4710. [CrossRef]

26. Bader, R. *Atoms in Molecules: A Quantum Theory*; Oxford University Press: Oxford, UK, 1990.

27. Popelier, P.L. *Atoms in Molecules: An Introduction*; Prentice Hall: London, UK, 2000.

28. Grabowski, S.J. What Is the Covalency of Hydrogen Bonding? *Chem. Rev.* **2011**, *111*, 2597–2625. [CrossRef] [PubMed]

29. Becke, A. *The Quantum Theory of Atoms in Molecules: From Solid State to DNA and Drug Design*; John Wiley & Sons: Hoboken, NJ, USA, 2007.

30. Reed, A.E.; Curtiss, L.A.; Weinhold, F. Intermolecular interactions from a natural bond orbital, donor-acceptor viewpoint. *Chem. Rev.* **1988**, *88*, 899–926. [CrossRef]

31. Weinhold, F.; Landis, C.R. *Discovering Chemistry with Natural Bond Orbitals*; Wiley-VCH: Hoboken NJ, USA, 2012; 319p.

32. Glendening, E.D.; Landis, C.R.; Weinhold, F. Natural bond orbital methods. *Wiley Interdiscip. Rev. Comput. Mol. Sci.* **2012**, *2*, 1–42. [CrossRef]

33. Weinhold, F.; Landis, C.; Glendening, E. What is NBO analysis and how is it useful? *Int. Rev. Phys. Chem.* **2016**, *35*, 399–440. [CrossRef]

34. Johnson, E.R.; Keinan, S.; Mori-Sánchez, P.; Contreras-García, J.; Cohen, A.J.; Yang, W. Revealing Noncovalent Interactions. *J. Am. Chem. Soc.* **2010**, *132*, 6498–6506. [CrossRef]

35. DiLabio, G.A.; Otero-de-la Roza, A., Noncovalent Interactions in Density Functional Theory. In *Reviews in Computational Chemistry*; John Wiley & Sons, Ltd.: Hoboken, NJ, USA, 2016; Chapter 1, pp. 1–97. [CrossRef]

36. Metropolis, N.; Rosenbluth, A.W.; Rosenbluth, M.N.; Teller, A.H.; Teller, E. Equation of State Calculations by Fast Computing Machines. *J. Chem. Phys.* **1953**, *21*, 1087–1092. [CrossRef]

37. Kirkpatrick, S.; Gelatt, C.D.; Vecchi, M.P. Optimization by Simulated Annealing. *Science* **1983**, *220*, 671–680. [CrossRef] [PubMed]

38. van Laarhoven, P.; Aarts, E. *Simulated Annealing: Theory and Applications*; Mathematics and Its Applications; Springer: Dordrecht, The Netherlands, 1987.

39. Pérez, J.; Restrepo, A. ASCEC V02: Annealing Simulado con Energía Cuántica. In *Property, Development, and Implementation: Grupo de Química–Física Teórica*; Instituto de Química, Universidad de Antioquia: Medellín, Colombia, 2008.

40. Pérez, J.F.; Hadad, C.Z.; Restrepo, A. Structural studies of the water tetramer. *Int. J. Quant. Chem.* **2008**, *108*, 1653–1659. [CrossRef]

41. Pérez, J.F.; Florez, E.; Hadad, C.Z.; Fuentealba, P.; Restrepo, A. Stochastic Search of the Quantum Conformational Space of Small Lithium and Bimetallic Lithium–Sodium Clusters. *J. Phys. Chem. A* **2008**, *112*, 5749–5755. [CrossRef] [PubMed]

42. Tomasi, J.; Persico, M. Molecular Interactions in Solution: An Overview of Methods Based on Continuous Distributions of the Solvent. *Chem. Rev.* **1994**, *94*, 2027–2094. [CrossRef]

43. Tomasi, J.; Mennucci, B.; Cammi, R. Quantum Mechanical Continuum Solvation Models. *Chem. Rev.* **2005**, *105*, 2999–3094. [CrossRef]

44. Gonzalez, C.; Restrepo-Cossio, A.; Márquez, M.; Wiberg, K.B. Ab Initio Study of the Stability of the Ylide-like Intermediate Methyleneoxonium in the Reaction between Singlet Methylene and Water. *J. Am. Chem. Soc.* **1996**, *118*, 5408–5411. [CrossRef]

45. Frisch, M.J.; Trucks, G.W.; Schlegel, H.B.; Scuseria, G.E.; Robb, M.A.; Cheeseman, J.R.; Scalmani, G.; Barone, V.; Mennucci, B.; Petersson, G.A.; et al. *Gaussian 09 Revision E.01*; Gaussian Inc.: Wallingford, CT, USA, 2009.

46. Gómez, S.; Rojas-Valencia, N.; Gómez, S.A.; Cappelli, C.; Merino, G.; Restrepo, A. A molecular twist on hydrophobicity. *Chem. Sci.* **2021**, *12*, 9233–9245. [CrossRef]

47. Gómez, S.A.; Rojas-Valencia, N.; Gómez, S.; Lans, I.; Restrepo, A. Initial recognition and attachment of the Zika virus to host cells: A molecular dynamics and quantum interaction approach. *ChemBioChem* **2022**, *23*, e202200351. [CrossRef]

48. Gómez, S.A.; Rojas-Valencia, N.; Gómez, S.; Egidi, F.; Cappelli, C.; Restrepo, A. Binding of SARS-CoV-2 to Cell Receptors: A Tale of Molecular Evolution. *ChemBioChem* **2021**, *22*, 724–732. [CrossRef]

49. Rojas-Valencia, N.; Gómez, S.; Guerra, D.; Restrepo, A. A detailed look at the bonding interactions in the microsolvation of monoatomic cations. *Phys. Chem. Chem. Phys.* **2020**, *22*, 13049–13061. [CrossRef]

50. Moreno, N.; Hadad, C.Z.; Restrepo, A. Microsolvation of electrons by a handful of ammonia molecules. *J. Chem. Phys.* **2022**, *157*, 134301. [CrossRef]

51. Correa, E.; Montaño, D.; Restrepo, A. Cation··· anion bonding interactions in 1–Ethyl–3–Methylimidazolium based ionic liquids. *Chem. Phys.* **2022**, *562*, 111648. [CrossRef]
52. Flórez, E.; Acelas, N.; Gómez, S.; Hadad, C.; Restrepo, A. To Be or Not To Be? that is the Entropic, Enthalpic, and Molecular Interaction Dilemma in the Formation of (Water) 20 Clusters and Methane Clathrate. *ChemPhysChem* **2022**, *23*, e202100716. [CrossRef]
53. Flórez, E.; Gómez, S.; Acelas, N.; Hadad, C.; Restrepo, A. Microsolvation versus encapsulation in mono, di, and trivalent cations. *ChemPhysChem* **2022**, e202200456. [CrossRef] [PubMed]
54. Lu, T.; Chen, F. Multiwfn: A multifunctional wavefunction analyzer. *J. Comput. Chem.* **2012**, *33*, 580–592. [CrossRef] [PubMed]
55. Glendening, E.D.; Badenhoop, J.K.; Reed, A.E.; Carpenter, J.E.; Bohmann, J.A.; Morales, C.M.; Landis, C.R.; Weinhold, F. *NBO 6.0*; Theoretical Chemistry Institute, University of Wisconsin: Madison, WI, USA, 2013.
56. Contreras-García, J.; Johnson, E.R.; Keinan, S.; Chaudret, R.; Piquemal, J.P.; Beratan, D.N.; Yang, W. NCIPLOT: A Program for Plotting Noncovalent Interaction Regions. *J. Chem. Theory Comput.* **2011**, *7*, 625–632. [CrossRef] [PubMed]
57. Herraez, A. Biomolecules in the computer: Jmol to the rescue. *Biochem. Mol. Biol. Educ.* **2006**, *34*, 255–261. [CrossRef] [PubMed]
58. Humphrey, W.; Dalke, A.; Schulten, K. VMD—Visual Molecular Dynamics. *J. Mol. Graph.* **1996**, *14*, 33–38. [CrossRef]
59. Reed, A.E.; Weinhold, F. Natural bond orbital analysis of near-Hartree–Fock water dimer. *J. Phys. Chem.* **1983**, *78*, 4066–4073. [CrossRef]
60. Copeland, C.; Menon, O.; Majumdar, D.; Roszak, S.; Leszczynski, J. Understanding the influence of low-frequency vibrations on the hydrogen bonds of acetic acid and acetamide dimers. *Phys. Chem. Chem. Phys.* **2017**, *19*, 24866–24878. [CrossRef]
61. Kalescky, R.; Kraka, E.; Cremer, D. Accurate determination of the binding energy of the formic acid dimer: The importance of geometry relaxation. *J. Chem. Phys.* **2014**, *140*, 084315. [CrossRef] [PubMed]
62. Farfán, P.; Echeverri, A.; Diaz, E.; Tapia, J.D.; Gómez, S.; Restrepo, A. Dimers of formic acid: Structures, stability, and double proton transfer. *J. Chem. Phys.* **2017**, *147*, 044312. [CrossRef] [PubMed]
63. Murillo, J.; David, J.; Restrepo, A. Insights into the structure and stability of the carbonic acid dimer. *Phys. Chem. Chem. Phys.* **2010**, *12*, 10963–10970. [CrossRef] [PubMed]
64. Knop, O.; Rankin, K.N.; Boyd, R.J. Coming to Grips with N–H··· N Bonds. 1. Distance Relationships and Electron Density at the Bond Critical Point. *J. Phys. Chem. A* **2001**, *105*, 6552–6566. [CrossRef]
65. Knop, O.; Rankin, K.N.; Boyd, R.J. Coming to Grips with N–H··· N Bonds. 2. Homocorrelations between Parameters Deriving from the Electron Density at the Bond Critical Point. *J. Phys. Chem. A* **2003**, *107*, 272–284. [CrossRef]
66. Ramírez, F.; Hadad, C.; Guerra, D.; David, J.; Restrepo, A. Structural studies of the water pentamer. *Chem. Phys. Lett.* **2011**, *507*, 229–233. [CrossRef]
67. Cremer, D.; Kraka, E. A description of the chemical bond in terms of local properties of electron density and energy. *Croat. Chem. Acta* **1984**, *57*, 1259–1281.
68. Rozas, I.; Alkorta, I.; Elguero, J. Behavior of ylides containing N, O, and C atoms as hydrogen bond acceptors. *J. Am. Chem. Soc.* **2000**, *122*, 11154–11161. [CrossRef]
69. Jenkins, S.; Morrison, I. The chemical character of the intermolecular bonds of seven phases of ice as revealed by ab initio calculation of electron densities. *Chem. Phys. Lett.* **2000**, *317*, 97–102. [CrossRef]
70. Arnold, W.D.; Oldfield, E. The chemical nature of hydrogen bonding in proteins via NMR: J-couplings, chemical shifts, and AIM theory. *J. Am. Chem. Soc.* **2000**, *122*, 12835–12841. [CrossRef]
71. Gómez, S.A.; Rojas-Valencia, N.; Gómez, S.; Cappelli, C.; Restrepo, A. The Role of Spike Protein Mutations in the Infectious Power of SARS-COV-2 Variants: A Molecular Interaction Perspective. *ChemBioChem* **2022**, *23*, e202100393. [CrossRef]
72. Flórez, E.; Acelas, N.; Ramírez, F.; Hadad, C.; Restrepo, A. Microsolvation of F^-. *Phys. Chem. Chem. Phys.* **2018**, *20*, 8909–8916. [CrossRef]
73. Weinhold, F.; Klein, R.A. Anti-Electrostatic Hydrogen Bonds. *Angew. Chem. Int. Ed.* **2014**, *53*, 11214–11217. [CrossRef]

Article

The Halogen Bond in Weakly Bonded Complexes and the Consequences for Aromaticity and Spin-Orbit Coupling

Ana V. Cunha [1,*,†], Remco W. A. Havenith [2,3,†], Jari van Gog [1], Freija De Vleeschouwer [4], Frank De Proft [4] and Wouter Herrebout [1,*]

[1] MolSpec, Departement Chemie, Universiteit Antwerpen, Groenenborgerlaan 171, 2020 Antwerpen, Belgium
[2] Stratingh Institute for Chemistry and Zernike Institute for Advanced Materials, Rijksuniversiteit Groningen, Nijenborgh 4, 9747 AG Groningen, The Netherlands
[3] Gent Quantum Chemistry Group, Faculteit Wetenschappen, Universiteit Gent, Krijgslaan 281 (S3), 9000 Gent, Belgium
[4] Algemene Chemie (ALGC), Vrije Universiteit Brussel (VUB), Pleinlaan 2, 1050 Brussel, Belgium
* Correspondence: ana.cunha@uantwerpen.be (A.V.C.); wouter.herrebout@uantwerpen.be (W.H.); Tel.: +32-326-533-73 (W.H.)
† These authors contributed equally to this work.

Abstract: The halogen bond complexes $CF_3X \cdots Y$ and $C_2F_3X \cdots Y$, with Y = furan, thiophene, selenophene and X = Cl, Br, I, have been studied by using DFT and CCSD(T) in order to understand which factors govern the interaction between the halogen atom X and the aromatic ring. We found that PBE0-dDsC/QZ4P gives an adequate description of the interaction energies in these complexes, compared to CCSD(T) and experimental results. The interaction between the halogen atom X and the π-bonds in perpendicular orientation is stronger than the interaction with the in-plane lone pairs of the heteroatom of the aromatic cycle. The strength of the interaction follows the trend Cl < Br < I; the chalcogenide in the aromatic ring nor the hybridization of the C–X bond play a decisive role. The energy decomposition analysis shows that the interaction energy is dominated by all three contributions, viz., the electrostatic, orbital, and dispersion interactions: not one factor dominates the interaction energy. The aromaticity of the ring is undisturbed upon halogen bond formation: the π-ring current remains equally strong and diatropic in the complex as it is for the free aromatic ring. However, the spin-orbit coupling between the singlet and triplet $\pi \rightarrow \pi^*$ states is increased upon halogen bond formation and a faster intersystem crossing between these states is therefore expected.

Keywords: halogen bonds; density functional theory; energy decomposition analysis; ring current analysis; spin-orbit coupling

Citation: Cunha, A.V.; Havenith, R.W.A.; van Gog, J.; De Vleeschouwer, F.; De Proft, F.; Herrebout, W. The Halogen Bond in Weakly Bonded Complexes and the Consequences for Aromaticity and Spin-Orbit Coupling. *Molecules* **2023**, *28*, 772. https://doi.org/10.3390/molecules28020772

Academic Editors: Qingzhong Li, Steve Scheiner and Zhiwu Yu

Received: 29 October 2022
Revised: 31 December 2022
Accepted: 5 January 2023
Published: 12 January 2023

1. Introduction

Van der Waals interactions are omnipresent in nature, and are crucial for understanding the dynamics and structure of a wide variety of systems. A particularly weak interaction is the halogen bond. The International Union of Pure and Applied Chemistry (IUPAC) defines a halogen bond as "a net attractive interaction between an electrophilic region associated with a halogen atom in a molecular entity, and a nucleophilic region in another, or the same, molecular entity" [1]. In 1961, Zingaro et al. [2] described complexes formed in solution by halogens and phosphine oxides and sulfides. This was the first time that the term "halogen bond" was used to describe interactions where halogens act as electrophilic species. However, it was much later that Glaser et al. [3] suggested to use the term halogen bond to describe an interaction between halogen atoms, regardless of their electrophilic or nucleophilic nature. It was in 2009 that the IUPAC gave a unified conceptual framework for the interactions involving halogens [1]. This interaction, which captured the attention of the scientific community for decades, can be discussed in terms of its unique features, such as directionality, tunability, hydrophobicity, and the atomic radius of the donor atom.

Halogen bonds (XB) of the form R–X$\cdots$Y, with X the halogen bond acceptor and Y the donor, are particularly directional interactions, due to the localization of the σ-hole exactly on the elongation of the covalent bond that the halogen atom is involved [4]. The effective atomic size along the extended R–X bond axis, in monovalent halogen atoms, is smaller than in the direction perpendicular to this axis [5–7]. This corresponds to a region of depleted electron density, the so-called σ-hole. Thus, shorter and stronger halogen bonds are more directional than longer and weaker counterparts. However, the directionality of the XB acceptor is along the axis of the donated lone pair on Y [8–19]. For the case in which the XB acceptor is an isolated π-system, the axis of the R–X bond is approximately along the axis perpendicular to the π-system. Furthermore, when the XB acceptor contains both lone pairs and aromatic π-pairs, the lone pairs are generally the ones involved in the XB formation. However, the two exceptions are furan and thiophene, for which both the lone pairs and the π-bonding orbitals are involved in the XB formation; the hydrogen bond interactions with the lone pairs or the π-bonds are of similar strength [20]. Tunability is another feature of this interaction, in which the XB ability increases on the order I > Br > Cl > F. This trend is often concordant with the positive character of the σ-holes [21], which decreases with the electronegativity of the halogen atom, but increases with the polarizability [22–25]. Thus, the donor ability of a given compound can be tuned by selecting a halogen atom as a donor site [26,27]. However, the tuning of the σ-hole magnitude can also be achieved by modifying the hybridization of the carbon bond to the XB donor site. Thus, the strength increases in the following order: C(sp)-X > C(sp^2)-X > C(sp^3)-X, and it has been observed for various systems [16,19,28–32]. Hence, there are three main possibilities to tune the XB interaction strength: (I) by single atom mutation; (II) by changing the hybridization; and (III) by modifying the electron withdrawing groups [33–37]. The size of the donor atoms influences the steric hindrance, as halogen atoms have large van der Waals radii. Thus, when compared to the hydrogen bond, the XB bond is more sensitive to steric effects [38–40]. This has been shown in the formation of DNA pairs where HB is replaced by XB. The latter shows that bromine gives more stable pairs when compared to iodine, due to the steric repulsions arising from the larger radius of the I atom [41–50]. Furthermore, these halogen bonds also impact optical transitions of supramolecular complexes, that can undergo singlet to triplet intersystem crossings [25,51–57]. The size of halogen atoms plays an important role in the photoluminescence of halogenated chromophores. This is shown in reference [58], in which the singlet to triplet intersystem crossing rate increases by a factor of 60 when using iodine corroles instead of fluorine ones. Hence, the heavy atom effect confers to XB-based materials particularly exciting promising applications. Such applications have been demonstrated in [52–55,59–61], in which halogen bonding was used to tune room-temperature phosphorescence in organic crystals. A large number of studies showed the use of halogen bonds for changing the electronic properties in numerous types of materials, ranging from crystalline solids to amorphous [56,57,60,62–66]. Furthermore, XB bonds have also proven to be suitable for changing the efficiency of solar cells, and perovskites. However, the modelling of these systems is still a bottleneck for computational chemistry.

Most of the electronic structure methods used for molecular modelling capture 99% of the total electronic energy [67]. However, the remaining missing fraction, which is crucial for molecular properties, such as relative energies [68–71] and binding properties [72–75], arise from correlated motion of electrons [76–78]. The main component of this energy is the long-range contribution, also known as van der Waals (vdW) or dispersion interactions [70,79]. These forces are dominant in weakly bonded complexes, and, thus, the modeling of these systems requires the introduction of dispersion corrections [80–83], or correlated wavefunction methods.

In this study, we present a detailed bonding analysis of halogen bonding in complexes consisting of a small molecule (CF$_3$X/C$_2$F$_3$X (X = Cl, Br, I)) and an aromatic molecule (furan/thiophene/selenophene) in various orientations. These complexes provide insights into the structure of halogenated crystals. The orientations that we consider are the halogen bonds with the lone pair (parallel orientation) and with the π-bonds (perpendicular orientation) of the chalcogenide of the aromatic ring (Figure 1). We also study the effect of a halogen bond on the electronic transitions of the aromatic molecule and possible intersystem crossing due to increased spin-orbit coupling as a consequence of the vicinity of a heavy atom. We use DFT and CCSD(T) calculations to retrieve the effects of correlation, and decompose the interaction energies into their different contributions. In this way, we are able to investigate several factors affecting the halogen bond, namely, the halogen atom of the donor, the aromatic ring as the acceptor, the hybridization of the carbon atom to which the halogen atom is attached, and the orientation of the halogen bond (parallel or perpendicular). Moreover, we are able to benchmark dispersion-corrected DFT and judge its suitability to describe these weakly interacting systems with halogen bonds. Finally, we investigate whether the formation of the halogen bond affects the aromaticity of the ring, and if it affects the $\pi \rightarrow \pi^*$ transition in the aromatic ring, and the spin-orbit coupling between the singlet and triplet $\pi \rightarrow \pi^*$ states.

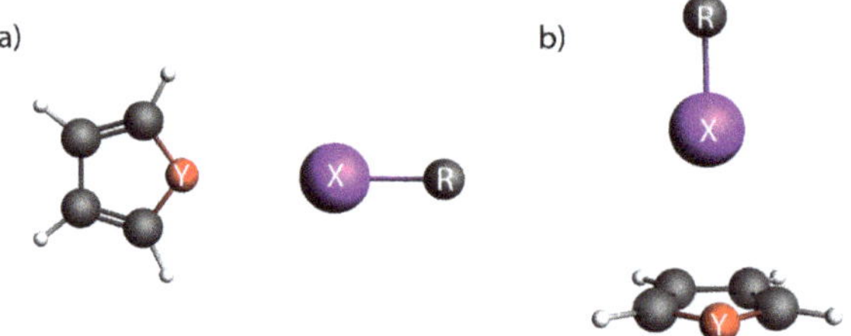

Figure 1. Schematic Yes you can move it. representation of geometries in (**a**) parallel ($\parallel$), and (**b**) perpendicular ($\perp$) orientation. X = Cl, Br, I; Y = O, S, Se; R = CF$_3$, C$_2$F$_3$.

2. Results and Discussion

2.1. Interaction Energies at Different Levels of Theory

The interaction energies obtained with different basis sets, for the complexes in the parallel orientation, are shown in Figure 2. The interaction energies (without zero-point vibrational energy correction) change significantly with the increase of the basis set size, with differences ranging between 0.1 to 13.6 kJ/mol (Figure 2). It shows, as expected, the importance of the use of a large basis set for weakly bonded complexes. Note that for C$_2$F$_3$Cl$\cdots$selenophene in parallel orientation, hardly any interaction is found with all three basis sets. However, the differences in the interaction energies obtained by using the TZ2P and QZ4P basis sets are considerably smaller and range only from 0 to 3.9 kJ/mol; this indicates that the interaction energy is close to convergence with a TZ2P basis set, showing that the QZ4P basis therefore is certainly sufficiently large enough.

For the perpendicular oriented complexes, the same trend is observed (Figure 3). The interaction energies calculated with the DZP basis set differ at most 10.2 kJ/mol and at least 6.5 kJ/mol. The differences between the interaction energies evaluated by using the TZ2P and QZ4P basis sets decrease again. Here, the largest difference is 2.1 kJ/mol and the minimum difference calculated is 0 kJ/mol (supplementary information). Taking into account the small differences between the interaction energies with a TZ2P and QZ4P basis set, we can conclude that the QZ4P basis set is of sufficient quality. Hence, this basis set will be used for the EDA calculations.

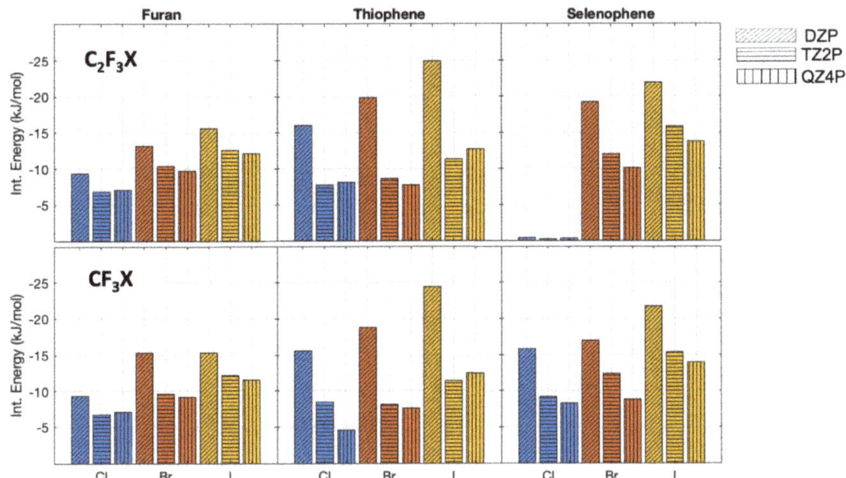

Figure 2. PBE0−dDsC We changed it interaction energies (without zero-point vibrational energy correction) with different basis sets for the complexes oriented in parallel. The different halogens are depicted with the following colormap: chlorine blue, bromine red, and iodine yellow. The DZP basis set is represented with slanted lines, the TZ2P with horizontal lines, and QZ4P with vertical lines.

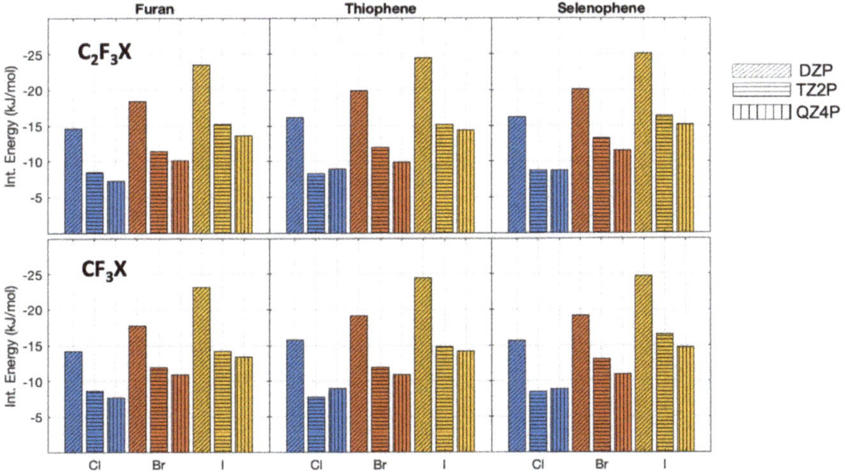

Figure 3. PBE0−dDsC interaction energies (without zero-point vibrational energy correction) with different basis sets for the complexes oriented in perpendicular. The halogen atoms are depicted with the following colormap: chlorine blue, bromine red, and iodine yellow. The DZP basis set is represented with slanted lines, the TZ2P with horizontal lines, and QZ4P with vertical lines.

Although the DFT results are robust with basis set size, an important question remains to be answered: "Does the PBE0 functional with dispersion corrections give a reasonable description of these weak interacting complexes?" Thus, we have calculated the interaction energies also by using the CCSD(T)-DLPNO/CBS approach. For both the parallel oriented complexes (Table 1) and the perpendicular oriented complexes (Table 2), the differences between the CCSD(T) interaction energies and the DFT ones are largest for the iodine halogen bonds but do not vary more than 5 kJ/mol. Hence, PBE0-dDsC/QZ4P is suitable to describe the weak bonding interactions in these complexes and captures most of the relevant physics.

Table 1. Interaction energies (kJ/mol) for the parallel oriented complexes calculated with PBE0-dDsC/QZ4P and CCSD(T)-DLPNO/CBS.

		Furan		Thiophene		Selenophene	
		CCSD(T) CBS	DFT QZ4P	CCSD(T) CBS	DFT QZ4P	CCSD(T) CBS	DFT QZ4P
	Cl	−7.2	−7.0	−9.1	−8.1	−0.5	−0.3
C_2F_3X	Br	−9.4	−9.7	−8.5	−7.7	−10.8	−10.1
	I	−13.9	−12.1	−16.2	−12.7	−17.2	−13.8
	Cl	−7.0	−7.1	−5.1	−4.6	−9.6	−8.3
CF_3X	Br	−9.1	−9.1	−8.0	−7.6	−8.7	−8.8
	I	−13.3	−11.5	−15.7	−12.4	−16.2	−13.9

Table 2. Interaction energies (kJ/mol) for the perpendicular oriented complexes calculated with PBE0-dDsC/QZ4P and CCSD(T)-DLPNO/CBS.

		Furan		Thiophene		Selenophene	
		CCSD(T) CBS	DFT QZ4P	CCSD(T) CBS	DFT QZ4P	CCSD(T) CBS	DFT QZ4P
	Cl	−7.7	−7.2	−10.1	−8.9	−10.0	−8.7
C_2F_3X	Br	−10.4	−10.1	−10.8	−9.9	−12.3	−11.5
	I	−15.0	−13.6	−17.7	−14.4	−17.8	−15.1
	Cl	−7.8	−7.7	−10.0	−9.0	−9.7	−8.8
CF_3X	Br	−11.0	−10.9	−11.8	−10.9	−11.3	−10.9
	I	−14.5	−13.4	−16.6	−14.2	−17.5	−14.7

A last benchmark of the theoretical data is a comparison of the calculated interaction enthalpies with experimentally determined enthalpies (Table 3) for $CF_3I\cdots$ furan. For the compounds for which experimental data are available, the agreement between theory and experiment is good, and only small deviations are found, which are within chemical accuracy. Thus, the PBE0-dDsC/QZ4P calculations are of sufficient quality and in the following sections only the results obtained with PBE0-dDsC/QZ4P are discussed and the remaining results can be found in the supplementary information.

Table 3. Calculated (PBE0-dDsC/QZ4P) and experimental interaction enthalpies (kJ/mol) [a] for the parallel ($\parallel$) and perpendicular ($\perp$) oriented $CF_3I\cdots$ furan complexes in the gas phase.

Compound	ΔH^{150K}	ΔH^{298K}	ΔH^{150K}_{exp}
$CF_3I\cdots$ furan ($\parallel$)	−12.8	−14.0	−14.0(8)
$CF_3I\cdots$ furan ($\perp$)	−12.0	−10.8	−14.4(9)

[a] Gas phase complexation enthalpy obtained by correcting the experimental value in LKr for solvent effects. Corrections were introduced by using MC-FEP simulations similar to those described in [20].

2.2. Interaction Energies and Trends

The hard–soft acid–base (HSAB) principle states that soft acids preferably interact with soft bases, while hard acids prefer interacting with hard bases, when all other factors are equal. Thus, when oxygen is switched to selenium in a molecule, the size of the atom increases as well the polarizability; however, the electronegativity decreases. Therefore, the oxygen can be considered as a hard base and selenium as a soft base, and the sulfur is in between. The same reasoning can be made for the halogen atoms, which results in the following ranking of their softness as acids: Cl < Br < I. The delocalized π-systems in the heteroaromatic molecules act as Lewis bases and are considered to be soft while the lone pairs present on the heteroatoms are harder. Based on the HSAB principle, it is expected that complexes having the lowest interaction energy contain the following combination of halogen and chalcogen atoms: chlorine (hard) and selenium (soft) or iodine (soft) and oxygen (hard). The strongest interaction, on the other hand, is expected to occur between complexes containing chlorine and oxygen or iodine and selenium. For the complexes corresponding to a perpendicular geometry, where the halogen atom interacts with the soft, delocalized π-system, the strongest interacting complex is expected to contain a soft iodine atom. The complex with the lowest interaction energy in perpendicular position, however, can be predicted to accommodate a hard chlorine atom. Based on the σ-hole theory, which considers only electrostatic interactions, a difference is expected between the sp^2 and sp^3 hybridized complexes: the carbon atom with sp^2 hybridization is expected to be more electronegative (higher s-character), and thus these complexes are expected to have a larger electrostatic interaction for the same halogen atom than their sp^3 hybridized counterparts. Furthermore, as the halogen atom is changed from chlorine to bromine, and further to iodine, the polarizability of the halogen atom increases, and its electronegativity will decrease. The increase in polarizability is expected to lead to an increased σ-hole and subsequently to a larger electrostatic contribution. Moreover, going down in the periodic table will also lead to an increased dispersion interaction. In addition to these longer-range interactions, Pauli repulsion and potentially charge transfer may also play their part, as demonstrated in a study of various CX_3I and halide anion interactions [21]. Hence, which effect will dominate is not clear based on qualitative arguments, as has been found previously [84].

These hypotheses can be tested by carrying out an EDA analysis on the PBE0-dDsC/QZ4P interaction energies (E_{bond}) for all complexes, presented in Tables 4 and 5. The total bonding (interaction) energies are decomposed in contributions of the Pauli repulsion term (E_{Pauli}), electrostatic interaction (E_{els}), the steric interaction ($E_{steric} = E_{Pauli} + E_{els}$), the orbital interaction term (E_{orb}), and the dispersion (E_{disp}) term.

A first glance at the interaction energies in parallel orientation (Table 4) shows that indeed the weakest interacting complex is $C_2F_3Cl\cdots$ selenophene, as predicted by applying the HSAB argumentation. The strongest interacting species are $C_2F_3I\cdots$ selenophene and $CF_3I\cdots$ selenophene, also inline with the HSAB rules. However, the interaction in $CF_3Cl\cdots$ selenophene, predicted to be weak, are on par with other species. Moreover, the interaction in $C_2F_3Cl\cdots$ furan, predicted to be strong, is only half as strong as the interaction in $C_2F_3I\cdots$ furan. For the perpendicular cases (Table 5), the $I\cdots Se$ interactions are indeed the strongest, but the $Cl\cdots Se$ interactions, predicted to be weak, are stronger than the $Cl\cdots O$ interactions, predicted to be strong. This indicates that the HSAB principle does not tell the full story, and predictions based solely on these considerations can be substantially in error. In addition, the prediction based on the σ-hole theory—that the sp^2 hybridized species interact more strongly than their sp^3 counterparts—is not immediately confirmed by the DFT interaction energies. The differences between the sp^2 and sp^3 hybridized species is in general very small, with only a few exceptions in one case confirming and in one case disproving the σ-hole predictions. It is evident that predicting the interaction strength cannot be based purely on qualitative arguments and that the interaction is a more subtle interplay of different contributions.

One trend that is generally followed for all species is that when X is varied from Cl to I, the interaction energy increases, as is predicted from considering the larger polarizability of X when going down in the periodic table. This effect is much less pronounced when Y of the aromatic ring is varied from O to Se: much smaller variations in the interaction energies are discernible and an increasing trend is not always found. This highlights again that these interactions are not solely determined by one factor.

Table 4. Energy decomposition analysis (EDA) in kJ/mol for the parallel oriented complexes.

Complex	E_{Pauli}	E_{els}	E_{steric}	E_{orb}	E_{disp}	E_{bond}
$C_2F_3Cl\cdots$ furan	7.0	−7.5	−0.5	−3.1	−3.0	−6.7
$C_2F_3Br\cdots$ furan	10.7	−11.2	−0.5	−5.1	−3.4	−9.0
$C_2F_3I\cdots$ furan	16.9	−16.7	0.2	−7.9	−4.2	−11.9
$C_2F_3Cl\cdots$ thiophene	7.0	−6.5	0.5	−3.5	−4.8	−7.8
$C_2F_3Br\cdots$ thiophene	10.2	−8.6	1.7	−5.4	−3.8	−7.5
$C_2F_3I\cdots$ thiophene	18.6	−14.0	4.6	−10.8	−6.1	−12.3
$C_2F_3Cl\cdots$ selenophene	−0.0	−0.2	−0.2	0.0	−0.3	−0.5
$C_2F_3Br\cdots$ selenophene	12.1	−10.1	2.0	−6.9	−4.8	−9.7
$C_2F_3I\cdots$ selenophene	23.6	−17.3	6.3	−13.3	−6.0	−13.1
$CF_3Cl\cdots$ furan	7.5	−7.9	−0.4	−3.4	−2.9	−6.8
$CF_3Br\cdots$ furan	9.6	−10.2	−0.7	−4.6	−3.2	−8.5
$CF_3I\cdots$ furan	17.6	−16.9	0.6	−7.9	−4.1	−11.4
$CF_3Cl\cdots$ thiophene	1.1	−2.5	−1.4	−1.2	−2.1	−4.7
$CF_3Br\cdots$ thiophene	10.0	−8.3	1.8	−5.7	−3.2	−7.1
$CF_3I\cdots$ thiophene	20.6	−15.1	5.6	−11.7	−5.8	−12.0
$CF_3Cl\cdots$ selenophene	7.5	−6.6	0.9	−4.1	−4.8	−8.0
$CF_3Br\cdots$ selenophene	14.5	−11.4	3.1	−7.8	−3.6	−8.2
$CF_3I\cdots$ selenophene	28.5	−20.4	8.1	−15.9	−5.8	−13.6

Table 5. Energy decomposition analysis (EDA) in kJ/mol for the perpendicular oriented complexes.

Complex	E_{Pauli}	E_{els}	E_{steric}	E_{orb}	E_{disp}	E_{bond}
$C_2F_3Cl\cdots$ furan	5.6	−5.6	0.0	−2.9	−4.0	−7.0
$C_2F_3Br\cdots$ furan	13.1	−11.3	1.8	−6.4	−4.7	−9.4
$C_2F_3I\cdots$ furan	22.0	−17.8	4.2	−11.9	−5.8	−13.5
$C_2F_3Cl\cdots$ thiophene	8.2	−7.1	1.1	−4.0	−5.3	−8.3
$C_2F_3Br\cdots$ thiophene	8.7	−8.3	0.5	−4.8	−5.0	−9.3
$C_2F_3I\cdots$ thiophene	23.7	−18.3	5.4	−12.3	−7.1	−14.1
$C_2F_3Cl\cdots$ selenophene	7.7	−6.9	0.7	−3.8	−5.2	−8.3
$C_2F_3Br\cdots$ selenophene	14.0	−11.7	2.3	−7.3	−5.7	−10.6
$C_2F_3I\cdots$ selenophene	23.5	−18.6	4.9	−12.7	−7.0	−14.9
$CF_3Cl\cdots$ furan	9.2	−8.0	1.2	−4.4	−4.1	−7.2
$CF_3Br\cdots$ furan	16.4	−12.8	3.6	−8.4	−5.0	−9.8
$CF_3I\cdots$ furan	24.0	−18.7	5.3	−13.1	−5.7	−13.4
$CF_3Cl\cdots$ thiophene	8.2	−7.2	1.0	−4.2	−5.1	−8.3
$CF_3Br\cdots$ thiophene	14.4	−11.6	2.8	−7.4	−5.4	−10.0
$CF_3I\cdots$ thiophene	25.9	−19.9	6.1	−13.7	−6.7	−14.3
$CF_3Cl\cdots$ selenophene	7.6	−7.1	0.5	−3.9	−5.0	−8.3
$CF_3Br\cdots$ selenophene	16.2	−12.6	3.5	−8.2	−5.3	−9.9
$CF_3I\cdots$ selenophene	27.6	−20.6	7.1	−14.3	−7.0	−14.2

If we look at the EDA analysis in more detail, we notice that the total steric interaction, which is a sum of the Pauli repulsion and the electrostatic interaction, is for most cases (parallel and perpendicular) rather small and in many cases repulsive. The electrostatic interactions do not dominate the final interaction energy. The orbital interaction energy and the dispersion energy also contribute significantly to the interaction between the two molecules in the complex, and both are equally important. The dispersion energy follows the expected trend of increasing when going down in the periodic table, if the halogen atom is changed. This trend is not clearly visible if the heteroatom in the aromatic ring is changed from O to Se. The differences in the dispersion contribution for sp^2 and sp^3 species are small, and more notable when the parallel geometries are compared to the perpendicular ones. For the perpendicular cases, where the halogen atom interacts with the π-system, the dispersion interactions are in general larger.

The orbital interaction term is in nearly all cases substantial. The orbital interaction term has its origin here in the relaxation of the orbitals of both fragments due to mutual polarization and to a lesser extent charge transfer. Thus, for the heavier halogen atoms, these orbital interactions are stronger, as they are more polarizable. Trends in the variations in the orbital contribution term due to the change from sp^2 to sp^3, due to the change from O to Se, or due to the change from parallel to perpendicular are harder to detect, and an a priori rule for predicting the magnitude of this term based on qualitative considerations proves to be unreliable.

We found that in general the interactions between the molecules in perpendicular orientation are larger than in parallel orientation, but not one factor is dominating in this increase. In some cases, the dispersion contribution is larger, in other cases, the orbital interaction term. The nature of the aromatic ring (Y = O, S, or Se) plays a much less decisive role, and no clear trends are visible. The hybridization of the carbon atom to which the halogen atom is attached also does not govern the interaction energy. The only clear trend is with halogen atom, when it becomes larger (X = Cl, Br, I), the interaction energy becomes stronger, mainly due to polarization of the fragments.

As we have seen that the interactions between the halogen bond donor and the aromatic π-system can be relatively strong, and caused by mutual polarization (as shown by the orbital interaction term), two remaining issues have to be resolved. One is whether the changes, induced by halogen bond formation, in the orbitals are such that the aromaticity of the five-membered ring is influenced, and the second one is whether the presence of the heavy halogen atom and the accompanying changes in orbitals affect the optical properties of the five-membered ring, and the spin-orbit coupling between the lowest electronic states. To answer the first question, the π-current density, induced by an external magnetic field has been calculated. Plots of the induced π-current density for $CF_3I\cdots$furan in parallel and perpendicular geometry are presented in Figure 4, together with a plot of the induced current density for the free aromatic ring. The induced current density is plotted in a plane, 1 a_0 above the molecular plane of the five-membered ring. The plots are visually indistinguishable from each other, indicating that no major changes in aromaticity occur upon complexation. This is further corroborated by the j_{max} values, the maximum strength of the current density. For furan, a j_{max} value of 0.0855 a.u. is found, whereas for $CF_3I\cdots$furan ($\parallel$) a value of 0.0824 a.u. is found and for $CF_3I\cdots$furan ($\perp$) 0.0827 a.u. Hence, upon formation of the halogen bonds, the induced π-current density is unaffected, and the aromatic character of the ring remains unchanged. This result is expected based on the symmetry selection rules in the ipsocentric formulation [85,86], as the nodal structure of the π-orbitals is not influenced by the halogen bond formation.

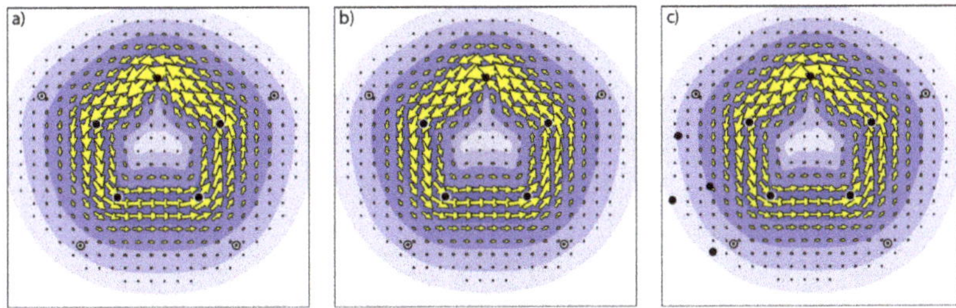

Figure 4. Plots of the π-current density for (**a**) furan, (**b**) $CF_3I\cdots$furan ($\parallel$), and (**c**) $CF_3I\cdots$furan ($\perp$).

For the exploration of the influence of the halogen bond on the excited states of the aromatic ring, we have selected several complexes (Table 6) with varying halogen (Br/I), orientation, and hybridization. The spin-orbit coupled excitation energies of the lowest, bright, $\pi \rightarrow \pi^*$ transition is barely influenced by the presence of the halogen bond. That also holds for the energy of the lowest triplet state that has $\pi \rightarrow \pi^*$ character. In all complexes, the three levels of the triplet remain degenerate, but in the perpendicular orientation, for the I$\cdots$furan complexes, a very small splitting in the three levels is visible, due to the close proximity of the iodine atom to the π-system.

The spin-orbit coupling between the bright $\pi \to \pi^*$ state and the lowest triplet state is small for all aromatic rings. A negligible increase is discernible if the halogen atom forms a halogen bond in parallel orientation, but in the perpendicular orientation, the spin-orbit coupling is significantly increased. The increase in spin-orbit coupling follows the trend selenophene > thiophene > furan, but, unexpectedly, the spin-orbit coupling in $C_2F_3I\cdots$furan is smaller than for the analogous Br complex, which may be caused by the closer proximity of the heavy atom to the π-system in the Br case than in the I case. Furthermore, despite the fact that in the I case a stronger mixing of the occupied π-system of the furan with the π-system of C_2F_3I occurs, a larger mixing in the unoccupied π-orbitals with σ-like C_2F_3Br orbitals occurs, which may enhance the spin-orbit coupling more in the Br case. An unexpectedly large spin-orbit interaction is observed for $C_2F_3I\cdots$thiophene complex, with concomitantly a strong mixing of the C_2F_3I orbitals with the occupied and unoccupied π-orbitals of thiophene, which further enhances the spin-orbit interaction.

Table 6. Excitation energies (eV) of the bright singlet and dark triplet $\pi \to \pi^*$ states and the spin-orbit coupling matrix element (cm^{-1}) for $CF_3I\cdots$furan in parallel ($\parallel$) and perpendicular ($\perp$) orientation.

Compound	$S(\pi \to \pi^*)$	$T(\pi \to \pi^*)$	$\langle S|H_{SO}|T\rangle$
furan	6.28	3.97/3.97/3.97	0.01
thiophene	5.85	3.74/3.74/3.74	0.06
selenophene	5.50	3.53/3.53/3.53	0.38
$CF_3Br\cdots$furan ($\parallel$)	6.29	3.97/3.97/3.97	1.05
$CF_3Br\cdots$furan ($\perp$)	6.35	4.00/4.00/4.00	14.32
$C_2F_3Br\cdots$furan ($\perp$)	6.22	4.00/4.00/4.00	11.77
$CF_3I\cdots$furan ($\parallel$)	6.31	4.00/4.00/4.00	0.09
$CF_3I\cdots$furan ($\perp$)	6.29	4.03/4.03/4.04	23.73
$CF_3I\cdots$thiophene ($\perp$)	5.80	3.80/3.80/3.80	41.17
$CF_3I\cdots$selenophene ($\perp$)	5.44	3.61/3.61/3.61	68.56
$C_2F_3I\cdots$furan ($\perp$)	6.24	4.03/4.03/4.04	6.84
$C_2F_3I\cdots$thiophene ($\perp$)	5.85	3.80/3.80/3.80	64.62
$C_2F_3I\cdots$selenophene ($\perp$)	5.49	3.61/3.61/3.61	67.56

In all perpendicular cases, an increase in spin-orbit coupling matrix element is observed due to formation of the halogen bond, hence, intersystem crossings from singlet to triplet can be accelerated by the formation of halogen bonds with heavy atoms. Thus, triplet formation in π-conjugated molecules can be accelerated by adding iodine-containing additives that form halogen bonds with the π-systems. This effect may find an application in organic electronic devices when it is desirable to increase the formation of triplets.

3. Materials and Methods

The starting geometries of the weakly bonded complexes were built by using the Amsterdam Modelling Suite–Graphical User Interface (AMS-GUI). The aromatic molecule (furan/thiophene/selenophene) was complexed with a small molecule with sp^2, and sp^3 hybridization (C_2F_3X or CF_3X, with X = Cl, Br, I), which were placed in a parallel and perpendicular orientation, Figure 1. This procedure rendered 36 different geometries.

Geometry optimizations were performed with the PBE0 functional (following the work of [87]), together with density-dependent dispersion corrections (dDsC) [88], and using the DZP, TZ2P, and QZ4P basis sets. Scalar relativistic effects were taken into consideration by using the ZORA formalism [89–91]. Frequency calculations were performed at the same level of theory for all basis sets to confirm that all stationary points are minima on the PES (see SI for a list of all frequencies for all molecules obtained with the QZ4P basis set). At the optimized geometries, the interaction energies were further analysed by using the energy decomposition analysis (EDA) [92]. These calculations were performed with the

AMS-2022 suite [82,93,94]. Single-point energy calculations were also performed with the CCSD(T) DLPNO method [95], with the ORCA-5.070 package [96,97], with extrapolation to the complete basis set limit by using the def2-TZVP and def2-QZVP basis sets on the PBE0-dDsC/QZ4P optimized geometries. The aromaticity of the weakly bonded complexes was studied by using the magnetic criterion, and the magnetically induced current density was calculated by using Gamess-UK [98,99] and SYSMO [100] by using the PBE0 functional, def2-TZVP basis set, and CTOCD-DZ method [85,86,101–103]. Time-dependent DFT calculations were performed with AMS (PBE0-dDsC/QZ4P) to explore the excited state properties. Spin-orbit coupling was taken into account by using the perturbational approach [104].

4. Conclusions

The halogen bond complexes $CF_3X\cdots Y$ and $C_2F_3X\cdots Y$, with Y = furan, thiophene, selenophene and X = Cl, Br, I, have been studied by using DFT and CCSD(T). It turns out that the PBE0-dDsC/QZ4P gives an adequate description of the interaction energies in these complexes, compared to CCSD(T) and experimental results. The energy decomposition analysis shows that all complexes are significantly stabilized by electrostatic, orbital, and dispersion interactions: not one factor dominates the interaction energy. In general, the interaction between the halogen atom and the π-bonds is stronger than with the lone pairs: the interaction is larger in the perpendicular orientation. The strength of the interaction follows the trend Cl < Br < I; the chalcogenide in the aromatic ring nor the hybridization plays a decisive role. Upon halogen bond formation, the aromaticity of the five-membered ring is unaffected: the π-ring current remains equally strong and diatropic in the complex as it is for the free aromatic ring. However, the photophysical properties of the complex are affected. The spin-orbit coupling between the singlet and triplet $\pi \rightarrow \pi^*$ states is increased, and a faster intersystem crossing is therefore expected. This effect of halogen bond formation can play a role in the formation of triplets in organic electronic devices when iodine containing additives are added.

Supplementary Materials: The supporting information can be downloaded at: https://www.mdpi.com/article/10.3390/molecules28020772/s1.

Author Contributions: The research was designed by W.H. The conceptualization and setting up the methodology was done by A.V.C., R.W.A.H. and W.H.; calculations and analysis were performed by A.V.C., R.W.A.H., J.v.G. and F.D.V. The original draft was written by A.V.C. and R.W.A.H., and draft was revised by all authors. A.V.C., W.H. and F.D.V. provided supervision. All authors have read and agreed to the published version of the manuscript.

Funding: This research was done with the support of a Strategic Research Program (SRP) from the VUB awarded to the General Chemistry research group.

Institutional Review Board Statement: Not applicable.

Informed Consent Statement: Not applicable.

Data Availability Statement: All inputs/outputs are available upon request.

Acknowledgments: The work of R.W.A.H. was sponsored by NWO Exact and Natural Sciences for the use of supercomputer facilities (contract no. 17197 7095) and R.W.A.H. and A.V.C. thank S. Dolas (SURF, NL) for allowing us to perform calculations on the experimental AMD platform kleurplaat maintained and operated by SURF Open Innovation Lab. F.D.V. acknowledges the VUB for the Strategic Research Program awarded to the ALGC research group. F.D.V. and F.D.P. wish to acknowledge the Vrije Universiteit Brussel for the support through a Strategic Research Program (SRP).

Conflicts of Interest: The authors declare no conflict of interest.

Abbreviations

The following abbreviations are used in this manuscript:

CCSD(T)	Single double coupled cluster with perturbative triples
EDA	Energy Decomposition Analysis
CDFT	Conceptual Density Functional Theory
DFT	Density Functional Theory
DZ	Double Zeta
TZ2P	Triple-Zeta with two polarization functions
QZ4P	Valence Quadruple-Zeta + 4 polarization function, relativistically optimized

References

1. Desiraju, G.R.; Ho, P.S.; Kloo, L.; Legon, A.C.; Marquardt, R.; Metrangolo, P.; Politzer, P.; Resnati, G.; Rissanen, K. Definition of the Halogen Bond (IUPAC Recommendations 2013). *Pure Appl. Chem.* **2013**, *85*, 1711–1713. [CrossRef]
2. Zingaro, R.A.; Hedges, M. Phosphine Oxide-Halogen Complexes: Effect on P-O and P-S Stretching Frequencies. *J. Phys. Chem.* **1961**, *65*, 1132–1138. [CrossRef]
3. Glaser, R.; Murphy, R.F. What s in a Name? Noncovalent Ar-Cl·(H-Ar)n Interactions and Terminology Based on Structure and Nature of the Bonding. *CrystEngComm* **2006**, *8*, 948–951. [CrossRef]
4. Clark, T.; Murray, J.S.; Lane, P.; Politzer, P. Why Are Dimethyl Sulfoxide and Dimethyl Sulfone Such Good Solvents? *J. Mol. Model.* **2008**, *14*, 689–697. [CrossRef]
5. Bianchi, R.; Forni, A.; Pilati, T. Experimental Electron Density Study of the Supramolecular Aggregation between 4,4-Dipyridyl-N,N'-Dioxide and 1,4-Diiodotetrafluorobenzene at 90 K. *Acta Crystallogr. Sect. B Struct. Sci.* **2004**, *60*, 559–568. [CrossRef]
6. Politzer, P.; Murray, J.S.; Clark, T. Halogen Bonding: An Electrostatically-Driven Highly Directional Noncovalent Interaction. *Phys. Chem. Chem. Phys.* **2010**, *12*, 7748–7757. [CrossRef]
7. Politzer, P.; Murray, J.S.; Clark, T. Halogen Bonding and Other σ-Hole Interactions: A Perspective. *Phys. Chem. Chem. Phys.* **2013**, *15*, 11178–11189. [CrossRef]
8. Hassel, O.; Hvoslef, J. The Structure of Bromine 1,4-Dioxanate. *Acta Chem. Scand.* **1954**, *8*, 873. [CrossRef]
9. Forni, A.; Metrangolo, P.; Pilati, T.; Resnati, G. Halogen Bond Distance as a Function of Temperature. *Cryst. Growth Des.* **2004**, *4*, 291–295. [CrossRef]
10. Khavasi, H.R.; Tehrani, A.A. Influence of Halogen Bonding Interaction on Supramolecular Assembly of Coordination Compounds; Head-to-Tail N··· X Synthon Repetitivity. *Inorg. Chem.* **2013**, *52*, 2891–2905. [CrossRef]
11. Crihfield, A.; Hartwell, J.; Phelps, D.; Walsh, R.B.; Harris, J.L.; Payne, J.F.; Pennington, W.T.; Hanks, T.W. Crystal Engineering through Halogen Bonding. 2. Complexes of Diacetylene-Linked Heterocycles with Organic Iodides. *Cryst. Growth Des.* **2003**, *3*, 313–320. [CrossRef]
12. Bjorvatten, T.; Hassel, O. Crystal Structure of the 1:3 Addition Compound Iodoform-Quinoline. *Acta Chem. Scand.* **1962**, *16*, 249–255. [CrossRef]
13. Pigge, F.C.; Vangala, V.R.; Swenson, D.C. Relative Importance of X . . . O=C vs. X . . . X Halogen Bonding as Structural Determinants in 4-Halotriaroylbenzenes. *Chem. Commun.* **2006**, 2123–2125. [CrossRef] [PubMed]
14. Evangelisti, L.; Feng, G.; Gou, Q.; Grabow, J.U.; Caminati, W. Halogen Bond and Free Internal Rotation: The Microwave Spectrum of CF3Cl-Dimethyl Ether. *J. Phys. Chem. A* **2014**, *118*, 579–582. [CrossRef] [PubMed]
15. Syssa-Magalé, J.L.; Boubekeur, K.; Palvadeau, P.; Meerschaut, A.; Schöllhorn, B. Self-Assembly via (N··· I) Non-Covalent Bonds between 1,4-Diiodo-Tetrafluoro-Benzene and a Tetra-Imino Ferrocenophane. *J. Mol. Struct.* **2004**, *691*, 79. [CrossRef]
16. Holmesland, O.; Romming, C. Crystal Structure of the (1:1) Addition Compounds of Diiodoacetylene with 1,4-Dithiane and 1,4-Dieselenane Respectively. *Acta Chem. Scand.* **1966**, *20*, 2601. [CrossRef]
17. Cinčić, D.; Friščić, T.; Jones, W. Experimental and Database Studies of Three-Centered Halogen Bonds with Bifurcated Acceptors Present in Molecular Crystals, Cocrystals and Salts. *CrystEngComm* **2011**, *13*, 3224–3231. [CrossRef]
18. Raatikainen, K.; Huuskonen, J.; Lahtinen, M.; Metrangolo, P.; Rissanen, K. Halogen Bonding Drives the Self-Assembly of Piperazine Cyclophanes into Tubular Structures. *Chem. Commun.* **2009**, *16*, 2160–2162. [CrossRef]
19. Nagels, N.; Geboes, Y.; pinter, B.; De Proft, F.; Herrebout, W.A. Tuning the Halogen/Hydrogen Bond Competition: A Spectroscopic and Conceptual DFT Study of Some Model Complexes Involving CHF2I. *Chem. Eur. J.* **2014**, *20*, 8433–8443. [CrossRef]
20. Herrebout, W. *Infrared and Raman Measurements of Halogen Bonding in Cryogenic Solutions*; Springer International Publishing: Cham, Switzerland, 2015; pp. 79–154.
21. Thirman, J.; Engelage, E.; Huber, S.M.; Head-Gordon, M. Characterizing the interplay of Pauli repulsion, electrostatics, dispersion and charge transfer in halogen bonding with energy decomposition analysis. *Phys. Chem. Chem. Phys.* **2018**, *20*, 905–915. [CrossRef]
22. Messina, M.T.; Metrangolo, P.; Panzeri, W.; Ragg, E.; Resnati, G. Perfluorocarbon-Hydrocarbon Self-Assembly. Part 3. Liquid Phase Interactions between Perfluoroalkylhalides and Heteroatom Containing Hydrocarbons. *Tetrahedron Lett.* **1998**, *39*, 9069–9072. [CrossRef]

23. Metrangolo, P.; Panzeri, W.; Recupero, F.; Resnati, G. Perfluorocarbon-Hydrocarbon Self-Assembly Part 16. 19F NMR Study of the Halogen Bonding between Halo-Perfluorocarbons and Heteroatom Containing Hydrocarbons. *J. Fluorine Chem.* **2002**, *114*, 27–33. [CrossRef]
24. Bjorvatten, T. Crystal Structures of Chloro and Bromo Cyanoacetylene. *Acta Chem. Scand.* **1968**, *22*, 410. [CrossRef]
25. Geboes, Y.; Nagels, N.; Pinter, B.; De Proft, F.; Herrebout, W.A. Competition of C(sp2)-X···O Halogen Bonding and Lone Pair···π Interactions: Cryospectroscopic Study of the Complexes of C2F3X (X = F, Cl, Br, and I) and Dimethyl Ether. *J. Phys. Chem. A* **2015**, *119*, 2502–2516. [CrossRef]
26. Metrangolo, P.; Murray, J.S.; Pilati, T.; Politzer, P.; Resnati, G.; Terraneo, G. The Fluorine Atom as a Halogen Bond Donor, Viz. a Positive Site. *CrystEngComm* **2011**, *13*, 6593-6596. [CrossRef]
27. Metrangolo, P.; Murray, J.S.; Pilati, T.; Politzer, P.; Resnati, G.; Terraneo, G. Fluorine-Centered Halogen Bonding: A Factor in Recognition Phenomena and Reactivity. *Cryst. Growth Des.* **2011**, *11*, 4238–4246. [CrossRef]
28. Rege, P.D.; Malkina, O.L.; Goroff, N.S. The Effect of Lewis Bases on the 13C NMR of Iodoalkynes. *J. Am. Chem. Soc.* **2002**, *124*, 370–371. [CrossRef]
29. Perkins, C.; Libri, S.; Adams, H.; Brammer, L. Diiodoacetylene: Compact, Strong Ditopic Halogen Bond Donor. *CrystEngComm* **2012**, *14*, 3033–3038. [CrossRef]
30. Bock, H.; Holl, S. Crystallization and Structure Determination of σ-Donor-Acceptor Complexes between 1,4-Dioxane and the Polyiodine Molecules I2, I2C=CI2, (IC)4S and (IC)4NR. *Z. Naturforsch. B J. Chem. Sci.* **2001**, *56*, 111–121. [CrossRef]
31. Gagnaux, P.; Susz, B.P. Etudes de Composés D'addition Des Acides de LEWIS. XII. Structure, Spectre Infrarouge et Polarisation Moléculaire Du Composé D'addition Dioxanne-1, 4–diiodacétylène. *Helv. Chim. Acta* **1960**, *43*, 948. [CrossRef]
32. Glaser, R.; Chen, N.; Wu, H.; Knotts, N.; Kaupp, M. 13C NMR Study of Halogen Bonding of Haloarenes: Measurements of Solvent Effects and Theoretical Analysis. *J. Am. Chem. Soc.* **2004**, *126*, 4412–4419. [CrossRef] [PubMed]
33. Raatikainen, K.; Cametti, M.; Rissanen, K. The Subtle Balance of Weak Supramolecular Interactions: The Hierarchy of Halogen and Hydrogen Bonds in Haloanilinium and Halopyridinium Salts. *Beilstein J. Org. Chem.* **2010**, *6*, 4. [CrossRef] [PubMed]
34. Logothetis, T.A.; Meyer, F.; Metrangolo, P.; Pilati, T.; Resnati, G. Crystal Engineering of Brominated Tectons: N-Methyl-3,5-Dibromo-Pyridinium Iodide Gives Particularly Short C-Br···I Halogen Bonding. *New J. Chem.* **2004**, *28*, 760. [CrossRef]
35. Liantonio, R.; Metrangolo, P.; Pilati, T.; Resnati, G. Fluorous Interpenetrated Layers in a Three-Component Crystal Matrix. *Cryst. Growth Des.* **2003**, *3*, 355–361. [CrossRef]
36. Aakeröy, C.B.; Fasulo, M.; Schultheiss, N.; Desper, J.; Moore, C. Structural Competition between Hydrogen Bonds and Halogen Bonds. *J. Am. Chem. Soc.* **2007**, *129*, 13772–13773. [CrossRef]
37. Saccone, M.; Cavallo, G.; Metrangolo, P.; Pace, A.; Pibiri, I.; Pilati, T.; Resnati, G.; Terraneo, G. Halogen Bond Directionality Translates Tecton Geometry into Self-Assembled Architecture Geometry. *CrystEngComm* **2013**, *15*, 3102–3105. [CrossRef]
38. Messina, M.T.; Metrangolo, P.; Panzeri, W.; Pilati, T.; Resnati, G. Intermolecular Recognition between Hydrocarbon Oxygen-Donors and Perfluorocarbon Iodine-Acceptors: The Shortest O···I Non-Covalent Bond. *Tetrahedron* **2001**, *57*, 8543. [CrossRef]
39. Syssa-Magalé, J.L.; Boubekeur, K.; Leroy, J.; Chamoreau, L.M.; Fave, C.; Schöllhorn, B. Directed Synthesis of a Halogen-Bonded Open Porphyrin Network. *CrystEngComm* **2014**, *16*, 10380–10384. [CrossRef]
40. Syssa-Magalé, J.L.; Boubekeur, K.; Schöllhorn, B. First Molecular Self-Assembly of 1,4-Diiodo-Tetrafluoro-Benzene and a Ketone via (O···I) Non-Covalent Halogen Bonds. *J. Mol. Struct.* **2005**, *737*, 103. [CrossRef]
41. Parisini, E.; Metrangolo, P.; Pilati, T.; Resnati, G.; Terraneo, G. Halogen Bonding in Halocarbon-Protein Complexes: A Structural Survey. *Chem. Soc. Rev.* **2011**, *40*, 2267–2278. [CrossRef]
42. Auffinger, P.; Hays, F.A.; Westhof, E.; Ho, P.S. Halogen Bonds in Biological Molecules. *Proc. Natl. Acad. Sci. USA* **2004**, *101*, 16789–16794. [CrossRef] [PubMed]
43. Babu, S.S.; Praveen, V.K.; Ajayaghosh, A. Functional π-Gelators and Their Applications. *Chem. Rev.* **2014**, *114*, 1973–2129. [PubMed]
44. Bartelena, L.; Robbins, J. Thyroid Hormone Transport Proteins. *Clin. Lab. Med.* **1993**, *13*, 583–598. [CrossRef]
45. Xu, Z.; Yang, Z.; Liu, Y.; Lu, Y.; Chen, K.; Zhu, W. Halogen Bond: Its Role beyond Drug-Target Binding Affinity for Drug Discovery and Development. *J. Chem. Inf. Model.* **2014**, *54*, 69–78. [CrossRef]
46. Howard, E.I.; Sanishvili, R.; Cachau, R.E.; Mitschler, A.; Chevrier, B.; Barth, P.; Lamour, V.; Van Zandt, M.; Sibley, E.; Bon, C. Ultrahigh Resolution Drug Design I: Details of Interactions in Human Aldose Reductase-Inhibitor Complex at 0.66 Å. *Proteins Struct. Funct. Genet.* **2004**, *55*, 792–804. [CrossRef]
47. Hays, F.A.; Vargason, J.M.; Ho, P.S. Effect of Sequence on the Conformation of DNA Holliday Junctions. *Biochemistry* **2003**, *42*, 9586–9597. [CrossRef]
48. El-Kabbani, O.; Ramsland, P.; Darmanin, C.; Chung, R.P.T.; Podjarny, A. Structure of Human Aldose Reductase Holoenzyme in Complex with Statil: An Approach to Structure-Based Inhibitor Design of the Enzyme. *Proteins Struct. Funct. Genet.* **2003**, *50*, 230–238. [CrossRef]
49. Hays, F.A.; Teegarden, A.; Jones, Z.J.R.; Harms, M.; Raup, D.; Watson, J.; Cavaliere, E.; Ho, P.S. How Sequence Defines Structure: A Crystallographic Map of DNA Structure and Conformation. *Proc. Natl. Acad. Sci. USA* **2005**, *102*, 7157–7162. [CrossRef] [PubMed]
50. Eichman, B.F.; Vargason, J.M.; Mooers, B.H.M.; Ho, P.S. The Holliday Junction in an Inverted Repeat DNA Sequence: Sequence Effects on the Structure of Four-Way Junctions. *Proc. Natl. Acad. Sci. USA* **2000**, *97*, 3971–3976. [CrossRef] [PubMed]

51. Bolton, O.; Lee, K.; Kim, H.J.; Lin, K.Y.; Kim, J. Activating Efficient Phosphorescence from Purely Organic Materials by Crystal Design. *Nat. Chem.* **2011**, *3*, 205–210. [CrossRef] [PubMed]

52. You, Y.; Park, S.Y. Phosphorescent iridium(III) Complexes: Toward High Phosphorescence Quantum Efficiency through Ligand Control. *Dalton Trans.* **2009**, *8*, 1267–1282. [CrossRef] [PubMed]

53. Zhao, Q.; Huang, C.; Li, F. Phosphorescent Heavy-Metal Complexes for Bioimaging. *Chem. Soc. Rev.* **2011**, *40*, 2508–2524. [CrossRef] [PubMed]

54. Yuan, W.Z.; Shen, X.Y.; Zhao, H.; Lam, J.W.Y.; Tang, L.; Lu, P.; Wang, C.; Liu, Y.; Wang, Z.; Zheng, Q. Crystallization-Induced Phosphorescence of Pure Organic Luminogens at Room Temperature. *J. Phys. Chem. C* **2010**, *114*, 6090–6099. [CrossRef]

55. Bolton, O.; Lee, D.; Jung, J.; Kim, J. Tuning the Photophysical Properties of Metal-Free Room Temperature Organic Phosphors via Compositional Variations in Bromobenzaldehyde/Dibromobenzene Mixed Crystals. *Chem. Mater.* **2014**, *26*, 6644–6649. [CrossRef]

56. Kwon, M.S.; Lee, D.; Seo, S.; Jung, J.; Kim, J. Tailoring Intermolecular Interactions for Efficient Room-Temperature Phosphorescence from Purely Organic Materials in Amorphous Polymer Matrices. *Angew. Chem. Int. Ed.* **2014**, *53*, 11177. [CrossRef]

57. Gao, H.Y.; Zhao, X.R.; Wang, H.; Pang, X.; Jin, W.J. Phosphorescent Cocrystals Assembled by 1,4- Diiodotetrafluorobenzene and Fluorene and Its Heterocyclic Analogues Based on C–I··· π Halogen Bonding. *Cryst. Growth Des.* **2012**, *12*, 4377–4387. [CrossRef]

58. Shi, L.; Liu, H.Y.; Shen, H.; Hu, J.; Zhang, G.L.; Wang, H.; Ji, L.N.; Chang, C.K.; Jiang, H.F. Fluorescence Properties of Halogenated Mono-Hydroxyl Corroles: The Heavy-Atom Effects. *J. Porphyr. Phthalocyanines* **2009**, *13*, 1221–1226. [CrossRef]

59. Hirata, S.; Totani, K.; Zhang, J.; Yamashita, T.; Kaji, H.; Marder, S.R.; Watanabe, T.; Adachi, C. Efficient Persistent Room Temperature Phosphorescence in Organic Amorphous Materials under Ambient Conditions. *Adv. Funct. Mater.* **2013**, *23*, 3386–3397. [CrossRef]

60. Lee, D.; Bolton, O.; Kim, B.C.; Youk, J.H.; Takayama, S.; Kim, J. Room Temperature Phosphorescence of Metal-Free Organic Materials in Amorphous Polymer Matrices. *J. Am. Chem. Soc.* **2013**, *135*, 6325–6329. [CrossRef]

61. Pinter, B.; Nagels, N.; Herrebout, W.A.; De Proft, F. Halogen Bonding from a Hard and Soft Acids and Bases Perspective: Investigation by Using Density Functional Theory Reactivity Indices. *Chem. Eur. J.* **2013**, *19*, 519–530. [CrossRef]

62. Gao, H.Y.; Shen, Q.J.; Zhao, X.R.; Yan, X.Q.; Pang, X.; Jin, W.J. Phosphorescent Co-Crystal Assembled by 1,4-Diiodotetrafluorobenzene with Carbazole Based on C–I··· π Halogen Bonding. *J. Mater. Chem.* **2012**, *22*, 5336. [CrossRef]

63. Zhu, Q.; Gao, Y.J.; Gao, H.Y.; Jin, W.J. Effect of N-Methyl and Ethyl on Phosphorescence of Carbazole in Cocrystals Assembled by C–I··· π Halogen Bond, π-Hole··· π Bond and Other Interactions Using 1,4-Diiodotetrafluorobenzene as Donor. *J. Photochem. Photobiol. A* **2014**, *289*, 31. [CrossRef]

64. Shen, Q.J.; Wei, H.Q.; Zou, W.S.; Sun, H.L.; Jin, W.J. Cocrystals Assembled by Pyrene and 1,2- or 1,4-Diiodotetrafluorobenzenes and Their Phosphorescent Behaviors Modulated by Local Molecular Environment. *CrystEngComm* **2012**, *14*, 1010–1015. [CrossRef]

65. Pang, X.; Wang, H.; Zhao, X.R.; Jin, W.J. Co-Crystallization Turned on the Phosphorescence of Phenanthrene by C–Br··· π Halogen Bonding, π–hole··· π Bonding and Other Assisting Interactions. *CrystEngComm* **2013**, *15*, 2722–2730. [CrossRef]

66. Zhu, Q.; Wang, H.; Zhao, X.R.; Jin, W.J. The Phosphorescent Behaviors of 9-Bromo- and 9-Iodophenanthrene in Crystals Modulated by π-π Interactions, C-H··· π Hydrogen Bond and C-I··· π Halogen Bond. *J. Photochem. Photobiol. A* **2014**, *274*, 98–107. [CrossRef]

67. Stöhr, M.; Van Voorhis, T.; Tkatchenko, A. Theory and practice of modeling van der Waals interactions in electronic-structure calculations. *Chem. Soc. Rev.* **2019**, *48*, 4118–4154. [CrossRef]

68. Otero-de-la Roza, A.; Johnson, E.R. Van der Waals interactions in solids using the exchange-hole dipole moment model. *J. Chem. Phys.* **2012**, *136*, 174109. [CrossRef]

69. Otero-de-la Roza, A.; Johnson, E.R. A benchmark for non-covalent interactions in solids. *J. Chem. Phys.* **2012**, *137*, 054103. [CrossRef]

70. Gobre, V.V.; Tkatchenko, A. Scaling laws for van der Waals interactions in nanostructured materials. *Nat. Commun.* **2013**, *4*, 1–6. [CrossRef]

71. Reilly, A.M.; Tkatchenko, A. van der Waals dispersion interactions in molecular materials: Beyond pairwise additivity. *Chem. Sci* **2015**, *6*, 3289–3301. [CrossRef]

72. Grimme, S.; Ehrlich, S.; Goerigk, L. Effect of the Damping Function in Dispersion Corrected Density Functional Theory. *J. Comput. Chem.* **2011**, *32*, 1456–1465. [CrossRef] [PubMed]

73. Klimes, J.; Michaelides, A. Perspective: Advances and challenges in treating van der Waals dispersion forces in density functional theory. *J. Chem. Phys.* **2012**, *137*, 120901. [CrossRef]

74. Bjorkman, T. Testing several recent van der Waals density functionals for layered structures. *J. Chem. Phys.* **2014**, *141*, 074708. [CrossRef] [PubMed]

75. Ambrosetti, A.; Reilly, A.M.; DiStasio, R.A.; Tkatchenko, A. Long-range correlation energy calculated from coupled atomic response functions. *J. Chem. Phys.* **2014**, *140*, 18A508. [CrossRef]

76. Sinanoglu, O. Many-Electron Theory of Atoms, Molecules and their interactions. *Adv. Chem. Phys.* **1964**, *6*, 315–412.

77. Buhmann, S.Y. *Dispersion Forces I*; Springer: Berlin/Heidelberg, Germany, 2012.

78. Woods, L.M.; Dalvit, D.A.R.; Tkatchenko, A.; Rodriguez-Lopez, P.; Rodriguez, A.W.; Podgornik, R. Materials perspective on Casimir and van der Waals interactions. *Rev. Mod. Phys.* **2016**, *88*, 045003. [CrossRef]

79. Dobson, J.F.; White, A.; Rubio, A. Asymptotics of the dispersion interaction: Analytic benchmarks for van der Waals energy functionals. *Phys. Rev. Lett.* **2006**, *96*, 073201. [CrossRef]

80. Blum, V.; Gehrke, R.; Hanke, F.; Havu, P.; Havu, V.; Ren, X.G.; Reuter, K.; Scheffler, M. Ab initio molecular simulations with numeric atom-centered orbitals. *Comp. Phys. Commun.* **2009**, *180*, 2175–2196. [CrossRef]
81. Chai, J.D.; Head-Gordon, M. Long-range corrected hybrid density functionals with damped atom-atom dispersion corrections. *Phys. Chem. Chem. Phys.* **2008**, *10*, 6615–6620. [CrossRef]
82. Fonseca Guerra, C.; Snijders, J.G.; te Velde, G.; Baerends, E.J. Towards an order-N DFT method. *Theor. Chem. Acc.* **1998**, *99*, 391–403. [CrossRef]
83. Clark, S.J.; Segall, M.D.; Pickard, C.J.; Hasnip, P.J.; Probert, M.J.; Refson, K.; Payne, M.C. First principles methods using CASTEP. *Z. Kristallogr.* **2005**, *220*, 567–570. [CrossRef]
84. Jiménez-Grávalos, F.; Gallegos, M.; Martín Pendás, A.; Novikov, A.S. Challenging the electrostatic σ-hole picture of halogen bonding using minimal models and the interacting quantum atoms approach. *J. Comput. Chem.* **2021**, *42*, 676–687. [CrossRef] [PubMed]
85. Steiner, E.; Fowler, P.W. Patterns of ring currents in conjugated molecules: A few-electron model based on orbital contributions. *J. Phys. Chem. A* **2001**, *105*, 9553–9562. [CrossRef]
86. Steiner, E.; Fowler, P.W. Four- and two-electron rules for diatropic and paratropic ring currents in monocyclic π systems. *Chem. Commun.* **2001**, *21*, 2220–2221. [CrossRef]
87. De Vleeschouwer, F.; De Proft, F.; Ergün, O.; Herrebout, W.; Geerlings, P. A Combined Experimental/Quantum-Chemical Study of Tetrel, Pnictogen, and Chalcogen Bonds of Linear Triatomic Molecules. *Molecules* **2021**, *26*, 6767. [CrossRef]
88. Steinmann, S.; Corminboeuf, C. Comprehensive Benchmarking of a Density-Dependent Dispersion Correction. *J. Chem. Theory Comput.* **2011**, *7*, 3567–3577. [CrossRef]
89. van Lenthe, E.; Baerends, E.J.; Snijders, J.G. Relativistic regular two-component Hamiltonians. *J. Chem. Phys.* **1993**, *99*, 4597–4610. [CrossRef]
90. van Lenthe, E.; Baerends, E.J.S.J.G. Relativistic total energy using regular approximations. *J. Chem. Phys.* **1994**, *101*, 9783–9792. [CrossRef]
91. van Lenthe, E.; van Leeuwen, R.; Baerends, E.J.; Snijders, J.G. Relativistic regular two-component Hamiltonians. *Int. J. Quantum Chem.* **1996**, *57*, 281–293. [CrossRef]
92. Bickelhaupt, F.M.; Baerends, E.J., Kohn-Sham Density Functional Theory: Predicting and Understanding Chemistry. In *Rev. Comput. Chem.*; Lipkowitz, K.B., Boyd, D.B., Eds.; Wiley-VCH: New York, NY, USA, 2000; pp. 1–86.
93. te Velde, G.; Bickelhaupt, F.M.; Baerends, E.J.; Fonseca Guerra, C.; van Gisbergen, S.J.A.; Snijders, J.G.; Ziegler, T. Chemistry with ADF. *J. Comput. Chem.* **2001**, *22*, 931–967. [CrossRef]
94. Baerends, E.J. AMS2022.01, SCM, Theoretical Chemistry, Vrije Universiteit, Amsterdam, The Netherlands. Available online: http://www.scm.com (accessed on 30 July 2022).
95. Liakos, D.; Guo, Y.; Neese, F. Comprehensive benchmark results for the domain based local pair natural orbital coupled cluster method (DLPNO-CCSD(T)) for closed- and open-shell systems. *J. Phys. Chem. A* **2020**, *124*, 90–100. [CrossRef]
96. Neese, F.; Wennmohs, F.; Becker, U.; Riplinger, C. The ORCA quantum chemistry program package. *J. Chem. Phys.* **2020**, *152*, 224108. [CrossRef] [PubMed]
97. Neese, F. Software update: The ORCA program system—Version 5.0. *WIREs Comput. Mol. Sci.* **2022**, *12*, e1606. [CrossRef]
98. Guest, M.F.; Bush, I.J.; van Dam, H.J.J.; Sherwood, P.; Thomas, J.M.H.; van Lenthe, J.H.; Havenith, R.W.A.; Kendrick, J. The GAMESS-UK electronic structure package: Algorithms, developments and applications. *Mol. Phys.* **2005**, *103*, 719–747. [CrossRef]
99. Havenith, R.W.A.; Fowler, P. Ipsocentric ring currents in density functional theory. *Chem. Phys. Lett.* **2007**, *449*, 347–353. [CrossRef]
100. Lazzeretti, P.; Zanasi, R. *SYSMO Package*; Additional Routines by P. W. Fowler, E. Steiner, R. W. A. Havenith, A. Soncini; University of Modena: Modena, Italy, 1980.
101. Keith, T.; Bader, R.F.W. Calculation of magnetic response properties using a continuous set of gauge transformations. *Chem. Phys. Lett.* **1993**, *210*, 223–231. [CrossRef]
102. Lazzeretti, P.; Malagoli, M.; Zanasi, R. Electronic current density induced by nuclear magnetic dipoles. *Chem. Phys. Lett.* **1994**, *220*, 299–304. [CrossRef]
103. Coriani, S.; Lazzeretti, P.; Malagoli, M.; Zanasi, R. On CHF calculations of second-order magnetic properties using the method of continuous transformation of origin of the current density. *Theor. Chim. Acta* **1994**, *89*, 181–192. [CrossRef]
104. Wang, F.; Ziegler, T. A simplified relativistic time-dependent density-functional theory formalism for the calculations of excitation energies including spin-orbit coupling effect. *J. Chem. Phys.* **2005**, *123*, 154102. [CrossRef]

Article

Stabilizing Halogen-Bonded Complex between Metallic Anion and Iodide

Fei Ying [1], Xu Yuan [2,3], Xinxing Zhang [2,3,*] and Jing Xie [1,*]

[1] Key Laboratory of Cluster Science of Ministry of Education, Beijing Key Laboratory of Photoelectronic/Electrophotonic Conversion Materials, School of Chemistry and Chemical Engineering, Beijing Institute of Technology, Beijing 100081, China

[2] College of Chemistry, Key Laboratory of Advanced Energy Materials Chemistry (Ministry of Education), Renewable Energy Conversion and Storage Center (ReCAST), Tianjin Key Laboratory of Biosensing and Molecular Recognition, Shenzhen Research Institute, Frontiers Science Center for New Organic Matter, Nankai University, Tianjin 300071, China

[3] Haihe Laboratory of Sustainable Chemical Transformations, Tianjin 300192, China

* Correspondence: zhangxx@nankai.edu.cn (X.Z.); jingxie@bit.edu.cn (J.X.)

Abstract: Halogen bonds (XBs) between metal anions and halides have seldom been reported because metal anions are reactive for XB donors. The pyramidal-shaped $Mn(CO)_5^-$ anion is a candidate metallic XB acceptor with a ligand-protected metal core that maintains the negative charge and an open site to accept XB donors. Herein, $Mn(CO)_5^-$ is prepared by electrospray ionization, and its reaction with CH_3I in gas phase is studied using mass spectrometry and density functional theory (DFT) calculation. The product observed experimentally at m/z = 337 is assigned as $[IMn(CO)_4(OCCH_3)]^-$, which is formed by successive nucleophilic substitution and reductive elimination, instead of the halogen-bonded complex (XC) $CH_3-I\cdots Mn(CO)_5^-$, because the $I\cdots Mn$ interaction is weak within XC and it could be a transient species. Inspiringly, DFT calculations predict that replacing CH_3I with CF_3I can strengthen the halogen bonding within the XC due to the electro-withdrawing ability of F. More importantly, in so doing, the nucleophilic substitution barrier can be raised significantly, ~30 kcal/mol, thus leaving the system trapping within the XC region. In brief, the combination of a passivating metal core and the introduction of an electro-withdrawing group to the halide can enable strong halogen bonding between metallic anion and iodide.

Keywords: halogen bond; metallic anion; nucleophilic substitution reaction; quantum chemistry calculation; reductive elimination

Citation: Ying, F.; Yuan, X.; Zhang, X.; Xie, J. Stabilizing Halogen-Bonded Complex between Metallic Anion and Iodide. *Molecules* **2022**, *27*, 8069. https://doi.org/10.3390/molecules27228069

Academic Editors: Qingzhong Li, Steve Scheiner and Zhiwu Yu

Received: 29 October 2022
Accepted: 18 November 2022
Published: 21 November 2022

Publisher's Note: MDPI stays neutral with regard to jurisdictional claims in published maps and institutional affiliations.

1. Introduction

The halogen bond (XB) is a type of non-covalent interaction that has attracted the interests of experimentalists and theoretical chemists in recent years [1–9]. According to the International Union of Pure and Applied Chemistry (IUPAC), "a halogen bond occurs when there is evidence of a net attractive interaction between an electrophilic region associated with a halogen atom in a molecular entity and a nucleophilic region in another, or the same, molecular entity" [10]. This definition states unambiguously that the halogen atom serves as an electrophile and interacts with a nucleophilic moiety. Typically, an XB is denoted as $R-X\cdots Y$ with three dots representing the bond, and X is a halogen atom (i.e., XB donor) that has an electrophilic region on its electrostatic potential surface, and Y is an XB acceptor. For the XB donor molecule (i.e., $R-X$ molecule), the electrophilic (or positive) region on X, named as "σ-hole" [11,12], is induced by the $R-X$ bond, which leaves an anisotropic distribution of electrons. The σ-hole magnitude, which represents the XB strength given by the same XB acceptor, scales with the polarizability of the halogen atom, that is, F < Cl < Br < I. Hence, changing the X atom can tune the XB's strength, and there are other methods as well, including modifying the R-functional group and the electro-withdrawing ability of Y.

The nature and tunability of XB make it useful in different fields spanning from material sciences to biomolecular recognition and drug design [13–18].

The common XB acceptors are nucleophiles, such as N, O, S, P, or halogen atoms/anions; metal anions are rarely seen. This is because atomic metal anions are usually too reactive towards organohalogens. For example, the reaction between Au^-/Ag^-/Cu^- anions and CH_3I in gas phase give rise to a Grignard reagent-like product $[CH_3{-}M{-}I]^-$, where a covalent M$-$I bond is formed [19,20]. This structure is calculated to be ~2.0–3.0 eV more stable than the XB complex $[CH_3{-}I\cdots M]^-$ [20,21]. To achieve the goal of forming metallic acceptor-containing halogen bonds, one of our authors proposed two strategies: one is to utilize a metal cluster anion with a high electron detachment energy; the other is to design a ligand-passivated/protected metal core that can maintain the negative charge [22]. The goal of this work is to check the feasibility of the second strategy experimentally. Hence, herein, we prepared a $Mn(CO)_5^-$ anionic compound by electrospray ionization and investigated its reactivity with CH_3I.

To test whether $Mn(CO)_5^-$ anion is a suitable candidate to form a halogen-bonded complex, we first investigated the properties of $Mn(CO)_5^-$ anion by density functional theory (DFT) calculation using M06-2X method [23] with aug-cc-pVTZ basis set [24–26]. As shown in Figure 1a, the structure of $Mn(CO)_5^-$ anion has a pyramidal shape, with one CO ligand in the horizontal direction and the other four CO ligands almost in the same plane (see Figure S1 for an illustration), leaving the left an open site to accept a XB donor. A Mulliken charge analysis [27] (Figure 1a) indicated that the Mn-atom core is the most negative, with a charge of -0.84 e, and this is clearly displayed in the electrostatic potential map (Figure 1b). In addition, the HOMO of $Mn(CO)_5^-$ anion (Figure 1c) comprising C p orbital and Mn d_{x^2} orbital has electrons evenly delocalized on four C atoms, thus stabilizing the compound. In brief, $Mn(CO)_5^-$ anion fulfills the two criteria of the second strategy: the metal core is negatively charged and has at least one open site to accept the XB.

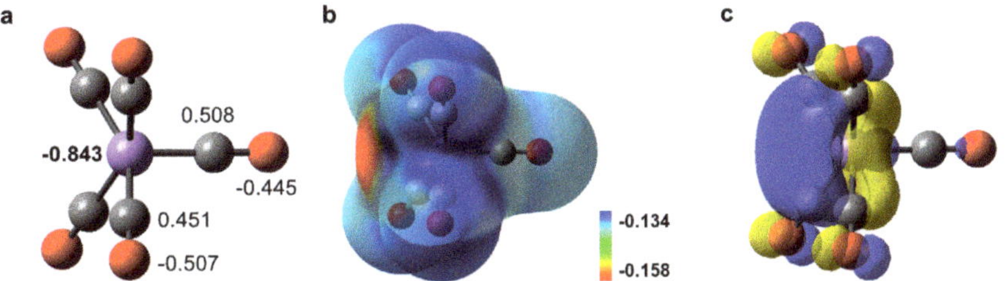

Figure 1. Calculated (**a**) structure and Mulliken charge, (**b**) electrostatic potential map in e/Bohr^3, and (**c**) HOMO of $Mn(CO)_5^-$ anion. The M06-2X/aug-cc-pVTZ level of theory is used [23–26]. Color code: C, grey; O, red; Mn, purple.

In this work, we will first study the reaction between $Mn(CO)_5^-$ anion and CH_3I in gas phase using a linear ion trap mass spectrometer. Then the products and mechanism will be analyzed with the help of DFT calculation. The stability of the halogen-bonded complex is evaluated, and a strategy is proposed to further stabilize it.

2. Methods

2.1. Experimental Methods

Mass spectra were acquired using a linear ion trap mass spectrometer (LTQ-XL, Thermo-Fisher, Waltham, MA, USA). The inlet capillary temperature of the mass spectrometer was maintained at 275 °C. The tube lens voltage on the LTQ-XL was set to be 0 V in order to avoid in-source fragmentation of the fragile species. The applied negative voltage was set at -4000 V in this study in order to trigger the electrospray ionization. A methanol solution of $Mn_2(CO)_{10}$ was sprayed to generate the $Mn(CO)_5^-$ anion. The collision-induced dissociation (CID) spectrum of the $Mn(CO)_5^-$ anion is presented in Figure S2. Gas-phase reaction between the $Mn(CO)_5^-$ anion and the neutral CH_3I molecule at room temperature was conducted in the linear ion trap by using the collision-induced dissociation (CID) mode that is, the MS^2 mode of the mass spectrometer in order to isolate the $Mn(CO)_5^-$ anion in the trap. The CH_3I molecules were introduced to the trap by putting a drop of CH_3I into a small stainless-steel reservoir that was connected to the pipeline of the He collision gas. The collision energy was set to be under 10 V in order to trigger the reactions.

2.2. Computational Methods

Geometry optimizations were performed using M06-2X functional [23], with aug-cc-pVTZ basis set [24–26] used for H, C, O, F, and Mn atoms, and aug-cc-pVTZ-PP basis set [28,29] used for I atoms. Various configurations were optimized for $CH_3I-Mn(CO)_3^-$, and the most stable structures were used for discussion (see Figure S3 for details). Harmonic vibrational frequencies were calculated to confirm the nature of the stationary points. Intrinsic reaction coordinate (IRC) calculations were performed on transition states to confirm that they connected the correct intermediates. The ground state of $Mn(CO)_4I^-$ is doublet, and the other metal-involved species are all singlet. The zero-point corrected energy is used in the potential energy profile. Gaussian 16 [30] package was used to perform all the calculations.

3. Results and Discussion

3.1. Mass Spectrometry

A typical mass spectrum showing the reaction products between $Mn(CO)_5^-$ and CH_3I is presented in Figure 2a. Three major peaks at m/z 281, 294, and 337 were observed, corresponding to the masses of $CH_3I-Mn(CO)_3^-$, $Mn(CO)_4I^-$, and $CH_3I-Mn(CO)_5^-$, among which the latter is the direct product from the reaction between $Mn(CO)_5^-$ and CH_3I, but the former two are the collision fragments of the latter. To obtain structural information for the m/z 337 peak, we further isolated it with the MS^3 mode of the mass spectrometer; its CID fragments with a collision energy of 5 V are presented in Figure 2b. If $CH_3I-Mn(CO)_5^-$ is a weakly bonded species of $Mn(CO)_5^-$ and CH_3I, its fragments should predominantly be $Mn(CO)_5^-$. However, two distinct fragments, $CH_3I-Mn(CO)_4^-$ and $MnIO_2^-$, were observed, suggesting that the m/z 337 peak is not a weakly bound species.

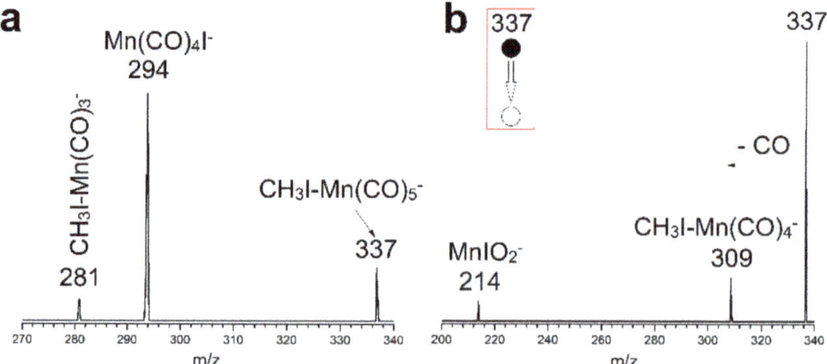

Figure 2. Mass spectrometric results. (**a**) A typical mass spectrum showing the reaction products between $Mn(CO)_5^-$ and CH_3I; (**b**) CID mass spectrum of $[Mn(CO)_5(CH_3I)]^-$ at m/z 337 taken with the MS^3 mode. The nominal applied CID voltage is 5 V.

3.2. Density Functional Theory Calculation

3.2.1. $Mn(CO)_5^-$ + CH_3I Reaction Mechanism

To identify the structure and understand the formation mechanism of the aforementioned observed products, we performed DFT calculations. Scheme 1 depicts the potential energy surfaces (PESs) for $Mn(CO)_5^- + CH_3I$, and selected structures are displayed in Figure 3. Enthalpy and free energy values at 298.15 K are listed in Table 1.

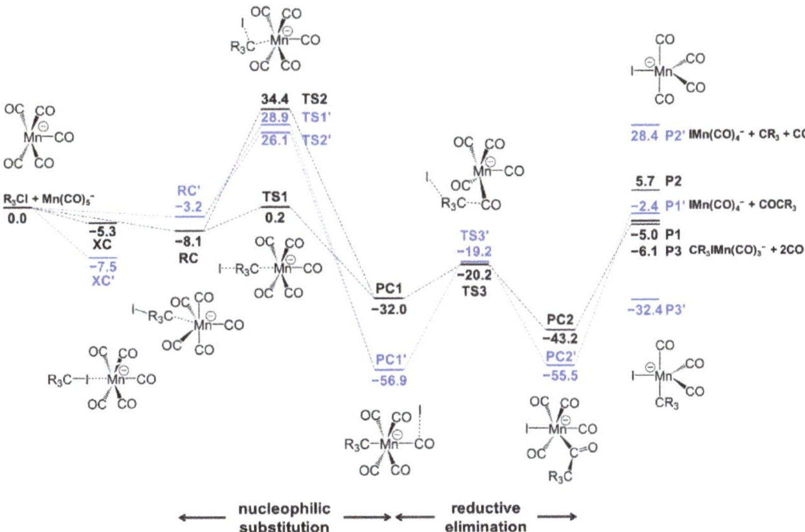

Scheme 1. Potential energy profile of $Mn(CO)_5^-$ reacting with CR_3I. The captions in black are for R = H, and the captions in blue are for R = F. Zero-point corrected energies (in kcal/mol) are given relative to the total energy of isolated R_3CI and $Mn(CO)_5^-$.

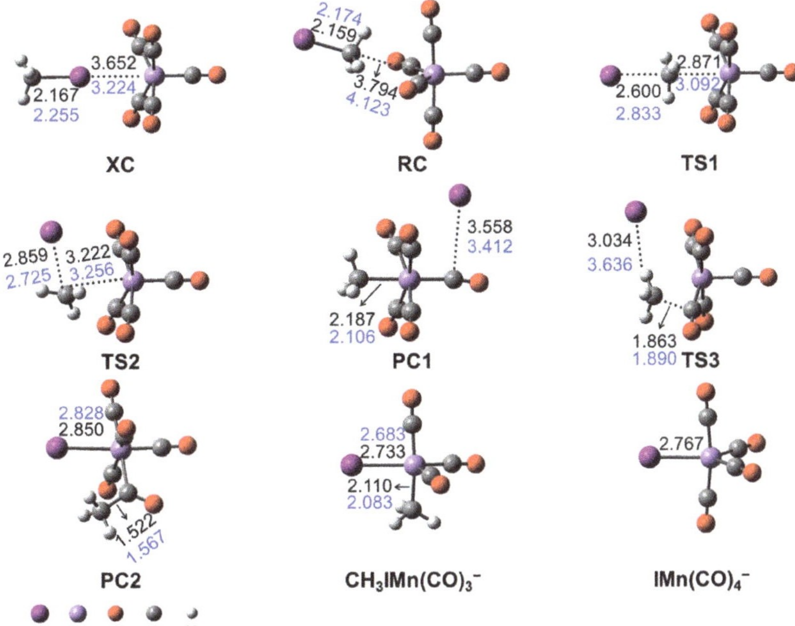

Figure 3. Structures of selected stationary points on the PES of $Mn(CO)_5^-$ reacting with CH_3I. The bond distances (Å) in black are for CH_3I, and those in blue are for CF_3I.

Table 1. Energetic values for stationary points on the PES of CR_3I reacting with $Mn(CO)_5^-$.

kcal/mol	ΔE_{elec}		$\Delta(E_{elec} + ZPE)$		$\Delta H_{298\cdot15K}$		$\Delta G_{298\cdot15K}$	
R	H	F	H	F	H	F	H	F
$Mn(CO)_5^- + CR_3I$	0.0	0.0	0.0	0.0	0.0	0.0	0.0	0.0
XC	−5.4	−17.5	−5.3	−17.5	−3.9	−16.5	1.4	−7.0
RC	−8.6	−3.6	−9.1	−3.2	−7.3	−1.8	−0.2	3.8
TS1	0.4	29.1	0.2	28.9	1.2	30.3	8.6	37.3
TS2	35.2	27.2	34.4	26.1	35.4	27.6	43.0	35.2
PC1	−31.7	−56.6	−32.0	−56.9	−31.1	−55.1	−22.4	−47.6
TS3	−20.3	−18.8	−20.2	−19.2	−19.5	−17.8	−10.4	−9.9
PC3	−44.4	−56.7	−43.2	−55.5	−42.4	−54.9	−33.3	−43.5
$Mn(CO)_4I^- + COCR_3$	−3.0	−0.7	−5.0	−2.4	−3.9	−1.4	−8.2	−5.8
$Mn(CO)_4I^- + CO + CR_3$	13.1	33.4	5.7	28.4	8.4	30.2	−4.7	16.3
$CR_3I-Mn(CO)_3^- + 2CO$	−1.6	−28.3	−6.1	−32.4	−4.4	−29.7	−15.5	−43.1

Figure 3 depicts that a halogen-bonded complex (XC) $CH_3-I\cdots Mn(CO)_5^-$ is formed by an I atom attacking the open site of Mn, and XC is 5.3 kcal/mol lower in energy than the reactants. The $I\cdots Mn$ distance within XC is 3.652 Å, which is 79% of the sum of the van der Walls radii of I (2.36 Å) and Mn (2.24 Å). At the same time, a slightly more stable pre-reaction complex (RC) is formed between a C atom interacting with Mn; it is lower in energy by 2.8 kcal/mol. Of note, additional conformers of RC that are higher in energy are localized: one has a linear I−C−Mn shape, and the other I−C−Mn angle is ~ 90°. These two structures are structural isomers of RC, which may appear when CH_3I attacks $Mn(CO_5)^-$ in a different direction. For clarity, they are omitted in Figure 3 and are instead present in Figure S4. After crossing a back-side attack nucleophilic substitution barrier (TS1) of 8.3 kcal/mol, it proceeds to post-reaction complex PC1. We also considered the front-side attack S_N2 transition state (TS2); however, it is too high (34.4 kcal/mol) to occur. Within PC1, CH_3 fragment and I fragment are located on the opposite side of Mn with a weak I-C interaction. Relative to the reactants, PC1 is −32.0 kcal/mol in energy. Then PC1 can undergo a reductive elimination barrier (TS3) of 11.8 kcal/mol and ends up with the formation of a C−C bond and the migration of I to bond with Mn. This resulted complex (PC2) is very stable, and is −43.2 kcal/mol relative to the reactants. Because TS1 is almost thermally neutral and TS3 is lower in energy than the reactants, the most stable PC2 can be formed under room temperature (the experimental condition). For this reason, in Figure 2, the signal at 337 m/z was assigned to be PC2, and it agrees with experimental results that this species is quite stable.

The calculated energy for $Mn(CO)_5^- + CH_3I \rightarrow Mn(CO)_4I^- + COCH_3$ reaction is −5.0 kcal/mol, and the calculated energy to form $Mn(CO)_4I^- + CO + CH_3$ is 5.7 kcal/mol. Therefore, we believe the experimentally observed $Mn(CO)_4I^-$ is more likely to be dissociated from PC2 and to generate $COCH_3$ at the same time, and is less likely to be caused by the collision-induced dissociation that forms CO and CH_3.

The calculated energy from generating $CH_3I-Mn(CO)_3^- + 2CO$ from reactants is −6.1 kcal/mol downhill. The most stable structure of $CH_3I-Mn(CO)_3^-$ has a pseudo-bipyramidal shape (Figure 3). Analyzing the PES indicates that it may dissociate from PC1 or, provided sufficient energy, dissociate from TS2.

In brief, although there is considerable halogen bonding between CH_3I and $Mn(CO)_5^-$, the passivated Mn center within $Mn(CO)_5^-$ is still reactive towards CH_3I, thus making the XC a transient species. The observed $CH_3I-Mn(CO)_5^-$ signal is PC2, which forms by nucleophilic substitution and the following reductive elimination.

3.2.2. Stabilizing Halogen-Bonded Complex by CF_3I

It is known that introducing an electron-withdrawing group, such as F, to the methyl group can increase the σ-hole magnitude, and thus the XB strength. Therefore, we changed CH_3I to CF_3I, which induces a greater positive region on the I atom, in the hope that it can stabilize the halogen-bonded complex when interacting with $Mn(CO)_5{}^-$. On the other hand, changing CH_3 to a heavier CF_3 group is expected to raise the inversion S_N2 barrier [31], thus preventing the S_N2 reaction. This may also help trap the system in a halogen-bonded complex well, so we computed the PES of $Mn(CO)_5{}^- + CF_3I$.

As shown in Scheme 1, the halogen-bonded complex (XC') $CF_3-I\cdots Mn(CO)_5{}^-$ well is 17.5 kcal/mol deep, where the pre-reaction complex RC' is 14.3 kcal/mol higher than it is. Within XC', the $I\cdots Mn$ distance is 3.224 Å, being 0.428 Å shorter than the corresponding value of XC. This is consistent with XC' being more stable than XC. To characterize the interaction between CR_3I and $Mn(CO_5)^-$ within the halogen-bonded complex $CR_3-I\cdots Mn(CO)_5{}^-$, natural bond orbital (NBO) [32,33] calculations were performed for R = H and F in order to analyze the donor–acceptor charge transfer properties. As shown in Figure 4, taking $CF_3-I\cdots Mn(CO)_5{}^-$ as an example, the donor orbital is the $Mn-C$ bonding σ orbital and Mn 3p orbital, and the acceptor orbital is the C–I antibonding σ^* orbital. The same scenario also applies for $CH_3-I\cdots Mn(CO)_5{}^-$. In comparison, when the halogen-bonded complex is composed of a main group nucleophile, such as F^- and CH_3I (i.e., $[CH_3-I\cdots F]^-$), the donor NBO is a 2p orbital; when the nucleophile is $Cu^-/Ag^-/Au^-$, the donor NBO is an s orbital [21,34].

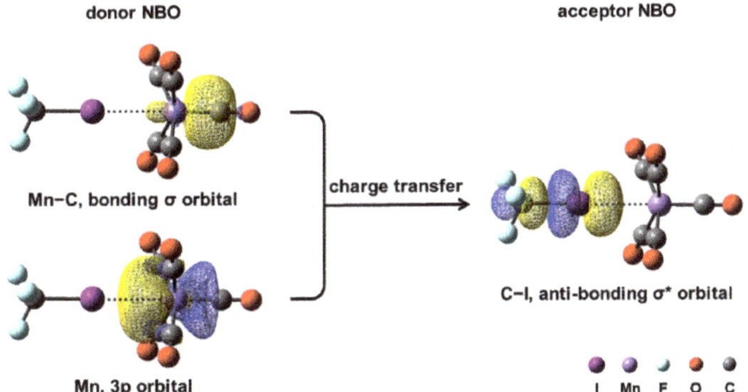

Figure 4. Donor and acceptor natural bond orbitals (NBOs) of halogen-bonded complex $CF_3-I\cdots Mn(CO)_5{}^-$ to illustrate the charge transfer interaction between I and Mn.

Additionally, the back-side attack S_N2 barrier (TS1') is largely raised to 28.9 kcal/mol relative to the reactants. Although the front-side attack transition state (TS2') is lower than TS2, TS2' is still 26.1 kcal/mol uphill. Of note, different from CH_3I, the front-side attack S_N2 barrier (TS2') is lower than the back-side attack S_N2 barrier (TS1') by 2.8 kcal/mol. If the reactants are cooled to room temperature or even lower, they are unlikely to cross these barriers or proceed to nucleophilic substitution and the following reductive elimination. Of note, the PC1' and PC2' complexes are even lower than PC1 and PC2, but PC1' needs to cross a barrier (TS3') of 37.7 kcal/mol. In another words, if the system crosses TS1', both PC1' and PC2' can be formed and stable.

To summarize, calculations show that replacing CH_3I with CF_3I stabilizes the halogen-bonded complex and raises the nucleophilic substitution barrier. This is an effective strategy to obtain strong halogen bonding between iodide and metallic anions.

4. Conclusions

To achieve the goal of constructing a stable halogen-bonded complex between metallic anionic species and halide, we adopted the strategy of passivating the reactive metallic anion by introducing protected ligands. Thus, we designed the $Mn(CO)_5{}^-$ anionic compound, and DFT calculation confirms that it maintains a negatively charged core and has an open site to accept halogen bond donors. Next, the $Mn(CO)_5{}^-$ species was prepared by electrospray ionization and then reacted with CH_3I in gas phase using a linear ion trap mass spectrometer. The major products were $CH_3I\text{-}Mn(CO)_3^-$, $Mn(CO)_4I^-$, and $CH_3I\text{-}Mn(CO)_5^-$. DFT calculations suggested that $CH_3I\text{-}Mn(CO)_5^-$ is a stable species (i.e., $[IMn(CO)_4(OCCH_3)]^-$) that forms by nucleophilic substitution and reductive elimination. The halogen-bonded complex $CH_3-I\cdots Mn(CO)_5^-$ could be a transient species because the interaction between I and Mn is weak.

By substituting CH_3I to CF_3I, calculations predicted that the resulted halogen-bonded complex $CF_3-I\cdots Mn(CO)_5^-$ is stabilized considerably. In addition, the barrier for nucleophilic substitution was greatly raised, allowing the system to trap in the XB complex well, given that the system is cool enough to avoid crossing the S_N2 barrier. This work presents an example of stabilizing the halogen bonding between a ligand-protected metal anion and halide with strong electro-withdrawing group. By adopting a similar strategy, it is anticipated that more metallic acceptor-containing XBs will be discovered.

Supplementary Materials: The following supporting information can be downloaded at: https://www.mdpi.com/article/10.3390/molecules27228069/s1, Figure S1: Optimized structure of $Mn(CO)_5^-$ anion. Figure S2: CID fragments of the $Mn(CO)_5^-$ anion. Figure S3: Optimized structures of $CH_3I^-Mn(CO)_3^-$. Figure S4: Additional conformers of RC. Coordinates of all computed structures.

Author Contributions: Conceptualization: X.Z. and J.X.; Funding acquisition: X.Z. and J.X.; Investigation: F.Y. and X.Y.; Resources: X.Z. and J.X.; Writing—original draft: F.Y. and X.Y.; Writing—review and editing: X.Z. and J.X. All authors have read and agreed to the published version of the manuscript.

Funding: J.X. acknowledges the Beijing Natural Science Foundation (No. 2222028), the National Natural Science Foundation of China (22273004, 21903004) and the Teli Fellowship from Beijing Institute of Technology, China. X.Z. acknowledges the National Natural Science Foundation of China (22174073 & 22003027), the NSF of Tianjin City (21JCJQJC00010), the National Key R&D Program of China (2018YFE0115000), Haihe Laboratory of Sustainable Chemical Transformations, Beijing National Laboratory for Molecular Sciences (BNLMS202106), and the Frontiers Science Center for New Organic Matter at Nankai University (63181206).

Institutional Review Board Statement: Not applicable.

Informed Consent Statement: Not applicable.

Data Availability Statement: Data is contained within the article or Supplementary Materials.

Conflicts of Interest: The authors declare no conflict of interest.

References

1. Cavallo, G.; Metrangolo, P.; Milani, R.; Pilati, T.; Priimagi, A.; Resnati, G.; Terraneo, G. The Halogen Bond. *Chem. Rev.* **2016**, *116*, 2478–2601. [CrossRef] [PubMed]
2. Gou, Q.; Feng, G.; Evangelisti, L.; Vallejo-López, M.; Spada, L.; Lesarri, A.; Cocinero, E.J.; Caminati, W. Internal Dynamics in Halogen-Bonded Adducts: A Rotational Study of Chlorotrifluoromethane–Formaldehyde. *Chem. Eur. J.* **2015**, *21*, 4148–4152. [CrossRef] [PubMed]
3. Inscoe, B.; Rathnayake, H.; Mo, Y. Role of Charge Transfer in Halogen Bonding. *J. Phys. Chem. A* **2021**, *125*, 2944–2953. [CrossRef] [PubMed]
4. Chen, J.; Wang, J.; Zheng, Y.; Feng, G.; Gou, Q. Halogen Bond in the Water Adduct of Chloropentafluoroethane Revealed by Rotational Spectroscopy. *J. Chem. Phys.* **2018**, *149*, 154307. [CrossRef] [PubMed]
5. Kolář, M.H.; Hobza, P. Computer Modeling of Halogen Bonds and Other σ-Hole Interactions. *Chem. Rev.* **2016**, *116*, 5155–5187. [CrossRef] [PubMed]
6. Robertson, C.C.; Wright, J.S.; Carrington, E.J.; Perutz, R.N.; Hunter, C.A.; Brammer, L. Hydrogen Bonding *vs.* Halogen Bonding: The Solvent Decides. *Chem. Sci.* **2017**, *8*, 5392–5398. [CrossRef] [PubMed]
7. Szabó, I.; Olasz, B.; Czakó, G. Deciphering Front-Side Complex Formation in S_N2 Reactions via Dynamics Mapping. *J. Phys. Chem. Lett.* **2017**, *8*, 2917–2923. [CrossRef] [PubMed]
8. Wang, H.; Wang, W.; Jin, W.J. σ-Hole Bond vs π-Hole Bond: A Comparison Based on Halogen Bond. *Chem. Rev.* **2016**, *116*, 5072–5104. [CrossRef] [PubMed]
9. Zheng, Y.; Herbers, S.; Gou, Q.; Caminati, W.; Grabow, J.U. Chlorine "Equatorial Belt" Activation of CF_3Cl by CO_2: The $C\cdots Cl$ Tetrel Bond Dominance in CF_3Cl–CO_2. *J. Phys. Chem. Lett.* **2021**, *12*, 3907–3913. [CrossRef] [PubMed]
10. Desiraju, G.R.; Ho, P.S.; Kloo, L.; Legon, A.C.; Marquardt, R.; Metrangolo, P.; Politzer, P.; Resnati, G.; Rissanen, K. Definition of the Halogen Bond (IUPAC Recommendations 2013). *Pure Appl. Chem.* **2013**, *85*, 1711–1713. [CrossRef]

11. Clark, T. σ-Holes. *WIREs Comput. Mol. Sci.* **2013**, *3*, 13–20. [CrossRef]
12. Clark, T.; Hennemann, M.; Murray, J.S.; Politzer, P. Halogen Bonding: The σ-Hole. *J. Mol. Model.* **2007**, *13*, 291–296. [CrossRef]
13. Auffinger, P.; Hays, F.A.; Westhof, E.; Ho, P.S. Halogen Bonds in Biological Molecules. *Proc. Natl. Acad. Sci. USA* **2004**, *101*, 16789–16794. [CrossRef] [PubMed]
14. Gilday, L.C.; Robinson, S.W.; Barendt, T.A.; Langton, M.J.; Mullaney, B.R.; Beer, P.D. Halogen Bonding in Supramolecular Chemistry. *Chem. Rev.* **2015**, *115*, 7118–7195. [CrossRef]
15. Mukherjee, A.; Tothadi, S.; Desiraju, G.R. Halogen Bonds in Crystal Engineering: Like Hydrogen Bonds yet Different. *Acc. Chem. Res.* **2014**, *47*, 2514–2524. [CrossRef]
16. Priimagi, A.; Cavallo, G.; Metrangolo, P.; Resnati, G. The Halogen Bond in the Design of Functional Supramolecular Materials: Recent Advances. *Acc. Chem. Res.* **2013**, *46*, 2686–2695. [CrossRef]
17. Sirimulla, S.; Bailey, J.B.; Vegesna, R.; Narayan, M. Halogen Interactions in Protein–Ligand Complexes: Implications of Halogen Bonding for Rational Drug Design. *J. Chem. Inf. Model.* **2013**, *53*, 2781–2791. [CrossRef] [PubMed]
18. Zhang, X.; Liu, G.; Ciborowski, S.; Bowen, K. Stabilizing Otherwise Unstable Anions with Halogen Bonding. *Angew. Chem. Int. Ed.* **2017**, *56*, 9897–9900. [CrossRef] [PubMed]
19. Muramatsu, S.; Koyasu, K.; Tsukuda, T. Formation of Grignard Reagent-like Complex CH_3-M-I^- via Oxidative Addition of CH_3I on Coinage Metal Anions M^- (M = Cu, Ag, Au) in the Gas Phase. *Chem. Lett.* **2017**, *46*, 676–679. [CrossRef]
20. Muramatsu, S.; Koyasu, K.; Tsukuda, T. Oxidative Addition of CH_3I to Au^- in the Gas Phase. *J. Phys. Chem. A* **2016**, *120*, 957–963. [CrossRef]
21. Wang, F.; Ji, X.; Ying, F.; Zhang, J.; Zhao, C.; Xie, J. Computational Studies of Coinage Metal Anion M^- + CH_3X (X = F, Cl, Br, I) Reactions in Gas Phase. *Molecules* **2022**, *27*, 307. [CrossRef]
22. Zhang, X.; Bowen, K. Designer Metallic Acceptor-Containing Halogen Bonds: General Strategies. *Chem. Eur. J.* **2017**, *23*, 5439–5442. [CrossRef] [PubMed]
23. Zhao, Y.; Truhlar, D.G. The M06 Suite of Density Functionals for Main Group Thermochemistry, Thermochemical Kinetics, Noncovalent Interactions, Excited States, and Transition Elements: Two New Functionals and Systematic Testing of Four M06-Class Functionals and 12 Other Functionals. *Theor. Chem. Acc.* **2008**, *120*, 215–241. [CrossRef]
24. Kendall, R.A.; Dunning, T.H., Jr.; Harrison, R.J. Electron Affinities of the First-Row Atoms Revisited. Systematic Basis Sets and Wave Functions. *J. Chem. Phys.* **1992**, *96*, 6796–6806. [CrossRef]
25. Dunning, T.H., Jr. Gaussian Basis Sets for Use in Correlated Molecular Calculations. I. The Atoms Boron through Neon and Hydrogen. *J. Chem. Phys.* **1989**, *90*, 1007–1023. [CrossRef]
26. Balabanov, N.B.; Peterson, K.A. Systematically Convergent Basis Sets for Transition Metals. I. All-Electron Correlation Consistent Basis Sets for the 3d Elements Sc–Zn. *J. Chem. Phys.* **2005**, *123*, 064107. [CrossRef] [PubMed]
27. Mulliken, R.S. Electronic Population Analysis on LCAO–MO Molecular Wave Functions. I. *J. Chem. Phys.* **1955**, *23*, 1833–1840. [CrossRef]
28. Peterson, K.A.; Figgen, D.; Goll, E.; Stoll, H.; Dolg, M. Systematically Convergent Basis Sets with Relativistic Pseudopotentials. II. Small-Core Pseudopotentials and Correlation Consistent Basis Sets for the Post-d Group 16–18 Elements. *J. Chem. Phys.* **2003**, *119*, 11113–11123. [CrossRef]
29. Peterson, K.A.; Shepler, B.C.; Figgen, D.; Stoll, H. On the Spectroscopic and Thermochemical Properties of ClO, BrO, IO, and Their Anions. *J. Phys. Chem. A* **2006**, *110*, 13877–13883. [CrossRef]
30. Frisch, M.J.; Trucks, G.W.; Schlegel, H.B.; Scuseria, G.E.; Robb, M.A.; Cheeseman, J.R.; Scalmani, G.; Barone, V.; Petersson, G.A.; Nakatsuji, H.; et al. *Gaussian 16, Revision A.03*; Gaussian, Inc.: Wallingford, CT, USA, 2016.
31. Mensa-Bonsu, G.; Tozer, D.J.; Verlet, J.R.R. Photoelectron Spectroscopic Study of $I^-\cdot ICF_3$: A Frontside Attack S_N2 Pre-Reaction Complex. *Phys. Chem. Chem. Phys.* **2019**, *21*, 13977–13985. [CrossRef]
32. Reed, A.E.; Curtiss, L.A.; Weinhold, F. Intermolecular Interactions from a Natural Bond Orbital, Donor-Acceptor Viewpoint. *Chem. Rev.* **1988**, *88*, 899–926. [CrossRef]
33. Reed, A.E.; Weinstock, R.B.; Weinhold, F. Natural Population Analysis. *J. Chem. Phys.* **1985**, *83*, 735–746. [CrossRef]
34. Ji, X.; Zhao, C.; Xie, J. Investigating the Role of Halogen-Bonded Complexes in Microsolvated $Y^-(H_2O)_n$ + CH_3I S_N2 reactions. *Phys. Chem. Chem. Phys.* **2021**, *23*, 6349–6360. [CrossRef] [PubMed]

Article

Interaction of Vinyl-Type Carbocations, $C_3H_5^+$ and $C_4H_7^+$ with Molecules of Water, Alcohols, and Acetone

Evgenii S. Stoyanov *, Irina Yu. Bagryanskaya and Irina V. Stoyanova

N.N. Vorozhtsov Institute of Organic Chemistry, Siberian Branch of Russian Academy of Sciences, 630090 Novosibirsk, Russia
* Correspondence: evgenii@nioch.nsc.ru

Abstract: X-ray diffraction analysis and IR spectroscopy were used to study the products of the interaction of vinyl cations $C_3H_5^+$ and $C_4H_7^+$ (Cat$^+$) (as salts of carborane anion $CHB_{11}Cl_{11}^-$) with basic molecules of water, alcohols, and acetone that can crystallize from solutions in dichloromethane and C_6HF_5. Interaction with water, as content increased, proceeded via three-stages. (1) adduct Cat$^+\cdot OH_2$ forms in which H_2O binds (through the O atom) to the C=C$^+$ bond of the cation with the same strength as seen in the binding to Na in $Na(H_2O)_6^+$. (2) H$^+$ is transferred from cation Cat$^+\cdot OH_2$ to a water molecule forming H_3O^+ and alcohol molecules (L) having the CH=CHOH entity. The O-atom of alcohols is attached to the H atom of the C=C$^+$-H moiety of Cat$^+$ thereby forming a very strong asymmetric H–bond, (C=)C$^+$-H$\cdots$O. (3) Finally all vinyl cations are converted into alcohol molecule L and H_3O^+ cations, yielding proton disolvates L-H$^+$-L with a symmetric very strong H-bond. When an acetone molecule (Ac) interacts with Cat$^+$, H$^+$ is transferred to Ac giving rise to a reactive carbene and proton disolvate Ac-H$^+$-Ac. Thus, the alleged high reactivity of vinyl cations seems to be an exaggeration.

Keywords: vinyl cations; vinyl cation adduct; very strong H-bond; proton disolvate; carborane salt

Citation: Stoyanov, E.S.; Bagryanskaya, I.Y.; Stoyanova, I.V. Interaction of Vinyl-Type Carbocations, $C_3H_5^+$ and $C_4H_7^+$ with Molecules of Water, Alcohols, and Acetone. *Molecules* **2023**, *28*, 1146. https://doi.org/10.3390/molecules28031146

Academic Editors: Qingzhong Li, Steve Scheiner and Zhiwu Yu

Received: 6 December 2022
Revised: 11 January 2023
Accepted: 19 January 2023
Published: 23 January 2023

1. Introduction

Carbocations as positively charged particles are strong electrophiles and may (1) react with a nucleophile thus yielding adducts (SN1 reaction), (2) act as a protonating agent turning into highly active species that enter into secondary reactions, or (3) get rearranged into other carbocations [1–3]. These reactions cannot be studied in liquid superacids, where carbocations have so far been mainly investigated by NMR [2], because nucleophiles are inevitably protonated under these conditions. Therefore, experimental studies on carbocation/nucleophile interactions are scarce.

The reactivity of unsaturated carbocations having C=C and C≡C bonds is more difficult to study than that of saturated ones. As a class of reactive intermediates, they have been the subject of extensive theoretical and experimental research in the past five decades [4–12]. Nonetheless, most reactivity studies have been focused on solvolysis reactions where the reactive vinyl cation is intercepted by heteroatom-containing solvent molecules [4,5,13], and these reactions have not been analyzed in detail. It follows from these works that vinyl cations are highly reactive and, therefore, uncontrollable intermediates. This point of view has been refuted by Mayr and coworkers [14], who found that the vinyl cation is even less reactive than diarylcarbenium cations, some of the most stable trisubstituted cations. An overestimation of the reactivity of the vinyl cation, as follows from the research on solvolysis reactions, is also evidenced by high stability of vinyl cation salts in solutions in dichloromethane, from which they can be isolated into a crystalline phase, which has made it possible to study them by X-ray diffraction [15–17]. Recently, the reason for the apparent high reactivity of the $C_6H_5CH_2^+$ benzyl cation toward such a nucleophile as benzene was established: it protonates benzene, turning it into a highly reactive carbene, $C_6H_5\overset{..}{C}H$, which enters into a secondary reaction with the available carbocation [18].

We are not aware of reports about reactions of vinyl carbocations with the simplest oxygen-containing nucleophiles L. It would be expected that such reactions would proceed by the mechanism of both attachment of the nucleophile and its protonation, with the formation of proton disolvates L-H$^+$-L as well, which have been studied by X-ray diffraction and IR spectroscopy (for L = H_2O, Et_2O, benzophenone, nitrobenzene, tetrahydrofuran, and others) [19,20].

In this work, we examined the interaction of unsaturated vinyl-type carbocations $C_3H_5{}^+$ and $C_4H_7{}^+$ (as carborane salts of the $CHB_{11}Cl_{11}{}^-$ anion) with the simplest nucleophiles: water, alcohols, and acetone. The reaction products that crystallized were studied by X-ray diffraction and IR spectroscopy. We chose carborane $CHB_{11}Cl_{11}{}^-$ as the counterion (hereafter denoted as ($Cl_{11}{}^-$), Figure 1) because of its exceptionally high stability at low basicity [21].

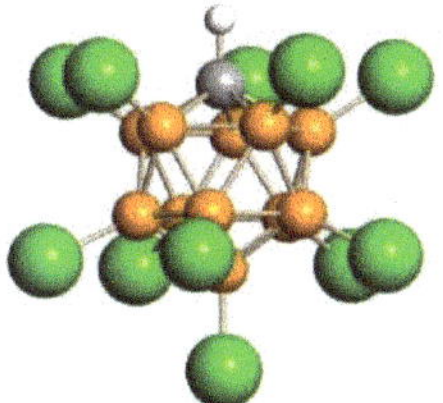

Figure 1. Undecachlorocarborane anion $CHB_{11}Cl_{11}{}^-$ (atoms: green Cl, brown B, grey C and white H).

2. Results

The bulk amount of salts of cations $C_3H_5{}^+$ and $C_4H_7{}^+$ can be easily obtained by adding to the acid $H(Cl_{11})$ powder such a small amount of 1,2-dichloropropane or 1,2-dichloro-2-methylpropane, respectively, with the sample remaining powdery. The quality of the resulting salts can be controlled by IR spectroscopy (their IR spectra should coincide with the reference spectra of salts [15,16], indicating the absence of impurities). These salts are soluble in pentafluorobenzene only in the presence of small amounts of water. Storage of such solutions for 1 day under ambient conditions led to a release of colorless crystals from it. The solubility of the salts and the yield of crystals increased with an increase in the content of water. X-ray diffraction analysis of crystals isolated from $C_4H_7{}^+(Cl_{11}{}^-)$ solutions showed that they are a salt of proton disolvate $L-H^+-L$ formed by two alcohol molecules L, representing 1-hydroxy-2-methylpropene (Figure 2).

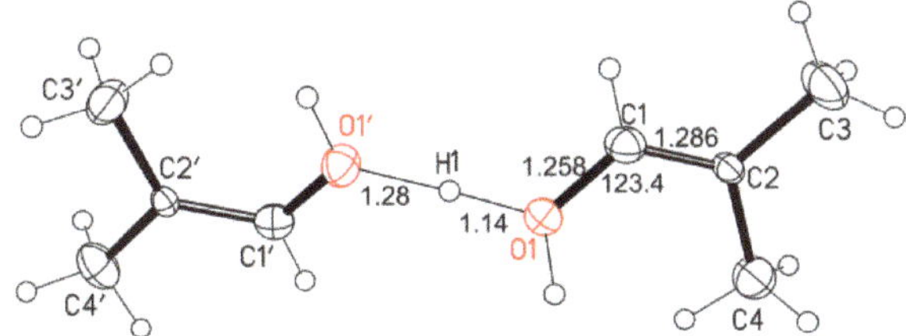

Figure 2. X-ray structure of the proton disolvate I*a*.

The O···O distance in the OH$^+$O moiety (2.420 Å) is typical for proton disolvates with very strong H-bonds [19,20]. The position of the bridging proton was determined by means of an electron density difference map (short O-H distance of 1.14 Å). Bond lengths and bond angles of the cation, which we will designate as I*a* (Figure 3), are given in Table 1 and Figure 2. Four carbon atoms C1-C4 and an oxygen atom of the alcohol molecule are in the same plane, and their CCC and CCO angles are close to 120°, i.e., C1 and C2 atoms have sp^2 hybridization and, therefore, are double bonded. The C-O···O angle is 118(3)°, which means that the O atom also has sp^2 hybridization and belongs to the alcohol OH group. The C=C double bond length of 1.286 Å was determined as the average of two isomeric alcohol molecules (Figure S1 in Supplementary Materials). It is shorter than that in molecular C=C(–OH) fragments of enol tautomers (1.362 Å) [22]. The C−O distance of 1.252 Å is also slightly shortened compared to that in a single C–O bond in the same (C=)C–OH fragment (1.333 Å) [22].

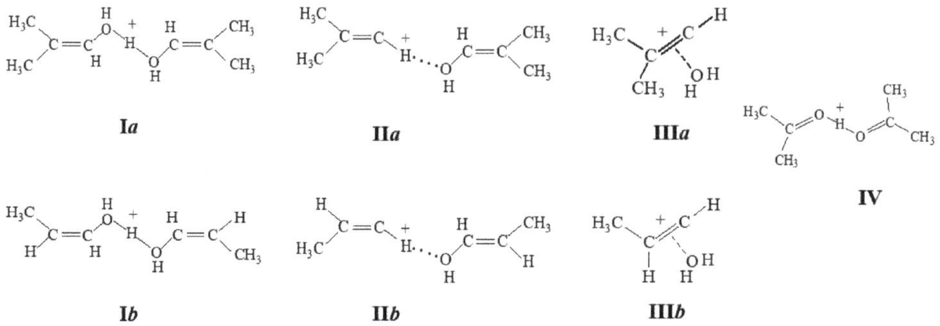

Figure 3. Schematic representation of the structures of the studied carbocation compounds.

Table 1. Selected geometric parameters of proton disolvate Ia (averaged from two solvate molecules $(CH_3)_2C=CHOH$) according to X-ray data.

Bond or Angle	Å or °
O1–C1	1.258(5)
C1–C2	1.286(5)
C2–CH$_3$	1.456(6)
O1–H1	1.14(7)
O1′–H1	1.28(7)
O1–C1–C2	123.4(4)
C1–C2–C3	121.3(4)
C1–C2–C4	121.6(3)
C3–C2–C4	117.1(4)
C1′–O1′⋯O1	118(3)

The IR spectrum of the crystals is characteristic of proton disolvates: it contains an intense absorption pattern of the OH$^+$O group, consisting of three broad bands at 905, 1297 and 1552 cm^{-1} (Figure 4, red). The band at 905 cm^{-1} belongs to ν_{as}(OH$^+$O), and bands 1297 and 1552 cm^{-1} to mixed stretching and bending vibrations of the OH$^+$O group [20]. The shape of the band at 905 cm^{-1} is distorted by the resonance effect leading to so-called transparent windows (Evans holes) [23], which appear as dips in the spectrum at 855 and 640 cm^{-1}.

The bands of CH stretching vibrations of CH$_3$ groups of cation I*a* (Figure 4, inset) are easily interpreted because they are similar to those of CH$_3$ groups of acetone in the previously studied disolvate Acetone−H$^+$−Acetone [24], (Table 2). A weak band at 3048 cm^{-1} may belong to the stretching vibration of the =C-H bond, the C atom of which is adjacent to the OH$^+$O group.

There are more uncertainties in the interpretation of the C-O and C=C stretching vibrations. It has been found that absorption bands of C−O stretching vibrations of alcohol molecules directly bound to H$^+$ in proton disolvates are so strongly broadened and weakened in intensity that they are not detectable in IR spectra [24]. Two bands at 1560 and 1683 cm^{-1} (Figure 4) are in the expected frequency range of C=C stretch vibrations [15–17] and can be attributed to them. The fact that the position of the bridging proton is determined by X-ray diffraction means that its two-well potential has a high enough potential barrier for the proton to be at the bottom of one of the wells for a sufficiently long time (at the time scale of IR spectroscopy) in order to demonstrate (in an IR spectrum) the non-equivalence of two alcohol molecules in I as two C=C stretching frequencies. One of them at 1560 cm^{-1} is close to the frequency (1555 cm^{-1}) of the cation in contact ion pair $(CH_3)_2C=C^+-H\ldots(Cl_{11}^-)$. Therefore, it can be assumed that the frequency 1560 cm^{-1} belongs to the isobutylene alcohol molecule, which has a shorter O^+-H bond (1.14 Å) and imitates a protonated alcohol molecule solvated by the second L alcohol molecule $(CH_3)_2C=CH-OH-^+H\ldots L$, which is less influenced by the positive charge and has an increased C=C stretch frequency: 1683 cm^{-1}. The CC and CH stretch frequencies are summarized in Table 2.

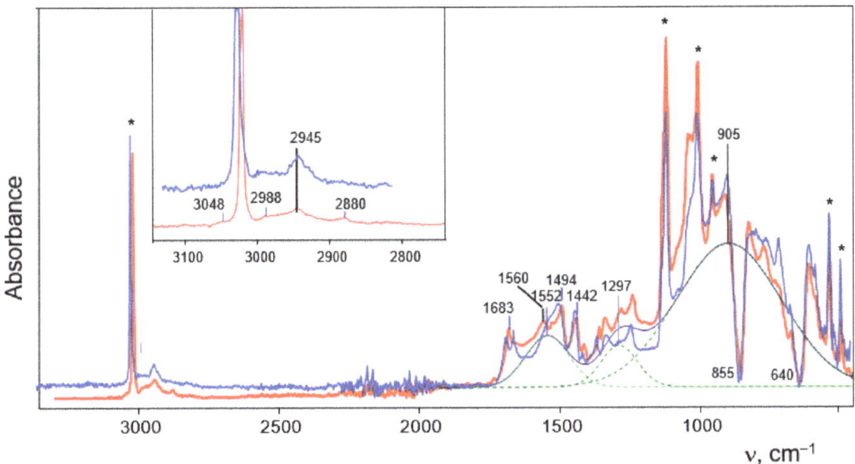

Figure 4. ATR IR spectra (with ATR correction) of crystals of salts of cations Ia (red) and Ib (blue). Anion signals are marked with asterisks.

Table 2. CH and C=C stretch frequencies (in cm^{-1}) of the vinyl cation in adducts I–IV and in $(Ac)_2H^+$.

Cation	CH Stretch Vibrations				νC=C in	
	=C-H	CH$_3$			$C_4H_7^+$	C_4H_7OH
		ν_{as}	ν_{as}	ν_s		
Ia	3048	2988	2945	2880	1560	1683
Ib	ND	2993	2945	2885	1560	1690 1664
IIa	~1630	2962	2932	2874	1633	1704
$C_4H_7^+$ [15]	3040	2960	2931	2875	1555	-
IV	-	3013	2962	2917	-	-
In salt $(Ac)_2H^+(FeCl_4^-)$ [24]	-	3016	2957	2915	-	-

ND: not determined.

It follows from the obtained results that disolvate Ia is generated by the interaction of the $C_4H_7^+$ carbocation with water molecules, as shown in Scheme 1. The weak spectrum of the H_3O^+ cation [25], which should arise simultaneously, actually manifests itself in the IR spectrum of the viscous phase, which precipitates concurrently with the crystalline phase.

$$2\ (CH_3)_2C=\overset{+}{C}H\ +\ 3H_2O\ \longrightarrow\ (CH_3)_2C=C(H)-O\cdots\overset{+}{H}\cdots O-C(H)=C(CH_3)_2$$

Ia

Scheme 1. Representation of the formation of cation Ia during the interaction of cation $C_4H_7^+$ with water molecules.

X-ray diffraction analysis of crystals—grown from a solution of the $C_3H_5^+(Cl_{11}^-)$ salt under the same conditions under which crystals of the salt of Ia were obtained—showed partially disordered structure. This property does not allow to see in detail the entire structure of the cation but enables us to determine its similarity with cation Ia. The similarity is confirmed by the identity of the IR spectrum of the crystals to that of the salt of disolvate Ia (Figure 4), which means that the crystals growing from solutions of $C_3H_5^+(Cl_{11}^-)$ are proton disolvates L_2-H^+-L_2, where L_2 is $CH_3CH=CHOH$. Hereafter, they will be referred to as cations Ib (Figure 2).

The solubility of salts of cations $C_3H_5^+$ and $C_4H_7^+$ in carefully dehydrated pentafluorobenzene is very low, and crystals cannot be grown from them. To increase the solubility of the salts, 1 vol% acetone was added to pentafluorobenzene containing trace amounts of water. This approach helped to obtain a solution with a heightened salt content and a reduced H_2O/cation$^+$ molar ratio. Keeping this solution over hexane vapor for 1 to 2 weeks led to the appearance of crystals. The X-ray diffraction analysis of the crystals obtained from solutions of the $C_4H_7^+(Cl_{11}^-)$ salts revealed that they contain the $C_4H_7^+$ cation solvated by one molecule of 1-hydroxy-2-methylpropane (Figure 5). The crystal lattice does not have an acetone molecule but contains one solvent molecule, C_6HF_5, per two salt molecules. The $C_4H_7^+$ cation is disordered over two positions differing in the location of the C2A atom slightly above or below the plane of three atoms: C1, C3, and C3A (Figure 5a and Figure S1 in Supplementary Materials). The C1-C2A$\cdots$O1 and C5-O1$\cdots$C2A angles are 137° and 121°, respectively (Figure 5b, Table 3), which means sp^2 hybridization of C2A and O1 atoms, i.e., an H atom is attached to each of them. The C$\cdots$O distance is 2.470 Å, which matches the maximum allowable distance between the O atoms of the symmetric O$\cdots$H$^+\cdots$O moiety in proton disolvates (2.47 Å for the most unstable proton disolvates obtained: H^+(nitrobenzene)$_2$) [19]. Therefore, with a high probability, a bridging proton is located between the C2A and O1 atoms, forming a short, strong, and low-barrier double-well H-bond, although compounds containing asymmetric moiety (C)C$-$H$^+\cdots$O with a strong H-bond are currently unknown. The short distance (C2A)H$\cdots$O1 of 1.97 Å, also points to the presence of a strong H- bond. The significant difference of the C2A-H$\cdots$O1 angle at 111° from the optimal one at 180° may be partly due to the inaccuracy of determining the localization of the H atom by the calculation method. The question of the presence of a strong H-bond with double-well proton potential is discussed below when the IR spectra of these crystals are examined. The $C_4H_7^+\cdot OHC_4H_7$ cation under consideration is hereafter denoted as IIa (Figure 2).

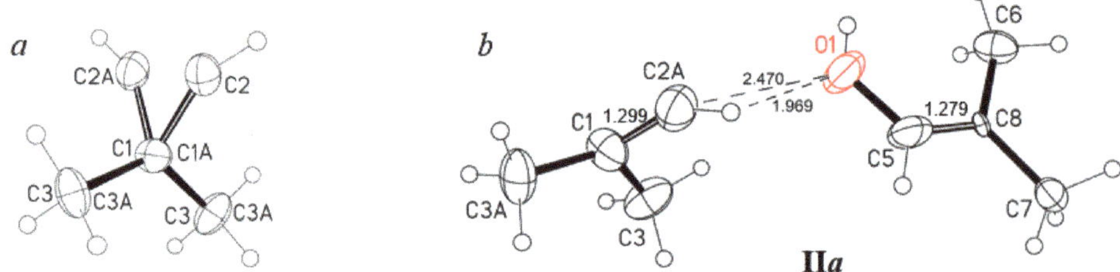

Figure 5. The $C_4H_7^+$ cation at two positions in the crystal lattice of salt $(C_4H_7^+\cdot OHC_4H_7)(Cl_{11}^-)$ (**a**) and the structure of the cationic adduct $C_4H_7^+\cdot OHC_4H_7$ as (**b**) determined by X-ray crystallography. Interatomic distances are given in Å.

X-ray diffraction analysis of crystals obtained from a solution of $C_3H_5^+(Cl_{11}^-)$ in C_6HF_5 + 1% acetone indicated that they contain the hydrocarbon cations and C_6HF_5 inclusion molecules of the solvents with significantly disordered C and F-atoms. We failed to localize the highly disordered structure of the cation. For this reason, we do not present or discuss these X-ray diffraction data. Nonetheless, they provide some useful information. For instance, in the crystal lattice, alcohol molecules C_3H_5OH with reliably fixed C and O atoms were identified, possibly indicating the solvation of the cation with the C_3H_5OH molecule in the same way as the C_4H_7OH molecule solvates the $C_4H_7^+$ cation in adduct II*a*. That is, we can assume the emergence of cationic adduct $C_3H_5^+ \cdot C_3H_5OH$, similar to II*a*, and designate it as II*b* (Figure 2). The IR spectra of crystals containing II*a* and II*b* adducts are identical (Figure 6), which confirms that II*b* is $C_3H_5^+ \cdot C_3H_5OH$.

Table 3. Selected geometric parameter of cationic adduct II*a* according to X-ray data.

Bond or Angle	Å or °
O1–C5	1.292(9)
C5–C8	1.236(9)
C8–CH$_3$	1.450(6)
C1–C2	1.350(11)
C1–CH$_3$	1.436(9)
O1–C5–C8	121.1(6)
C5–C8–C6	121.6(5)
C5–C8–C7	119.7(5)
C5–C8–C6	118.7(5)
C2–C1–C3	118.3(5)
C2–C1–C3a	110.1(5)
C2 … O1	2.47(1)

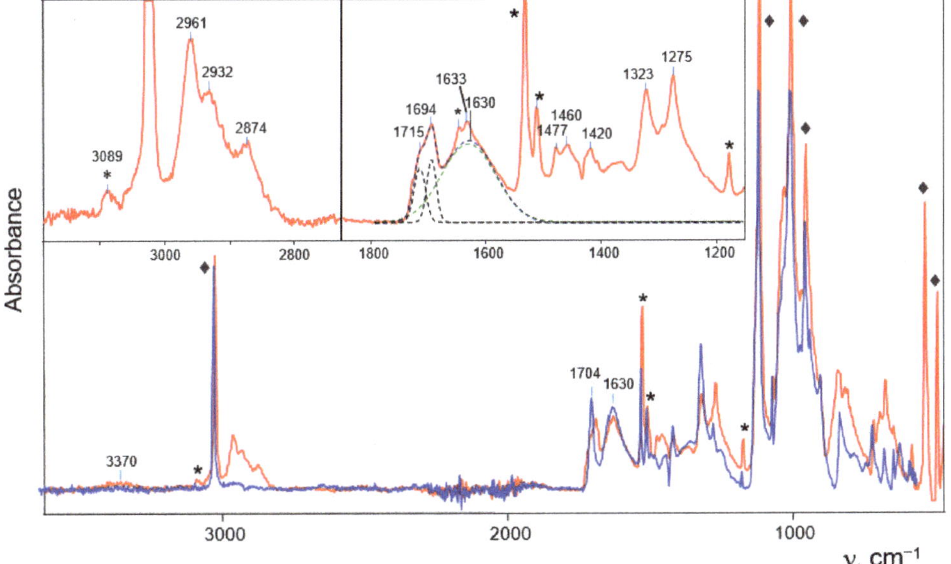

Figure 6. ATR IR spectra (with ATR correction) of crystals of $(C_4H_7^+ \cdots C_4H_7OH)(Cl_{11}^-)$ (II*a*, red) and $(C_3H_5^+ \cdots C_3H_5OH)(Cl_{11}^-)$ (II*b*, blue). Characteristic absorption bands of the C_6HF_5 inclusion molecule are indicated by asterisks, and the strongest absorption bands of the (Cl_{11}^-) anion are marked with ♦. The dotted line shows separation of contours of the bands of stretching C=C vibrations (in the inset).

In the IR spectra of cationic adducts II*a* and II*b*, specific absorption of a strong $C–H^+\cdots O$ H-bond should be present; however, this type of strong H-bond has not yet been found. Well-studied strong hydrogen bonds are symmetric $O\cdots H^+\cdots O$ or $N\cdots H^+\cdots N$ H-bonds in proton disolvates L_2H^+ with double- well proton potential separated by a low potential barrier that transforms it into flat-bottom potential for vibrational transitions. In these cases, the proton vibrations appear in the IR spectra as an intense and broad absorption pattern with a maximum at 850–1000 cm^{-1} [25], as observed in the spectra of cations Ia and Ib (Figure 4). In proton disolvates with an asymmetric moiety, for example, $N-H^+\cdots O$, the bottom of the double-well potential is asymmetric and the maximum of the broad and intense absorption shifts to 1400–1700 cm^{-1} [25]. In the spectra of the analyzed adducts IIa and IIb, a broad band at 1630 cm^{-1} is observed, as is absorption in the region of 1200–1500 cm^{-1}, which is not described by a single Gaussian (Figure 6, inset). They correspond fairly well to the asymmetric $X_1-H^+\cdots X_2$ fragment, in our case $=C-H^+\cdots O$.

The C=C bond of cations $C_3H_5^+$ and $C_4H_7^+$ is affected by the bridging proton, and this effect should specifically influence the absorption of its stretch vibrations. For example, it has been found [24] that if a single C-O bond is attached to a bridged proton its absorption band broadens and decreases in intensity so much that it is undetectable in the IR spectrum. If the C=O bond is attached to H^+, as in proton disolvate Acetone-H^+-Acetone (discussed below), it still appears in the spectrum as a weak-to-moderately intense broadened band [24]. Therefore, absorption of the C=C stretch vibration of $C_3H_5^+$ and $C_4H_7^+$ may look like a weak band at 1633 cm^{-1} (Figure 6). It is higher in frequency than νC=C at ~1560–1590 cm^{-1} in the contact ion pairs (CIPs) formed with the (Cl_{11}^-) anion in solutions and in a solid phase [15–17]. This means that the interaction of $C_3H_5^+$ and $C_4H_7^+$ cations with an alcohol molecule in II is stronger than the interaction with the (Cl_{11}^-) anion in the CIP.

The C=C band of the alcohol molecule in adducts IIa and IIb is subject to a weaker influence of the bridging proton. Its C=C stretching vibration is observed at 1704 cm^{-1} (Figure 6). It is a single band for adduct IIb, whereas in the spectrum of IIa it is split into two components at 1715 and 1694 cm^{-1} of equal intensity (Figure 6, inset). Obviously, in the crystal lattice of the II*a* adduct salt, there are two weakly nonequivalent C_4H_7OH molecules.

In the CH stretch frequency region, the three bands at 2961, 2932 and 2874 cm^{-1} can be unambiguously interpreted as vibrations of the CH_3 group (Table 2). The IR spectra of the crystals also contain absorption bands of the captured pentafluorobenzene molecule. They are easily identified because of the finding that when a crystal is crushed on an ATR accessory and its crystal lattice is destroyed, pentafluorobenzene is released and slowly evaporates when the sample is kept in a glove box atmosphere. Therefore, the intensity of its absorption decreases over time. The most intense C_6F_5H bands in Figure 6 are marked with asterisks.

The band at 3370 cm^{-1} may belong to OH groups of the C_4H_7OH alcohol molecule, which are engaged in weak H bonds with the (Cl_{11}^-) anion.

The solution from which crystals of the salt of II*b* grew was kept in a sealed ampoule, and after a few days a small number of tiny crystals grew from it. It was possible to find one crystal of sufficient size for X-ray analysis. It turned out that this was a salt of the monohydrate of the propylene cation ($C_3H_5^+\cdot OH_2$)(Cl_{11}^-) (Figure 7). The $C_3H_5^+\cdot OH_2$ species has two localizations in the unit cell with slightly different positions of C and O atoms, but they can be distinguished (Figure 7a). Similarly, an anion can be disordered over two positions, as indicated by the presence of electron density peaks in the difference map in the region of the anion. Nonetheless, we failed to localize the second position of the anion. This may be the reason for substantial deterioration of the R_f factor, but does not interfere with the determination of coordinates of the C and O atoms of the cation with accuracy sufficient to establish the topology of the cation qualitatively and its main geometric parameters (Table 4).

The main feature of the structure of the $C_3H_5^+\cdot OH_2$ cationic adduct, which is denoted below as III*b* (Figure 2), is as follows. The H_2O molecule is attached to the $C_3H_5^+$ cation through the O atom in the direction perpendicular to the C=C double bond at a distance of 2.32 Å, which is very close to the average $Na\cdots O(H_2)$ distance of 2.333 Å in the first hydration shell of hexahydrate $Na(OH_2)_6$ as determined by X-ray diffraction analysis for 13 structures (retrieval was made according to the Cambridge structural data base, ConQuest 2021.3.0) [26]. Therefore, the nature of the interaction of the H_2O molecule with the C=C$^+$ bond of the vinyl cation is similar to that of the interaction of water molecules with the alkali metal cation in its first hydration shell.

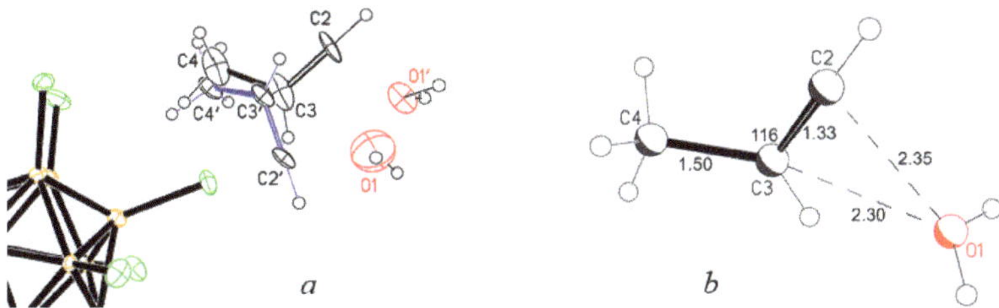

Figure 7. X-ray structure of the $(C_4H_7^+ \cdot OH_2)(Cl_{11}^-)$ (III*b*) with two locations of the cation (**a**), and more detailed representation of the $C_4H_7^+ \cdot OH_2$ adduct (**b**).

Table 4. Selected geometric parameters of cationic adduct III*b* (averaged over two positions) according to X-ray data.

Bond or Angle	Å or °
C2–C3	1.34(5)
C3–C4	1.51(8)
C2–C3–C4	115(5)
O1$\cdots$C2	2.35(6)
O1$\cdots$C3	2.30(6)

Unfortunately, the insufficient amount of the obtained crystals of $(C_3H_5^+ \cdot OH_2)(Cl_{11}^-)$ did not allow us to register their IR spectrum. Nevertheless, the spectra of these compounds have been obtained by us earlier, when we characterized crystalline salts $C_3H_5^+(Cl_{11}^-)$ and $C_4H_7^+(Cl_{11}^-)$ by X-ray diffraction and IR spectroscopy [15,16]. In the IR spectra of individually selected single crystals of these salts subjected to X-ray analysis, no traces of water absorption were found. On the other hand, during the recording of IR spectra of an aggregate of small crystals that arose simultaneously with larger ones, a weak spectrum of water molecules was observed (Figure 8). Its frequencies ν_{as}, ν_s and δ at 3612, 3544, and 1606 cm^{-1}, respectively, are similar to those of monomeric molecules dissolved in organic solvents or hydrated alkali metal cations (Table 5). This result was surprising and could not be explained. Now it is clear that the observed bands belong to non–H-bonded water molecules of monohydrates $(C_4H_7^+ \cdot OH_2)(Cl_{11}^-)$ and $(C_3H_5^+ \cdot OH_2)(Cl_{11}^-)$ designated subsequently as IIIa and IIIb (Figure 2), which are formed and co-crystallize with salts of anhydrous vinyl cations. They are typical of the H_2O molecules that hydrate the H_3O^+ cation in $H_3O^+(H_2O)_3(Cl_{11}^-)$ crystals [27], or alkali metal cations (Table 5). Such mostly ionic interaction $Cat^+\cdots OH_2$ leads to a slight decrease in the frequencies of OH stretching, compared to those of the dissolved monomer molecule. In adducts IIIa/b, OH stretching frequencies are somewhat lower than those in hydrates of alkali metal cations and H_3O^+; hence the strength of bond $Cat^+\cdots OH_2$ in IIIa/b is higher.

Table 5. IR frequencies of H_2O molecules bonded to cations through O-atom.

	ν_{as}	ν_s	δ
Monomeric H_2O [a]	3675	3591	1607
$H_3O^+(H_2O)_3(Cl_{11}^-)$ [b]	3637	3578	1605
$Cs^+(H_2O)_6(Cl_{11}^-)$ [c]	3656	3578	1611
i-$C_4H_7 \cdot H_2O(Cl_{11}^-)$	3612	3544	1606

[a] In diluted solutions in dichloroethane; [b] in the crystal phase studied by X-ray diffraction [27]; [c] in a dichloroethane extract from aqueous solutions of $Cs^+(Cl_{11}^-)$.

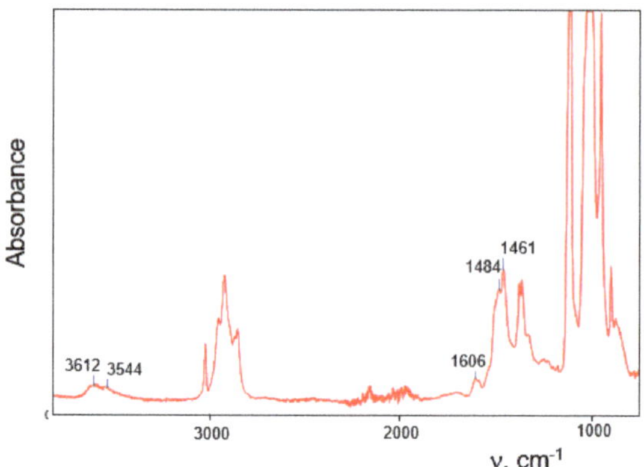

Figure 8. The IR spectrum of a set of i-$C_4H_7^+(Cl_{11}^-)$ crystals and $(i$-$C_4H_7^+ \cdot OH_2)(Cl_{11}^-)$ crystals cocrystallized with them.

As mentioned above, crystals of compounds IIa and IIb were obtained from saturated solutions of vinyl cation salts in C_6HF_5 containing 1 vol% acetone. It could be expected that after an increase in the acetone content, crystals containing acetone could be obtained. Incubation of saturated solutions of the $C_3H_5^+(Cl_{11}^-)$ or $C_4H_7^+(Cl_{11}^-)$ salts in C_6HF_5 + 5% acetone over hexane vapor gave rise to needle-like crystals. X-ray diffraction analysis indicated that this was proton disolvate salt Ac-H$^+$-Ac (IV), where Ac is the acetone (Figures 2 and 9a). The same salt was obtained in a different way by reacting the $H(Cl_{11})$ acid with acetone vapor and its subsequent dissolution in dichloromethane. Slow evaporation of the solution in the glove box led to the formation of crystals. Their X-ray diffraction analysis showed that this was also proton disolvate salt $(Ac)_2H^+(Cl_{11}^-)$ but with other crystal lattice parameters, that is, it is a different polymorph (Figure 9b). The O···O distance in cations for both polymorphs is almost the same, 2.429 and 2.423 Å. Selected geometric parameters of IV and IV′ are given in Table 6.

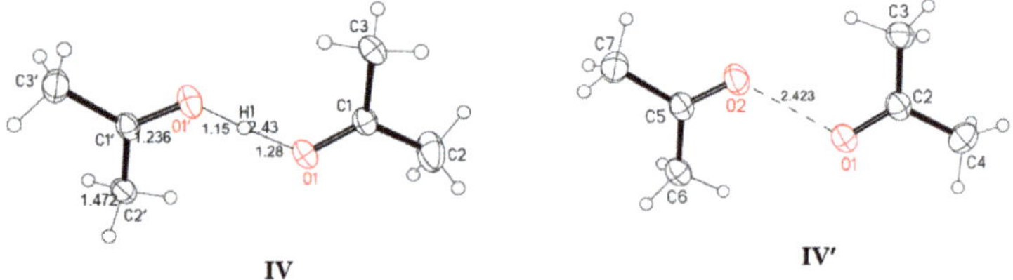

Figure 9. The proton disolvate in two polymorphs of salt $(Ac$-H^+-$Ac)(Cl_{11}^-)$. In polymorph IVb, it was not possible to localize the bridging proton by X-ray diffraction analysis.

Table 6. Selected geometric parameters of proton disolvates IVa and IVb (averaged over two independent fragments (acetone molecules) of the adduct) according to X-ray data.

IV		IV′	
Bond or Angle	**Å or °**	**Bond or Angle**	**Å or °**
O1–C1	1.237(3)	O1–C2	1.228(8)
C1–CH₃	1.473(4)	C–CH₃	1.473(8)
O1–H1	1.28(5)		
O1′–H1	1.15(5)		
O1–C1–C2	120.1(2)	O1–C2–C3	122.2(5)
O1–C1–C3	119.8(2)	O1–C2–C4	117.5(5)
C2–C1–C3	120.0(2)	C3–C2–C4	120.2(6)
O1 … O1′	2.429(3)	O1 … O2	2.423(9)

The IR spectra of the $(Ac)_2H^+$ cation (Figure S2 in Supplementary Materials) match its known IR spectra [24,28]; interpretation of the spectrum is given in ref. [28].

3. Discussion

The results make it possible to establish the sequence of the interaction of vinyl cations with water molecules. Initially, an H_2O molecule is attached to the $C=C^+$ bond of cation $RCH=C^+H$ or $R_2C=C^+H$ (where R is CH_3) via the O- atom (Equation (1), in it and in subsequent equations, the $R_2C=C^+H$ cation is used).

$$R_2C=C^+H \; + \; H_2O \; \longrightarrow \; R_2C=C^+H \atop H\overset{O}{\cdots}H \tag{1}$$

The nature of this bond is close to that of bonds formed by water molecules with alkali metal cations during their hydration or with H_3O^+ in crystal salts of $H_3O^+\cdot 3H_2O$, i.e., the bond is strongly ionic.

With an increase in the concentration of the salt of cationic adduct $R_2C=C^+H\cdots H_2O$ in solutions, the self-association of ion pairs increases (which is typical for salts of carbocations in solutions [29]). This enhances the contact interaction of H_2O with the cation and promotes the transfer of a proton to a water molecule with the formation of H_3O^+ and cationic adduct II:

$$2\left(\begin{array}{c} R_2C=C^+H \\ \vdots \\ O \\ H\;H \end{array}\right) \longrightarrow R_2C=C\overset{\textstyle +}{\underset{H}{\diagup}}H\cdots\cdot OCH=CR_2 \; + \; H_3O^+ \tag{2}$$

With a further increase in the contents of water and of the carbocation salt, the concentration of the resulting adducts II and of the H_3O^+ cation increases. The latter can protonate an alcohol molecule that is more basic than H_2O, thereby producing a proton disolvate and regenerating cation $R_2C=C^+H$ (Equation (3)),

$$2\left(\begin{array}{c} R_2C=C^+H \\ \vdots \\ HOCH=CR_2 \end{array}\right) \xrightarrow{+H_3O} R_2C=CO\overset{H}{\underset{H}{-}}H\overset{+}{-}OC=CR_2 \; + \; 2\,R_2C=C^+H \; + \; H_2O \tag{3}$$

which next interacts with water, closing the cycle. The alcohol molecule of adduct II can also be protonated directly by cation III, thus by passing the stage of H_3O^+ formation:

$$R_2C=C^+H \atop H\overset{O}{\cdots}H \quad + \quad \overset{H}{\underset{H}{\overset{\mid}{C}=CR_2}} \longrightarrow R_2C=CO\overset{H}{\underset{H}{-}}H\overset{+}{-}OC=CR_2 \tag{4}$$

The studied cationic adducts enable one to trace the change in the nature of the interaction of vinyl cations with basic molecules of H_2O, alcohol, and acetone as their basicity increases. The H_2O molecule is attached to the $C=C$ bond of the vinyl cation (Scheme 2, III) with a strength similar to that of an H_2O molecule attached to an alkali metal cation or to the H_3O^+ cation in $H_3O^+(H_2O)_3$. As the basicity of the O atom increases from H_2O to alcohol molecule, its interaction with the cation shifts to the H atom, thereby producing a strong H-bond, with a partially covalent character [30], and an increase in the $C\cdots\cdots O$ distance to 2.47 Å (Scheme 2, II). In this case, the asymmetric double-well potential of the bridging proton has a deeper minimum at the C atom. A further increase in the basicity of the oxygen atom (acetone molecule) causes a shift of the minimum of the double-well potential to the O atom with a high probability of proton transfer to the acetone molecule. The loss of a proton by the vinyl cation results in an extremely reactive carbene molecule (with $C=C$: as the active site), which then reacts with the components of the mixture, forming non-crystallizing products (wax phase). The released protonated acetone adds a second acetone molecule thereby generating proton disolvate IV.

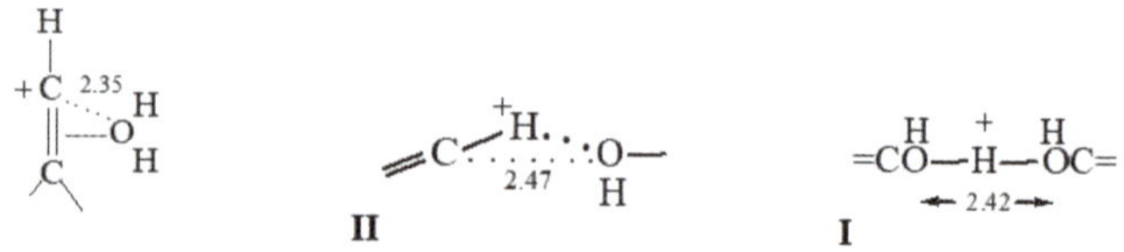

Scheme 2. Schematic representation of the charged moiety in adducts I–III.

Thus, the interaction of the vinyl cation with nucleophile L proceeds: (1) through its addition to a charged double bond (if L = H_2O) or H-bonded to the =C-H group (if L = alcohol molecule), and (2) through the protonation of L with the transition of the vinyl cation to the neutral carbene molecule. The second mechanism of interaction was proposed in ref. [31] and proved for the benzyl carbocation [32].

The finding that adducts II and III with strongly ionic $Cat^+ - L$ interaction exist is surprising. They can exist if the basicity of L slightly exceeds that of counterion (Cl_{11}^-). This means that vinyl cations behave like rather chemically inert particles, contradicting predictions of quantum chemical calculations. For example, if the crystallographic structures of adducts II and III are optimized, then the covalent interaction of the cation with the H_2O or alcohol molecule results in the emergence (in the case of $C_4H_7^+$) of molecules of protonated isobutenyl alcohol and diisobutenyl ether, respectively (Figure 10), with a significant gain in energy.

Figure 10. A comparison of two structures: IIa with optimized H atomic coordinates and fixed atomic C and O coordinates equal to those, determined from X-ray crystallography and optimized structure for all atomic coordinates for vacuum (at the UB3LYP/6- 311++G(d,p) level of theory).

For example, if we compare energies of the structures calculated at the UB3LYP/6-311++G(d,p) level of theory: (1) energy with optimized H atomic coordinates and fixed coordinates of C and O that are equal to the coordinates that follow from the X-Ray data for IIa, and (2) energy with fully optimized structure (including the C, O and H atomic coordinates), energies (1) and (2) are related as 69.71 and 0 kcal/mol with the transformation of structure IIa into protonated diisobutenyl ether. The environment cannot have a stronger stabilizing effect on the IIa structure because the purely ionic interaction of C and O atoms with Cl atoms of the anionic environment is weak (C⋯Cl and O⋯Cl distances exceed the sum of van der Waals atomic radii). Thus, it follows from quantum chemical calculations that adducts II and III should not exist, which contradicts the experimental findings. Therefore, the use of quantum chemical calculations requires caution in studies on mechanisms of the interaction of vinyl cations with neutral molecules.

4. Materials and Methods

The salts of vinyl cations $C_3H_5^+$ and $C_4H_7^+$ were obtained as described previously [15–17]. The pentafluorobenzene (Sigma-Aldrich, Saint Louis, MO, USA) used as a solvent was thoroughly dried with molecular sieves and was not purified further.

All sample handling was carried out in an atmosphere of argon (H_2O, $[O_2] < 0.5$ ppm) in a glove box. ATR IR spectra were recorded on a Shimadzu IRAffinity-1S spectrometer housed inside the glove box in the $4000-400$ cm^{-1} frequency range using an ATR accessory with a diamond crystal. The spectra were processed in the GRAMMS/A1 (7.00) software from Thermo Scientific, Waltham, MA, USA.

X-ray diffraction data were collected on a Bruker Kappa Apex II CCD diffractometer using φ,ω-scans of narrow (0.5°) frames with Mo Kα radiation (λ = 0.71073 Å) and a graphite monochromator at temperature 200 K. The structures were solved by direct methods with the help of SHELXT-2014/5 [33], and refined by the full-matrix least-squares method against all F2 in an anisotropic-isotropic (for H atoms) procedure using SHELXL-2018/3 [33]. Absorption corrections were applied by the empirical multiscan method using SADABS software [34]. Hydrogen atom positions were calculated using the riding model. The hydrogen atom positions for OH -groups were located by means of a difference Fourier map. The crystallographic data and details of the refinements for all structures are summarized in Table S1 in Supplementary Materials. We were unable to obtain good single crystals of the $(C_3H_7^+ \cdot H_2O)(Cl_{11}^-)$ salt (of adduct III) for X-ray diffraction analyses. Nonetheless, we conducted an X-ray diffraction experiment and localized non-hydrogen atoms, but could not refine the structure to obtain a good R-factor. Crystals of the salt of cationic adduct II contain strongly disordered C_6HF_5 solvent molecules. We were not able to find their atomic coordinates. The accessible volume of free solvent molecules in these crystals, as determined by routine PLATON analysis, was 15.0% (846 Å^3). The highly disordered C_6HF_5 molecules occupying this volume could not be modeled as a set of discrete atomic positions. We employed the PLATON/SQUEEZE procedure to calculate the contribution to the diffraction from the solvent region, and thereby produced a series of solvent-free diffraction intensities. The independent part of the unit cell of salt $[(CH_3)_2C{=}CHOH]_2H^+(Cl_{11}^-)$ contains two anions and two cationic proton disolvates I. In Table 1, the geometry of disolvate I is given as an average over two solvate molecules $(CH_3)_2C{=}CHOH$. A similar averaging of geometric parameters over two independent positions was performed for adduct III and proton disolvates IVa and IVb (Tables 4 and 6).

CCDC 2223519, 2223520, 2223521, 2223522 and 2223523 contain the supplementary crystallographic data for this paper. These data can be obtained free of charge from The Cambridge Crystallographic Data Center at http://www.ccdc.cam.ac.uk/data_request/cif (accessed on 9 January 2023).

The obtained crystal structures were analyzed for short contacts between non-bonded atoms using PLATON [35,36] and MERCURY software packages [37].

5. Conclusions

Vinyl cations $C_3H_5^+$ and $C_4H_7^+$ (Cat$^+$) in solutions of their salts in dichloromethane and C_6HF_5 interact with O-containing nucleophiles as follows.

An H_2O molecule attaches to the C=C bond of Cat$^+$ in a similar manner to the hydration of alkali metal cations, thereby yielding Cat$^+\cdot H_2O$ adducts with a strongly ionic bond. Its strength only slightly exceeds the strength of the interaction of the (Cl_{11}^-) anion with the vinyl cation in contact ion pairs Cat$^+(Cl_{11}^-)$.

With an increase in the content of Cat$^+\cdot H_2O$ adducts in solutions, the adduct self-associates and interacts with a transfer of a proton to one water molecule and attachment of the second one to the C=C bond, thus forming H_3O^+ and an alcohol molecule, respectively.

The alcohol molecule interacts predominantly with the H atom of the C=C$^+$−H moiety of the vinyl cation, thereby producing a proton disolvate with strong asymmetric H-bond =C−$^+$H$\cdots$O having double-well proton potential with a deeper minimum near the C atom.

A further increase in the water content in the solutions leads to complete conversion of vinyl cations into alcohol molecules with the formation of symmetric proton disolvates LH$^+$L containing strong and partially covalent O−H$^+$−O hydrogen bonds [30] and H_3O^+ cations.

The interaction of the vinyl cations with acetone molecules, which are more basic than H_2O or alcohol molecules, causes the formation of only symmetrical proton disolvates, LH$^+$L, in the absence of acetone-containing cationic adducts. The vinyl cation is converted into carbene containing a highly reactive C=C: moiety.

Summing up, we can say that the interaction of the vinyl cation with base L proceeds through two mechanisms: via the formation of adducts (SN1 reaction), and via the mechanism where vinyl cation acts as a protonating agent. When the basicity of L is close to that of a single water molecule, L attaches to the double C=C bond thereby producing an adduct. As the basicity of L increases, the interaction with the $C=C^+-H$ moiety of the vinyl cation strengthens and is shifted to the H atom, thus forming a solvate having a strong asymmetrical $=C-^+H\cdots O$ hydrogen bond. Further strengthening of the basicity of L leads to the transfer of a proton to L and to the emergence of the eventually symmetric LH^+L cation. The loss of a proton by the vinyl cation converts it into a neutral reactive carbene molecule containing a C=C: moiety.

The formation of adducts with water and alcohol molecules by vinyl cations is unexpected, because according to quantum chemical calculations, they are energetically unfavorable and should not exist.

The very existence of these adducts means that the alleged high reactivity of vinyl carbocations is an overestimation.

Supplementary Materials: The following supporting information can be downloaded at: https://www.mdpi.com/article/10.3390/molecules28031146/s1, Figure S1: Two locations of the $C_4H_7^+\cdot C_4H_7OH$ adduct in the crystal lattice of its salt with the $\{Cl_{11}^-\}$ anion (not shown); Figure S2: The ATR IR spectrum of proton disolvate IVa; Table S1: Crystallographic data and details of the X-ray diffraction experiment.

Author Contributions: Conceptualization, E.S.S. and I.V.S.; methodology, E.S.S.; validation, I.Y.B.; formal analysis, I.Y.B. and I.V.S.; writing—review and editing, E.S.S.; supervision, E.S.S.; project administration, E.S.S. All authors have read and agreed to the published version of the manuscript.

Funding: This work was supported the Ministry of Science and Higher Education of the Russian Federation (state registration No. 1021052806375-6-1.4.3).

Institutional Review Board Statement: Not applicable.

Informed Consent Statement: Not applicable.

Data Availability Statement: Not applicable.

Conflicts of Interest: The authors declare no conflict of interest.

Sample Availability: Samples of the compounds are available from the authors.

References

1. Olah, G. Carbocations and Electrophilic Reactions. *Angew. Chem. Int. Ed.* **1973**, *12*, 173–212. [CrossRef]
2. Olah, G.A.; Prakash, G.K.S. *Carbocation Chemistry*; John Willey & Sons: Hoboken, NJ, USA, 2004.
3. Mayr, H.; Lang, G.; Ofial, A.R. Reactions of Carbocations with Unsaturated Hydrocarbons: Electrophilic Alkylation or Hydride Abstraction? *J. Am. Chem. Soc.* **2002**, *124*, 4076–4083. [CrossRef] [PubMed]
4. Stang, P.J.; Rappoport, Z. *Dicoordinated Carbocations*; Wiley: Hoboken, NJ, USA, 1997.
5. Hanack, M. Mechanistic and Preparative Aspects of Vinyl Cation Chemistry. *Angew. Chem. Int. Ed.* **1978**, *17*, 333–341. [CrossRef]
6. Rappoport, Z.; Gal, A. Vinylic cations form solvolysis. I. Trianisylvinyl halide system. *J. Am. Chem. Soc.* **1969**, *91*, 5246–5254. [CrossRef]
7. Radom, L.; Hariharan, P.C.; Pople, J.A.; Schleyer, P.V.R. Molecular orbital theory of the electronic structure of organic compounds. XIX. Geometries and energies of C_3H_5 cations. Energy relations among allyl, vinyl, and cyclopropyl cations. *J. Am. Chem. Soc.* **1973**, *95*, 6531–6544. [CrossRef]
8. Sherrod, S.A.; Bergman, R.G. Synthesis and solvolysis of 1-cyclopropyl-1-iodoethylene. Generation of an unusually stable vinyl cation. *J. Am. Chem. Soc.* **1969**, *91*, 2115–2117. [CrossRef]
9. Müller, T.; Juhasz, M.; Reed, C.A. The X-ray Structure of a Vinyl Cation. *Angew. Chem. Int. Ed.* **2004**, *43*, 1543–1546. [CrossRef]
10. Jones, W.M.; Miller, F.W. The Generation of Vinyl Cations from Vinyltriazenes. *J. Am. Chem. Soc.* **1967**, *89*, 1960–1962. [CrossRef]
11. Hinkle, R.J.; McNeil, A.J.; Thomas, Q.A.; Andrews, M.N. Primary Vinyl Cations in Solution: Kinetics and Products of β,β-Disubstituted Alkenyl(aryl)iodonium Triflate Fragmentations. *J. Am. Chem. Soc.* **1999**, *121*, 7437–7438. [CrossRef]
12. Bagdasarian, A.L.; Popov, S.; Wigman, B.; Wei, W.; Lee, W.; Nelson, H.M. Urea-Catalyzed Vinyl Carbocation Formation Enables Mild Functionalization of Unactivated C−H Bonds. *Org. Lett.* **2020**, *22*, 7775–7779. [CrossRef]
13. Stang, P.J.; Rappaport, Z.; Hanack, M.; Subramanian, L.R. *Vinyl Cations*; Academic Press: Cambridge, MA, USA, 1979; 513p.
14. Byrne, P.A.; Kobayashi, S.; Würthwein, E.-U.; Ammer, J.; Mayr, H. Why Are Vinyl Cations Sluggish Electrophiles? *J. Am. Chem. Soc.* **2017**, *139*, 1499–1511. [CrossRef] [PubMed]

15. Stoyanov, E.S.; Bagryanskaya, I.Y.; Stoyanova, I.V. Unsaturated Vinyl-Type Carbocation $[(CH_3)_2C=CH]^+$ in Its Carborane Salts. *ACS Omega* **2021**, *6*, 15834–15843. [CrossRef] [PubMed]

16. Stoyanov, E.S.; Bagryanskaya, I.Y.; Stoyanova, I.V. Isomers of the Allyl Carbocation $C_3H_5^+$ in Solid Salts: Infrared Spectra and Structures. *ACS Omega* **2021**, *6*, 23691–23699. [CrossRef]

17. Stoyanov, E.S.; Bagryanskaya, I.Y.; Stoyanova, I.V. IR-Spectroscopic and X-ray- Structural Study of Vinyl-Type Carbocations in Their Carborane Salts. *ACS Omega* **2022**, *7*, 27560–27572. [CrossRef] [PubMed]

18. Stoyanov, E.S.; Stoyanova, I.V. Cloronium Cations in Dichloromethane Solutions as Catalysts for the Conversion of CH_2Cl_2 to $CHCl_3/CCl_4$ and CH_3Cl/CH_4. *ChemistrySelect* **2018**, *3*, 12181–12185. [CrossRef]

19. Stasko, D.; Hoffmann, S.P.; Kim, K.-C.; Fackler, N.L.P.; Larsen, A.S.; Drovetskaya, T.; Tham, F.S.; Reed, C.A.; Rickard, C.E.F.; Boyd, P.D.W.; et al. Molecular Structure of the Solvated Proton in Isolated Salts. Short, Strong, Low Barrier (SSLB) H-bonds. *J. Am. Chem. Soc.* **2002**, *124*, 13869–13876. [CrossRef]

20. Stoyanov, E.S.; Reed, C.A. The IR Spectrum of the $H_5O_2^+$ Cation in the Context of Proton Disolvates $L-H^+-L$. *J. Phys. Chem. A.* **2006**, *110*, 12992–13002. [CrossRef]

21. Reed, C. Carborane Acids. New "Strong yet Gentle" Acids for Organic and Inorganic Chemistry. *Chem. Com.* **2005**, 1669–1677. [CrossRef]

22. Allen, F.H.; Kenard, O.; Watson, D.G.; Bramer, L.; Orpen, A.G.; Taylor, R. Tables of bond lengths determined by X-ray and neutron diffraction. Part 1. Bond lengths in organic compounds. *J. Chem. Soc. Perkin Trans. II* **1987**, S1–S19. [CrossRef]

23. Videnova-Adrabinska, V. Polarized IR and Raman study of the $Na_3H(SO_4)_2$ single crystal. *J. Mol. Struct.* **1990**, *237*, 367–388. [CrossRef]

24. Stoyanov, E.S. A distinctive feature in the IR spectra of proton disolvates $[L_2H^+]$ and polysolvates $[(L_2H^+)\cdot nL]$: Unusual strong broadening of some absorption bands of ligands L bound with H^+. *Phys. Chem. Chem. Phys.* **2000**, *2*, 1137–1145. [CrossRef]

25. Iohansen, A.V. Infrared Spectroscopy and Spectral Determination of Hydrogen Bond Energy. In *Hydrogen Bond*; Sokolov, N.D., Ed.; Nauka: Moscow, Russia; pp. 112–155.

26. Bruno, I.J.; Cole, J.C.; Edgington, P.R.; Kessler, M.; Macrae, C.F.; McCabe, P.; Pearson, J.; Taylor, R. New software for searching the Cambridge Structural Database and visualizing crystal structures. *Acta Cryst.* **2002**, *B58*, 389–397. [CrossRef] [PubMed]

27. Stoyanov, E.S.; Stoyanova, I.V.; Tham, F.S.; Reed, C.A. The Nature of the Hydrated Proton $H_{(aq)}^+$ in Organic Solvents. *J. Am. Chem. Soc.* **2008**, *130*, 12128–12138. [CrossRef] [PubMed]

28. Douberly, G.E.; Ricks, A.M.; Ticknor, B.W.; Duncan, M.A. The structure of protonated acetone and its dimer: Infrared photodissociation spectroscopy from 800 to 4000 cm^{-1}. *Phys. Chem. Chem. Phys.* **2008**, *10*, 77–79. [CrossRef]

29. Stoyanov, E.S.; Stoyanova, I.V. Spontaneous transition of alkyl carbocations to unsaturated vinyl-type carbocations in organic solutions. *Int. J. Mol. Sci.* **2023**, *24*, 1802. [CrossRef]

30. Grabowski, S.J. What Is the Covalency of Hydrogen Bonding? *Chem. Rev.* **2011**, *111*, 2597–2625. [CrossRef]

31. Niggemann, M.; Gao, S. Are Vinyl Cations Finally Coming of Age? *Angew. Chem. Int. Ed.* **2018**, *57*, 2–5. [CrossRef]

32. Stoyanov, E.S.; Stoyanova, I.V. The Mechanism of High Reactivity of Benzyl Carbocation, $C_6H_5CH_2^+$, during Interaction with Benzene. *ChemistrySelect* **2020**, *5*, 9277–9280. [CrossRef]

33. Sheldrick, G.M. Crystal structure refinement with SHELXT. *Acta Crystallogr.* **2015**, *C71*, 3–8.

34. SADABS. v. 2008-1. Bruker AXS: Madison, WI, USA, 2008.

35. Spek, A.L. *PLATON, A Multipurpose Crystallographic Tool (Version 10M)*; Utrecht University: Utrecht, The Netherlands, 2003.

36. Spek, A.L. Single-crystal structure validation with the programPLATON. *J. Appl. Crystallogr.* **2003**, *36*, 7–13. [CrossRef]

37. Macrae, C.F.; Edgington, P.R.; McCabe, P.; Pidcock, E.; Shields, G.P.; Taylor, R.; Towler, M.; van de Streek, J. Mercury: Visualization and analysis of crystal structures. *J. Appl. Crystallogr.* **2006**, *39*, 453–457. [CrossRef]

Article

How the Position of Substitution Affects Intermolecular Bonding in Halogen Derivatives of Carboranes: Crystal Structures of 1,2,3- and 8,9,12-Triiodo- and 8,9,12-Tribromo *ortho*-Carboranes

Kyrill Yu. Suponitsky [1,2], Sergey A. Anufriev [1] and Igor B. Sivaev [1,2,*]

[1] A. N. Nesmeyanov Institute of Organoelement Compounds, Russian Academy of Sciences, 28 Vavilov Str., Moscow 119334, Russia

[2] Basic Department of Chemistry of Innovative Materials and Technologies, G.V. Plekhanov Russian University of Economics, 36 Stremyannyi Line, Moscow 117997, Russia

* Correspondence: sivaev@ineos.ac.ru

Abstract: The crystal structures of two isomeric triiodo derivatives of *ortho*-carborane containing substituents in the three most electron-withdrawing positions of the carborane cage, $1,2,3\text{-}I_3\text{-}1,2\text{-}C_2B_{10}H_9$, and the three most electron-donating positions, $8,9,12\text{-}I_3\text{-}1,2\text{-}C_2B_{10}H_9$, as well as the crystal structure of $8,9,12\text{-}Br_3\text{-}1,2\text{-}C_2B_{10}H_9$, were determined by single-crystal X-ray diffraction. In the structure of $1,2,3\text{-}I_3\text{-}1,2\text{-}C_2B_{10}H_9$, an iodine atom attached to the boron atom (position 3) donates its lone pairs simultaneously to the σ-holes of both iodine atoms attached to the carbon atoms (positions 1 and 2) with the I···I distance of 3.554(2) Å and the C-I···I and B-I···I angles of 169.2(2)° and 92.2(2)°, respectively. The structure is additionally stabilized by a few B-H···I-shortened contacts. In the structure of $8,9,12\text{-}I_3\text{-}1,2\text{-}C_2B_{10}H_9$, the I···I contacts of type II are very weak (the I···I distance is 4.268(4) Å, the B8-I8···I12 and B12-I12···I8 angles are 130.2(3)° and 92.2(3)°) and can only be regarded as dihalogen bonds formally. In comparison with the latter, the structure of $8,9,12\text{-}Br_3\text{-}1,2\text{-}C_2B_{10}H_9$ demonstrates both similarities and differences. No Br···Br contacts of type II are observed, while there are two Br···Br halogen bonds of type I.

Keywords: *ortho*-carborane; iodo derivatives; X-ray structure; I···I dihalogen bond

Citation: Suponitsky, K.Y.; Anufriev, S.A.; Sivaev, I.B. How the Position of Substitution Affects Intermolecular Bonding in Halogen Derivatives of Carboranes: Crystal Structures of 1,2,3- and 8,9,12-Triiodo- and 8,9,12-Tribromo *ortho*-Carboranes. *Molecules* **2023**, *28*, 875. https://doi.org/10.3390/molecules28020875

Academic Editors: Qingzhong Li, Steve Scheiner and Zhiwu Yu

Received: 31 October 2022
Revised: 10 January 2023
Accepted: 12 January 2023
Published: 15 January 2023

1. Introduction

The ability of halogens to form complexes with various electron pair donors was discovered over two hundred years ago [1–3], and the Nobel Prize laureate Odd Hassel provided crystallographic proof for the existence of such a bond, interpreting it as a charge-transfer interaction more than fifty years ago [4,5]. However, only at the beginning of the 21st century has halogen bonding grown from a scientific curiosity to one of the most interesting and actively studied non-covalent interactions for the construction of supramolecular assemblies [6–10].

This progress has been largely due to a better understanding of the principles on which the strength of the halogen bond depends. The performance of the halogen bond largely depends on the degree of polarization of the halogen atom; that is, the greater the positive electrostatic potential of the σ-hole, the more efficient the halogen bond donor will be [10–12]. The value of the positive potential of the σ-hole depends on the ability of the halogen atom to be polarized, which decreases in the following order: I > Br > Cl >> F [13,14]. The value of the positive potential of the σ-hole can be enhanced due to the electron-withdrawing ability of the fragment to which the halogen atom is attached.

For a halogen atom to be an electron acceptor in order to form a halogen bond, it must be bonded to an electron-withdrawing atom or group. Therefore, the *sp* hybridization of carbon atoms bearing a halogen is favored over sp^2 followed by sp^3 hybridization [15,16]. The hybridization of the carbon atom can be compensated by the electron-withdrawing effect of fluorine atoms, as evidenced by the close values of the σ-hole potential of the corresponding iodine atoms in 1-iodoethynyl-4-iodobenzene and 1,4-diiodoperfluoro-benzene (172 and 169 kJ/mol, respectively) [16]. The strength of the halogen bond is highly correlated with the degree of iodobenzene fluorination [17]. Therefore, it is not surprising that 1,4-diiodoperfluorobenzene and its analogs are widely used in the design of halogen-bonded supramolecular systems [18–28], although arylacetylene iodides also play an important role [16,29–35]. In the absence of other electron density donors, the iodine atoms in these compounds are also able to play this role, which leads to the formation of I· · ·I dihalogen bonds [36–39], and the number of such bonds, as a rule, increases with the number of iodine atoms in the molecule [40].

Icosahedral carboranes $C_2B_{10}H_{12}$ are another class of compounds whose derivatives are promising as halogen bond donors. The predicted strength of the halogen bonds with the same electron donor (based on the σ-hole potential) is larger for C-vertex halogen-substituted carboranes than for their organic aromatic counterparts [41–43]. In contrast to the iodo aromatics, wherein all iodine atoms are equivalent, in the iodo derivatives of *ortho*-carborane iodine atoms, depending on their position, they can act preferentially as an acceptor or a donor of a halogen bond. A typical example is 1,12-diiodo-*ortho*-carborane, in which one of the iodine atoms is bonded to the most electron-withdrawing position of the carborane cage (position 1), and the second to the most electron-donating position (position 12) (Figure 1) [44]. The first of them is an electron acceptor, and the last one is a donor, which form an ideal intermolecular I· · ·I dihalogen bond of type II [45].

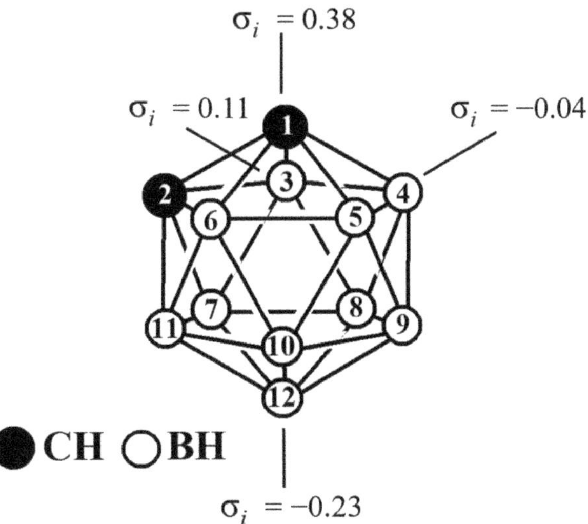

Figure 1. Atom numbering atoms and Hammett constants σ_i in *ortho*-carborane.

In this contribution, we studied intermolecular bonding in two isomers of triiodo-*ortho*-carborane containing substituents in the three most electron-withdrawing positions of the carborane cage (1,2,3) and the three most electron-donating positions (8,9,12); in addition, a comparative analysis of the crystal packings of the 8,9,12-triiodo and 8,9,12-tribromo derivatives of *ortho*-carborane was performed.

2. Results and Discussion

To date, a number of iodo derivatives of *ortho*-carborane have been synthesized, and the structures of a dozen of them have been established by single-crystal X-ray diffraction. The derivatives with a high degree of substitution such as 8,9,10,12-I_4-1,2-$C_2B_{10}H_8$ [46], 4,5,7,8,9,10,11,12-I_8-1,2-$C_2B_{10}H_4$ [47], and 3,4,5,6,7,8,9,10,11,12-I_{10}-1,2-$C_2B_{10}H_2$ [47], as in the case of iodo-aromatics, are characterized by the formation of numerous intermolecular I$\cdots$I dihalogen bonds varying from 3.74 to 4.05 Å. In contrast to the polyiodo derivatives, no intermolecular dihalogen bonds were found in any of the isomeric monoiodo derivatives of *ortho*-carborane 1-I-1,2-$C_2B_{10}H_{11}$ [45], 3-I-1,2-$C_2B_{10}H_{11}$ [48], 8-I-1,2-$C_2B_{10}H_{11}$ [49], and 9-I-1,2-$C_2B_{10}H_{11}$ [47].

As for the diiodo derivatives of *ortho*-carborane, the presence of intermolecular I$\cdots$I dihalogen bonds inside them depends on the position of the substituents. In addition to 1,12-diiodo-*ortho*-carborane 1,12-I_2-1,2-$C_2B_{10}H_{10}$, which is characterized by the presence of strong intermolecular I$\cdots$I dihalogen bonds (3.57 Å) [45], weak I$\cdots$I dihalogen bonds (4.09 Å) were found in the 3,6-diiodo derivative 3,6-I_2-1,2-$C_2B_{10}H_{10}$ [50], while the 3,10-I_2-1,2-$C_2B_{10}H_{10}$ [49], 4,7-I_2-1,2-$C_2B_{10}H_{10}$ [51], and 9,12-I_2-1,2-$C_2B_{10}H_{10}$ [52] isomers do not form dihalogen bonds. Therefore, we were interested in studying the possibility of the formation of intermolecular I$\cdots$I dihalogen bonds in triiodo-*ortho*-carboranes containing substituents in the three most electron-withdrawing positions of the caborane cage 1,2,3-I_3-1,2-$C_2B_{10}H_9$ and the three most electron-donating positions of 8,9,12-I_3-1,2-$C_2B_{10}H_9$.

The formation of 8,9,12-I_3-1,2-$C_2B_{10}H_9$ (**1**) was previously reported in the iodination of *ortho*-carborane with molecular iodine in acetic acid in the presence of a mixture of concentrated sulfuric and nitric acids [53]. We isolated the 8,9,12-triiodo derivative as a by-product of the reaction of *ortho*-carborane with iodine in dichloromethane in the presence of AlCl₃ [54]. It should be noted that the unit cell parameters of 8,9,12-I_3-1,2-$C_2B_{10}H_9$ (**1**) have been reported [55]; however, its structure has not been yet solved.

The crystal structure of 8,9,12-I_3-1,2-$C_2B_{10}H_9$ was determined by single-crystal X-ray diffraction. A general view of **1** is presented in Figure 2. All the B-I distances in 8,9,12-I_3-1,2-$C_2B_{10}H_9$ are nearly equal (B8-I8 is 2.165(7) Å, B9-I9 is 2.160(7) Å, and B12-I12 is 2.160(7) Å) and are only slightly longer than the B-I distances in 8,9,10,12-I_4-1,2-$C_2B_{10}H_8$ (for which the average value is 2.151 Å) [46].

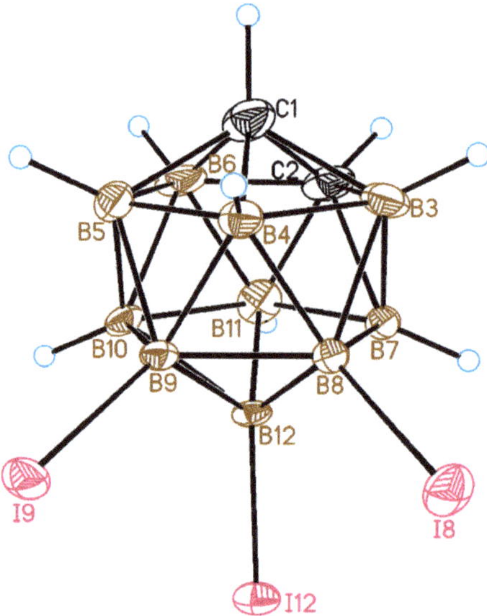

Figure 2. General view of 8,9,12-I_3-1,2-$C_2B_{10}H_9$ (**1**) showing atomic numbering. Thermal ellipsoids are given at 50% probability level.

A crystal-packing fragment of **1** is depicted in Figure 3. Only weak intermolecular interactions are observed in the crystal structure. From a formal point of view, four types of intermolecular interactions are observed in the crystal of **1**. Halogen atoms participate in both types (I and II) of halogen bonding, and I$\cdots$H-C(B) hydrogen bonds as well as B-H$\cdots$H-B contacts are formed. It should be noted that all intermolecular contacts except for one are somewhat longer than the sum of the van-der-Waals radii. For instance, the type II halogen bond is very weak (the I$\cdots$I distance is 4.268(4) Å, the B8-I8$\cdots$I12 angle is 130.2(3)°, and the B12-I12$\cdots$I8 angle is 92.2(3)°) (Figure 3) and can only be regarded as a type II halogen bond formally. At the same time, the I9$\cdots$I9 halogen bond of type I demonstrates an interhalogen distance (4.002(4) Å) shorter than the sum of the van-der-Waals radii (4.14 Å) [56]; however, halogen bonds of this type are usually relatively weak.

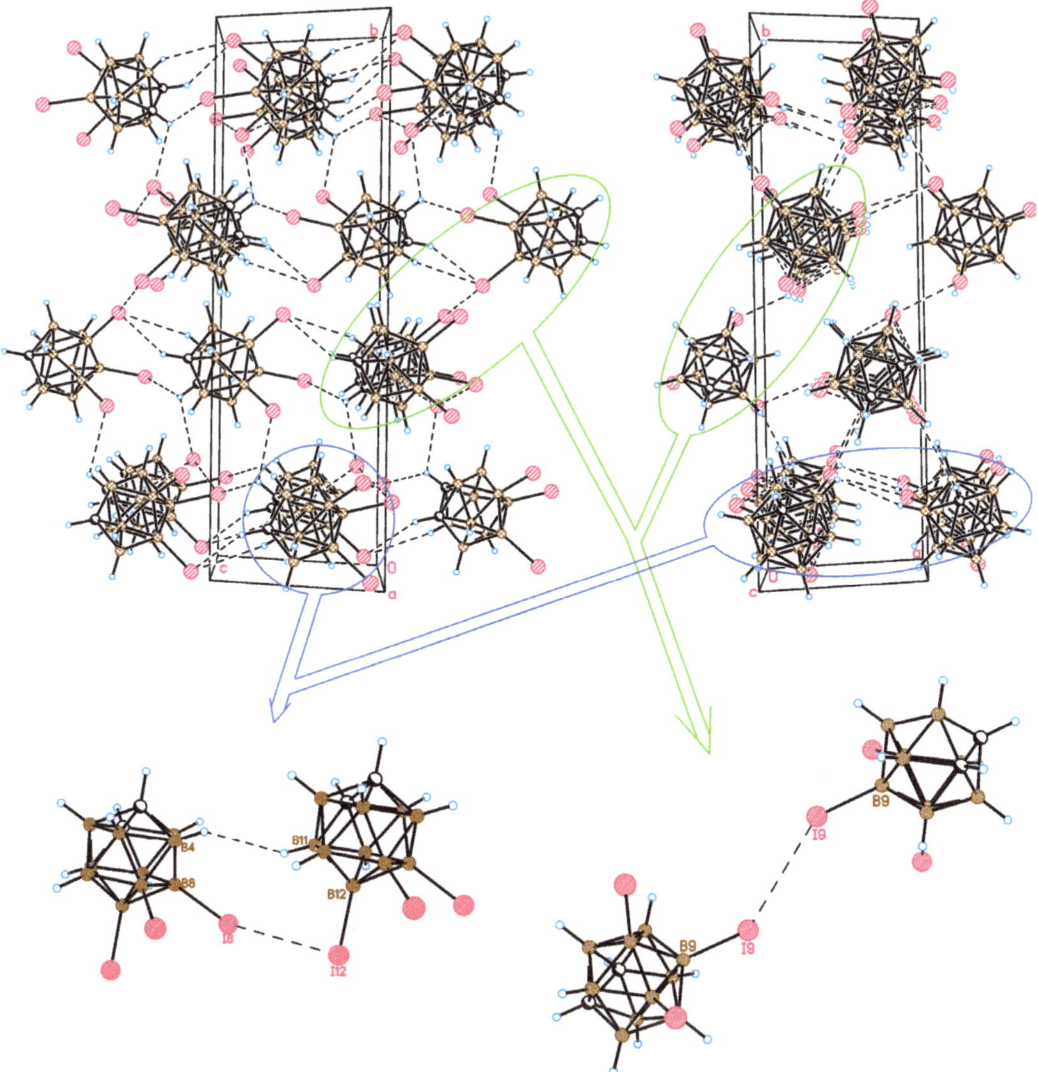

Figure 3. Crystal-packing fragment of 8,9,12-I$_3$-1,2-C$_2$B$_{10}$H$_9$. **Top-left**: view in the *bc* crystallographic plane; **top-right**: view in the *bc* crystallographic plane. Detailed view of type II weak halogen bonding (**bottom-left**) and type I halogen bonding (**bottom-right**).

Therefore, it is impossible to choose one or two of the most important contacts that can be considered to be structure-forming. Interactions in the *bc* crystallographic plane are due to I$\cdots$H-C(B) and H$\cdots$H contacts, while in the crystallographic direction, *a*, molecules are linked mostly by I$\cdots$I interactions. As a result, the crystal packing of 8,9,12-I$_3$-1,2-C$_2$B$_{10}$H$_9$ can be considered to be nearly isotropic.

It would be interesting to compare the crystal packing of 8,9,12-triiodo-*ortho*-carborane with that of its closest analog, 8,9,12-tribromo-*ortho*-carborane 8,9,12-Br$_3$-1,2-C$_2$B$_{10}$H$_9$ (**2**). Despite the fact that the bromination of *ortho*-carborane was first described as early as the mid-1960s [57], the chemistry of the bromo-derivatives of carborane has been studied to a much lesser extent compared to its iodo-derivatives due to the difficulty in isolating pure products. Recently, we published the synthesis and characterization of the 9,12-dibromo derivative of *ortho*-carborane [58]. Since the 8,9,12-tribromo derivative was one of the side-products of that reaction, we decided to increase the ratio of bromine to *ortho*-carborane (up to 3:1) and the reaction time. This allowed us to isolate the desired compound 8,9,12-Br$_3$-1,2-C$_2$B$_{10}$H$_9$ (**2**) at a 17% yield (see Section 3.3) It should be noted that the signal of the *CH* carborane groups of in the ^{1}H NMR spectrum in CDCl$_3$, which is a convenient indicator of the *CH*-acidity of carboranes [59,60], for compound **2** appears in a higher field at 3.87 ppm. compared to compound **1** (4.13 ppm). This indicates a lower acidity of the *CH*-carborane groups in the 8,9,12-tribromo derivative compared to the 8,9,12-triiodo derivative.

The crystal structure of 8,9,12-Br$_3$-1,2-C$_2$B$_{10}$H$_9$ was determined by single-crystal X-ray diffraction. A general view of **2** is presented in Figure 4.

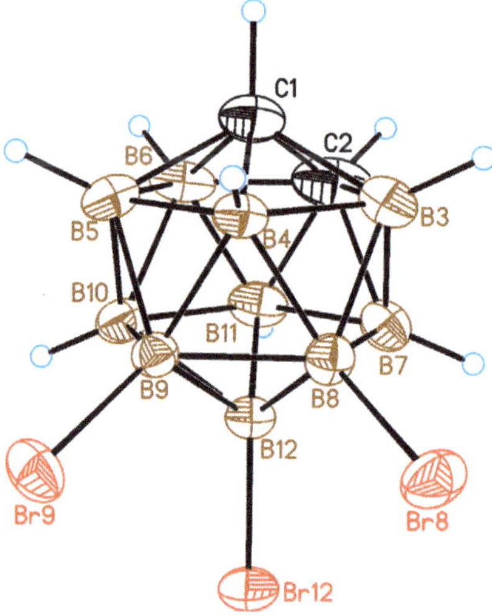

Figure 4. General view of 8,9,12-Br$_3$-1,2-C$_2$B$_{10}$H$_9$ (**2**) showing atomic numbering. Thermal ellipsoids are given at 50% probability level.

It should be noted that the structure of 8,9,12-tibromo-*ortho*-carborane was determined in 1966 [61] at room temperature. The quality of that experiment was evidently low, and the experiment itself mostly concentrated on the description of the compound's molecular geometry. Therefore, in the present study, we redetermined its structure at a low temperature (120 K), focusing on both its molecular structure and, especially, crystal-packing properties. Prior to the description of its crystal structure and comparison with that of **1**, it is interesting to mention some other studied bromo- and iodo-derivatives of *ortho*-carborane. For instance, the c.rystal structures of 1,2-Me$_2$-8,9,10,12-I$_4$-1,2-C$_2$B$_{10}$H$_6$ [47] and 1,2-Me$_2$-8,9,10,12-Br$_4$-1,2-C$_2$B$_{10}$H$_6$ [62] are isostructural. At the same time, the crystal structures of 1,12-I$_2$-1,2-C$_2$B$_{10}$H$_{10}$ [45] and 1,12-Br$_2$-C$_2$B$_{10}$H$_{10}$ [42] do not show any similarity. Only partial similarity in terms of crystal packing was observed for 9,12-I$_2$-1,2-C$_2$B$_{10}$H$_{10}$ [52] and 9,12-Br$_2$-1,2-C$_2$B$_{10}$H$_{10}$ [58]; however, the latter appeared to be isostructural to its chloro analog 9,12-Cl$_2$-1,2-C$_2$B$_{10}$H$_{10}$ [63] (see Figure S10 in SI).

A comparison of the crystal structures of **2** and **1** studied in this work demonstrates both similarities and differences. As in compound **1**, a Br9$\cdots$Br9 halogen bond of type I is observed in the crystal structure of 8,9,12-Br$_3$-1,2-C$_2$B$_{10}$H$_9$ (the Br$\cdots$Br distance is 3.586(2) Å, which is shorter than the sum of the van-der-Waals radii 3.79 Å) (Figure 5). At the same time, there are no type II halogen bonds; however, one more halogen bond of type I is found between Br8 atoms, wherein the Br$\cdots$Br distance (3.969(2) Å) is somewhat longer than the sum of the van-der-Waals radii. As in compound **1**, all the other intermolecular interactions are Br$\cdots$H-C(B) and H$\cdots$H. The differences in the crystal-packing properties described above result in some redistribution of the contact types (Figure 6): the contribution of Hal$\cdots$Hal contacts increases, which leads to a decrease in the number of Hal$\cdots$H contacts and to an increase in H$\cdots$H ones.

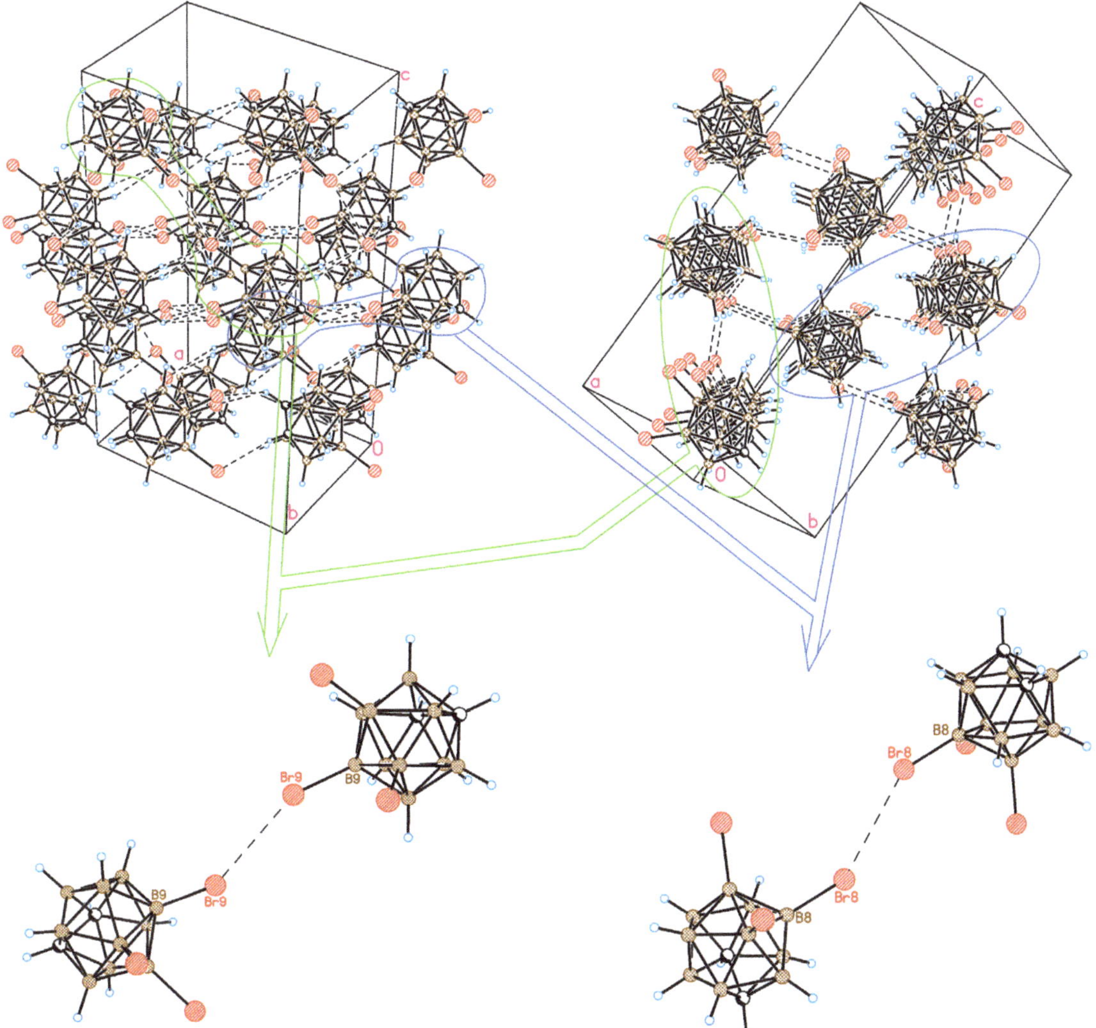

Figure 5. Crystal-packing fragment of 8,9,12-Br$_3$-1,2-C$_2$B$_{10}$H$_9$. Top-left: view down [1,−1,0] direction; top-right: view down [0,1,1] direction. Bottom: detailed view of type I halogen bonding (Br9$\cdots$Br9, left; Br8$\cdots$Br8, right). Projections of crystal packing are chosen to be consistent with those in Figures 3 and 7 (orientations of molecules are nearly the same).

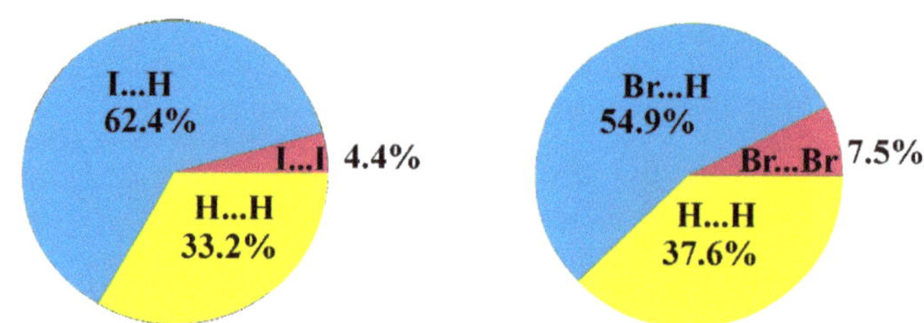

Figure 6. Distribution of intermolecular contacts in the crystal structures of 8,9,12-I_3-1,2-$C_2B_{10}H_9$ (**left**) and 8,9,12-Br_3-1,2-$C_2B_{10}H_9$ (**right**) as obtained using the Crystal Explorer program package [64].

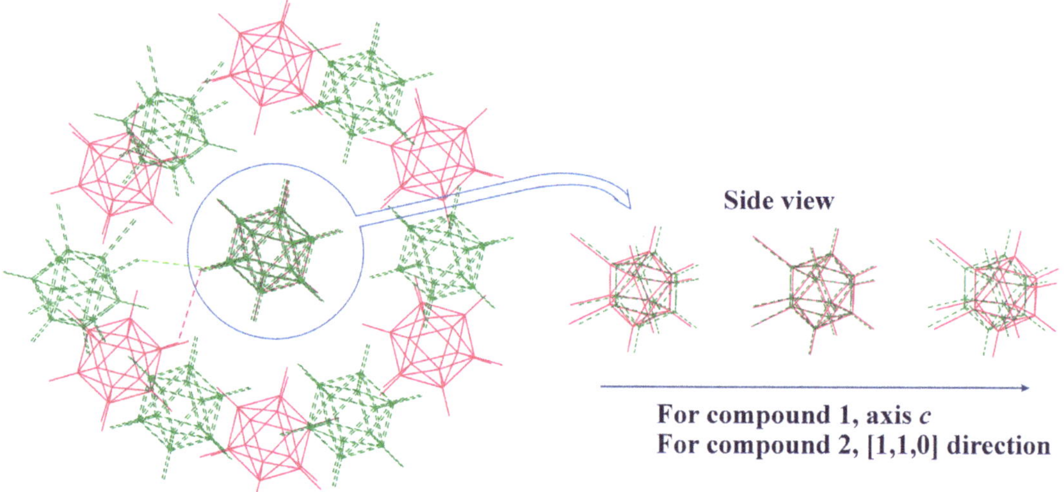

Figure 7. Superimposition of the closest environment of the crystal structure of compounds **1** (magenta) and **2** (green). The I9···I9 and Br9···Br9 halogen bonds (of type I) are shown by dashed lines on the left-side view.

The observed similarities and dissimilarities in the crystal packing of 8,9,12-I_3-1,2-$C_2B_{10}H_9$ and 8,9,12-Br_3-1,2-$C_2B_{10}H_9$ can be clearly seen in Figure 7. Similar C-H···I(Br)-bonded chains are formed in one direction, while in the perpendicular plane, the relative orientation of molecules is somewhat different.

The 1,2,3-isomer 1,2,3-I_3-1,2-$C_2B_{10}H_9$ (**3**) was prepared by the deprotonation of 3-iodo-*ortho*-carborane followed by a treatment of molecular iodine (see below). The crystal structure of 1,2,3-I_3-1,2-$C_2B_{10}H_9$ was determined by single-crystal X-ray diffraction. A general view of **3** is presented in Figure 8. The molecule in the crystal occupies a special position, as it is located at the two-fold symmetry axis. The C-I distances are the same (due to symmetry) and equal to 2.103(4) Å, while the B-I bond is somewhat longer at 2.160(5) Å. These lengths are slightly shorter than the C1-I1 (2.121(2) Å) and B12-I12 (2.179(2) Å) bonds in 1,12-I_2-*closo*-$C_2B_{10}H_{10}$ [45].

Contrary to 8,9,12-I_3-1,2-$C_2B_{10}H_9$, the crystal packing of 1,2,3-I_3-1,2-$C_2B_{10}H_9$ is formed by halogen-bonded planes parallel to the *bc* crystallographic plane (Figure 9). In the planes, the I2 atom (attached to the boron atom) donates its lone pairs simultaneously to the σ-holes of both iodine atoms attached to the carbon atoms (the I1···I2 distance is 3.554(2) Å, the C1-I1···I2 angle is 169.2(2)°, and the B3-I2···I1 angle is 92.2(2)°). Therefore, the main structure-forming unit is the trimeric halogen-bonded associate.

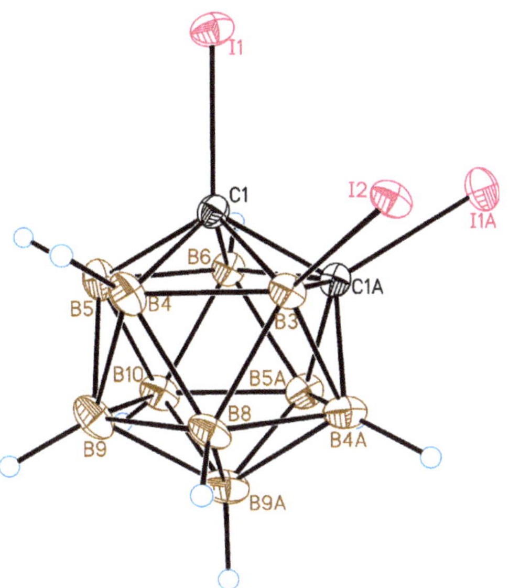

Figure 8. General view of 1,2,3-I$_3$-1,2-C$_2$B$_{10}$H$_9$ (**2**) showing atomic numbering. Thermal ellipsoids are given at 50% probability level.

In our recent study, we theoretically compared the dimer formation of 1,12- and 1,3-diiodo-*ortho*-carboranes [45]. According to our calculations, it appeared that both dimers are stabilized by a type II halogen bond and B-H···I hydrogen bonds. The role of the halogen bond is more pronounced in both dimers; however, in the 1,3-isomer, the halogen bond is weaker (but only by 2.5 kJ/mol), while the hydrogen bonds are stronger (in total by 0.4 kcal/mol). This means that the probability of the formation of a type II halogen bond in a real crystal of 1,3-I$_2$-1,2-C$_2$B$_{10}$H$_{10}$ is somewhat low. Nevertheless, it is formed and is a structure-forming interaction in the crystal structure of 1,2,3-I$_3$-1,2-C$_2$B$_{10}$H$_9$. Indeed, there are no H···H shortened contacts. The structure is additionally stabilized by a few B-H···I shortened contacts. However, some of them are formed between molecules already linked by halogen bonds. For a better understanding of the intermolecular connection in the trimers, we optimized its structure using density functional theory (DFT) at the PBE0/def2tzvp level followed by a topological analysis of the calculated electron density in terms of the "Atoms in Molecules" theory [65]. The intermolecular interaction energies were estimated from their correlation with the potential energy density at the bond critical point [66,67] using the AIMAll program [68].

This method of investigating structural details was successfully utilized in our recent studies on noncovalent interactions [69–71]. Good agreement was obtained between the calculated and experimental structures. The interhalogen distances are nearly the same (Figure 9), and the calculated angles C1-I1···I2 (168.2°) and B3-I2···I1 (90.5°) also strongly agree with the experiment. The H···I distances are somewhat shorter, as predicted by theory. According to the calculations, the energy of the halogen bond is equal to 8.8 kJ/mol, while the energies of the H4···I1 and H4···I2 contacts are 2.5 and 2.1 kJ/mol, respectively. Therefore, the attraction energy of each two molecules in the layer is equal to (8.8 + 2.5 + 2.1) 13.4 kJ/mol, while only weak B-H···I contacts are observed between layers. This allows us to consider the crystal packing of compound **3** as anisotropic unlike the 8,9,12-isomer. It is interesting to note that the crystal density of the latter is somewhat higher than that of the 1,2,3-isomer. This can be explained by the increased role of the I···I interactions (Figure 10). The presence of relatively strong I···I intermolecular interactions does not allow molecules to adjust their orientations to obtain closer packing.

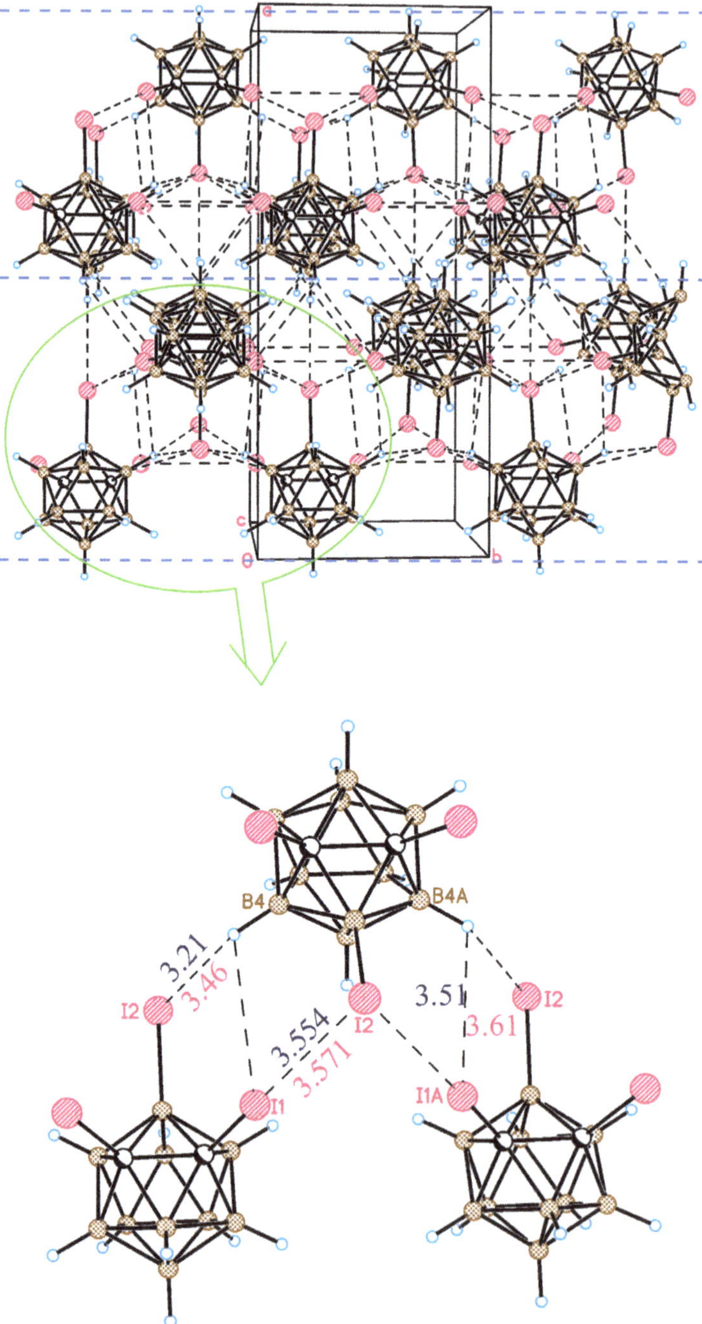

Figure 9. (**Top**) Crystal-packing fragment of 1,2,3-I$_3$-1,2-C$_2$B$_{10}$H$_9$. Blue, dashed lines separate halogen-bonded planes. (**Bottom**) Halogen-bonded trimer as a structure-forming unit of 1,2,3-I$_3$-1,2-C$_2$B$_{10}$H$_9$. Blue and red values correspond to the experimental and calculated I$\cdots$I and H$\cdots$I distances, respectively (given in Å).

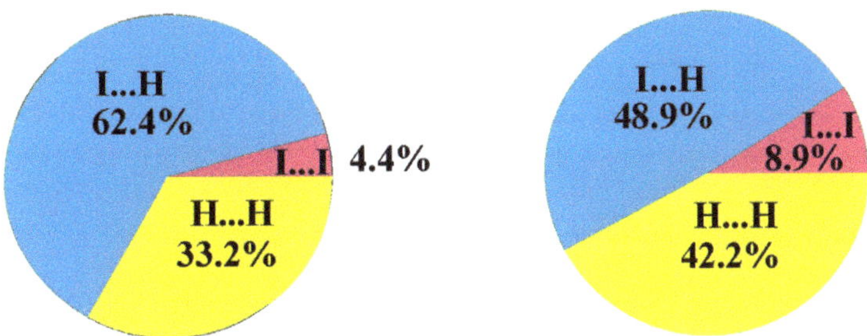

Figure 10. Distribution of intermolecular contacts in the crystal structures of 8,9,12-I_3-1,2-$C_2B_{10}H_9$ (**left**) and 1,2,3-I_3-1,2-$C_2B_{10}H_9$ (**right**).

The same reasons can be used to explain the higher density of water in comparison to ice, and have also been used to explain the differences in the crystal-packing density of polynitro compounds [72,73].

3. Materials and Methods

3.1. General Methods

The reactions were carried under an inert atmosphere. 3-Iodo-*ortho*-carborane was prepared according to a procedure from the literature [74]. 1,2-Dimethoxyethane was dried using standard procedures [75]. All other chemical reagents were purchased from Sigma Aldrich, Acros Organics, and ABCR and used without purification. The reaction progress was monitored by thin-layer chromatography (Merck F254 silica gel on aluminum plates) and visualized using 0.5% $PdCl_2$ in 1% HCl in aq. MeOH (1:10). Acros Organics silica gel (0.060–0.200 mm) was used for column chromatography. The NMR spectra at 400 MHz (^{1}H) and 128 MHz (^{11}B) were recorded with Varian Inova 400 spectrometer. The residual signal of the NMR solvent relative to Me_4Si was taken as the internal reference for ^{1}H spectra. ^{11}B NMR spectra were referenced using $BF_3 \cdot Et_2O$ as external standard.

3.2. Preparation of 8,9,12-Triiodo-ortho-Carborane 8,9,12-I₃-1,2-C₂B₁₀H₉

8,9,12-I_3-*ortho*-$C_2B_{10}H_9$ was isolated as a by-product from the di-iodination reaction of *ortho*-carborane under standard conditions [51]. Iodine (3.553 g, 14.00 mmol) and anhydrous $AlCl_3$ (0.400 g) were added to a solution of *ortho*-carborane (1.009 g, 7.00 mmol) in dichloromethane (30 mL) and heated under reflux for 8 h. Then, the reaction mixture was cooled and treated with a solution of $Na_2S_2O_3 \cdot 5H_2O$ (3.000 g) in water (50 mL). The organic phase was separated, and the aqueous fraction was extracted with dichloromethane (3×50 mL). The organic phases were combined, dried over Na_2SO_4, filtered, and concentrated under reduced pressure. The crude product was purified by column chromatography on silica using diethyl ether as eluent to yield 1.900 g (69%) of 9,12-I_2-1,2-$C_2B_{10}H_{10}$ and 0.102 g (3%) of 8,9,12-I_3-1,2-$C_2B_{10}H_9$ as white powders.

8,9,12-I_3-1,2-$C_2B_{10}H_9$: ^{1}H NMR ($CDCl_3$, ppm): 4.13 (2H, br s, CH_{carb}), 3.8–2.0 (7H, br m, B*H*). ^{11}B NMR ($CDCl_3$, ppm): δ −6.1 (1B, d, *J* = 157 Hz), −11.5 (4B, s + d), −13.1 (2B, d, *J* = 171 Hz), −14.7 (1B, d, *J* = 220 Hz), −16.4 (1B, d, *J* = 220 Hz), and −17.2 (1B, s, *B(8)*).

3.3. Preparation of 8,9,12-Tribromo-ortho-Carborane 8,9,12-Br₃-1,2-C₂B₁₀H₉

Bromine (1.08 mL, 3.356 g, and 21.00 mmol) and anhydrous aluminum chloride (0.400 g) were added to a solution of *ortho*-carborane (1.009 mg and 7.00 mmol) in 1,2-dichloroethane (30 mL) and heated under reflux for 40 h. Then, the reaction mixture was cooled and treated with a solution of $Na_2S_2O_3 \cdot 5H_2O$ (5.000 g) in water (50 mL). The organic phase was separated, and the aqueous fraction was extracted with dichloromethane (3×50 mL). The organic phases were combined, dried over Na_2SO_4, filtered, and concentrated under reduced pressure. The crude product was purified by column chromatography on silica using chloroform as eluent to yield 0.450 g (17%) of $8,9,12\text{-Br}_3\text{-}1,2\text{-}C_2B_{10}H_9$ as a white powder.

$8,9,12\text{-Br}_3\text{-}1,2\text{-}C_2B_{10}H_9$: ^{1}H NMR (CDCl$_3$, ppm): 3.87 (2H, br s, CH_{carb}), 3.5−1.7 (7H, br m, BH). ^{11}B NMR (CDCl$_3$, ppm): 0.4 (2B, s, $B(9,12)$), −5.3 (1B, s, $B(8)$), −8.4 (1B, d, J = 161 Hz, $B(10)$), −13.9 (2B, d, J = 176 Hz, $B(4,7)$), −15.6 (2B, d, J = 178 Hz, $B(5,11)$), −17.1 (1B, d, J = 188 Hz, $B(3)$), and −20.4 (1B, d, J = 185 Hz, $B(6)$). ^{13}C NMR (CDCl$_3$, ppm): 45.5 (CH_{carb}).

3.4. Preparation of 1,2,3-Triiodo-ortho-Carborane $1,2,3\text{-}I_3\text{-}1,2\text{-}C_2B_{10}H_9$

A 2.25 M hexane solution of *n*-butyllithium (390 μL; 0.88 mmol) was added to a solution of 3-iodo-*ortho*-carborane (110 mg; 0.41 mmol) in 1,2-dimethoxyethane (10 mL) and stirred at room temperature for 1 h. Iodine (244 mg; 0.96 mmol) was added in one portion and the reaction mixture was stirred at room temperature overnight; then, it was treated with a solution of $Na_2S_2O_3 \cdot 5H_2O$ (250 mg) in water (50 mL). The organic phase was separated, and the aqueous fraction was extracted with dichloromethane (3 × 50 mL). The organic phases were combined, dried over Na_2SO_4, filtered, and concentrated under reduced pressure. The crude product was purified by column chromatography on silica using diethyl ether as eluent to yield 137 mg of mixture of $3\text{-}I\text{-}1,2\text{-}C_2B_{10}H_{11}$ and $1,2,3\text{-}I_3\text{-}1,2\text{-}C_2B_{10}H_9$.

3.5. Single-Crystal X-ray Diffraction Study

Single-crystal X-ray diffraction experiments of **1, 2,** and **3** (see Supplementary Materials) were carried out using SMART APEX2 CCD diffractometer (λ(Mo-Kα) = 0.71073 Å; graphite monochromator; ω-scans) at 120 K. Collected data were processed by the SAINT and SADABS programs incorporated into the APEX2 program package [76]. The structures were determined by direct methods and refined by the full-matrix-least-squares procedure against F^2 in anisotropic approximation. The refinement was carried out with the SHELXTL program [77]. The CCDC numbers (2216663 for **1**, 2234154 for **2**, and 2216664 for **3**) contain the supplementary crystallographic data for this paper. These data can be obtained free of charge via www.ccdc.cam.ac.uk/data_request/cif.

Crystallographic data for $8,9,12\text{-}I_3\text{-}1,2\text{-}C_2B_{10}H_9$ (**1**): $C_2H_9B_{10}I_3$ are monoclinic; space group $P2_1/n$: a = 7.5776(6) Å, b = 24.0030(18) Å, c = 7.7535(6) Å, β = 109.487(2)°, V = 1329.46(18) Å^3, and Z = 4, M = 521.89, d_{cryst} = 2.607 g·cm^{-3}. $wR2$ = 0.0794 calculated on F^2_{hkl} for all 2611 independent reflections with $2\theta < 52.1°$ (GOF = 1.117, R = 0.0343 calculated on F_{hkl} for 2331 reflections with $I > 2\sigma(I)$).

Crystallographic data for $8,9,12\text{-}Br_3\text{-}1,2\text{-}C_2B_{10}H_9$ (**2**): $C_2H_9B_{10}Br_3$ are monoclinic; space group $C2/c$: a = 12.1453(6) Å, b = 8.4794(5) Å, c = 23.0632(11) Å, β = 90.089(2)°, V = 2375.2(2) Å^3, Z = 8, M = 380.92, d_{cryst} = 2.131 g·cm^{-3}. $wR2$ = 0.0797 calculated on F^2_{hkl} for all 2349 independent reflections with $2\theta < 52.1°$ (GOF = 1.024, R = 0.0347 calculated on F_{hkl} for 1908 reflections with $I > 2\sigma(I)$).

Crystallographic data for $1,2,3\text{-}I_3\text{-}1,2\text{-}C_2B_{10}H_9$ for (**3**): $C_2H_9B_{10}I_3$ are orthorhombic; space group *Pnma*: a = 19.1157(8) Å, b = 8.0014(3) Å, c = 8.7287(4) Å, V = 1335.08(10) Å^3, Z = 4, M = 521.89, d_{cryst} = 2.596 g·cm^{-3}. $wR2$ = 0.0580 calculated on F^2_{hkl} for all 1739 independent reflections with $2\theta < 56.2°$ (GOF = 1.143, R = 0.0224 calculated on F_{hkl} for 1613 reflections with $I > 2\sigma(I)$).

3.6. Quantum Chemical Calculation

Quantum chemical optimization of halogen-bonded trimeric associate of $1,2,3\text{-}I_3\text{-}1,2\text{-}C_2B_{10}H_9$ was carried out using the Gaussian program [78]. The initial geometry for optimization was taken from the X-ray data. Optimization was carried out using PBE0 functional and triple-zeta basis set def2tzvp. For better agreement with experimental geometry, calculation was carried out within polarizable continuum model (PCM) using SCRF keyword in the Gaussian program and highly polar water molecule. It has recently been shown that such a method of calculation results in better agreement of the geometry for noncovalent interactions [45,69].

Supplementary Materials: The following supporting information can be downloaded at: https://www.mdpi.com/article/10.3390/molecules28020875/s1; Crystallographic data for compounds **1**, **2**, and **3**.

Author Contributions: X-ray diffraction experiment and manuscript writing, K.Y.S.; Synthesis, S.A.A.; General manuscript concept and manuscript writing, I.B.S. All authors have read and agreed to the published version of the manuscript.

Funding: This research was supported by the Russian Science Foundation (Grant No. 21-13-00345).

Data Availability Statement: Crystallographic data for the structures of $8,9,12$-I_3-$1,2$-$C_2B_{10}H_9$ (**1**), $8,9,12$-Br_3-$1,2$-$C_2B_{10}H_9$ (**2**), and $1,2,3$-I_3-$1,2$-$C_2B_{10}H_9$ (**3**) were deposited in the Cambridge Crystallographic Data Centre as supplementary publications CCDC 2216663 (for **1**), 2234154 (for **2**), and 2216664 (for **3**). The Supplementary Information contains crystallographic data for compounds **1**, **2**, and **3**.

Acknowledgments: The single-crystal X-ray diffraction data were obtained using equipment from the Center for Molecular Structure Studies at A.N. Nesmeyanov Institute of Organoelement Compounds, operating with financial support from the Ministry of Science and Higher Education of the Russian Federation (agreement no. 075-00697-22-00).

Conflicts of Interest: The authors declare no conflict of interest. The funders had no role in the design of the study; in the collection, analyses, or interpretation of data; in the writing of the manuscript; or in the decision to publish the results.

References

1. Colin, J.J. Sur quelques combinaisons de l'iode. *Ann Chim.* **1814**, *91*, 252–272.
2. Wisniak, J. Jean-Jacques Colin. *Rev. CENIC Cienc. Biol.* **2017**, *48*, 112–120.
3. Guthrie, F. On the iodide of iodammonium. *J. Chem. Soc.* **1863**, *16*, 239–244. [CrossRef]
4. Hassel, O. Structural aspects of interatomic charge-transfer bonding. *Science* **1970**, *170*, 497–502. [CrossRef]
5. Turunen, L.; Hansen, J.H.; Erdélyi, M. Halogen bonding: An Odd chemistry? *Chem. Rec.* **2021**, *21*, 1252–1257. [CrossRef] [PubMed]
6. Metrangolo, P.; Resnati, G. (Eds.) *Halogen Bonding: Fundamentals and Applications*; Springer: Berlin/Heidelberg, Germany, 2010. [CrossRef]
7. Gilday, L.C.; Robinson, S.W.; Barendt, T.A.; Langton, M.J.; Mullaney, B.R.; Beer, P.D. Halogen bonding in supramolecular chemistry. *Chem. Rev.* **2015**, *115*, 7118–7195. [CrossRef] [PubMed]
8. Cavallo, G.; Metrangolo, P.; Milani, R.; Pilati, T.; Priimagi, A.; Resnati, G.; Terraneo, G. The halogen bond. *Chem. Rev.* **2016**, *116*, 2478–2601. [CrossRef] [PubMed]
9. Costa, P.J. The halogen bond: Nature and applications. *Phys. Sci. Rev.* **2017**, *2*, 20170136. [CrossRef]
10. Frontera, F.; Bauzá, A. On the importance of σ-hole interactions in crystal structures. *Crystals* **2021**, *11*, 1205. [CrossRef]
11. Politzer, P.; Murray, J.S.; Clark, T. Halogen bonding: An electrostatically-driven highly directional noncovalent interaction. *Phys. Chem. Chem. Phys.* **2010**, *12*, 7748–7757. [CrossRef]
12. Politzer, P.; Murray, J.S.; Clark, T. Halogen bonding and other σ-hole interactions: A perspective. *Phys. Chem. Chem. Phys.* **2013**, *15*, 11178–11189. [CrossRef] [PubMed]
13. Lommerse, J.P.M.; Stone, A.J.; Taylor, R.; Allen, F.H. The nature and geometry of intermolecular interactions between halogens and oxygen or nitrogen. *J. Am. Chem. Soc.* **1996**, *118*, 3108–3116. [CrossRef]
14. Metrangolo, P.; Murray, J.S.; Pilati, T.; Politzer, P.; Resnati, G.; Terraneo, G. The fluorine atom as a halogen bond donor, *viz.* a positive site. *CrystEngComm* **2011**, *13*, 6593–6596. [CrossRef]
15. Le Questel, Y.-J.; Laurence, C.; Graton, J. Halogen-bond interactions: A crystallographic basicity scale towards iodoorganic compounds. *CrystEngComm* **2013**, *15*, 3212–3221. [CrossRef]
16. Askeröy, C.B.; Baldrighi, M.; Desper, J.; Metrangolo, P.; Resnati, G. Supramolecular hierarchy among halogen-bond donors. *Chem. Eur. J.* **2013**, *19*, 16240–16247. [CrossRef] [PubMed]
17. Präsang, C.; Whitwood, A.C.; Bruce, D.W. Halogen-bonded cocrystals of 4-(*N,N*-dimethylamino)pyridine with fluorinated iodobenzenes. *Cryst. Growth Des.* **2009**, *9*, 5319–5326. [CrossRef]
18. Roper, L.C.; Präsang, C.; Kozhevnikov, V.N.; Whitwood, A.C.; Karadakov, P.B.; Bruce, D.W. Experimental and theoretical study of halogen-bonded complexes of DMAP with di- and triiodofluorobenzenes. A complex with a very short N···I halogen bond. *Cryst. Growth Des.* **2010**, *10*, 3710–3720. [CrossRef]

19. Wang, H.; Jin, W.J. Cocrystal assembled by 1,4-diiodotetrafluorobenzene and phenothiazine based on C-I···π/N/S halogen bond and other assisting interactions. *Acta Cryst. B* **2017**, *73*, 210–216. [CrossRef]

20. Christopherson, J.C.; Topic, F.; Barret, C.J.; Friscic, T. Halogen-bonded cocrystals as optical materials: Next generation control over light-matter interactions. *Cryst. Growth Des.* **2018**, *18*, 1245–1259. [CrossRef]

21. Bedeković, N.; Stilinović, V.; Friščić, T.; Cinčić, D. Comparison of isomeric *meta*- and *para*-diiodotetrafluorobenzene as halogen bond donors in crystal engineering. *New J. Chem.* **2018**, *42*, 10584–10591. [CrossRef]

22. Li, L.; Wu, W.X.; Liu, Z.F.; Jin, W.J. Effect of geometry factors on the priority of σ-hole···π and π-hole···π bond in phosphorescent cocrystals formed by pyrene or phenanthrene and trihaloperfluorobenzenes. *New J. Chem.* **2018**, *42*, 10633–10641. [CrossRef]

23. Lin, J.; Chen, Y.; Zhao, D.; Lu, X.; Lin, Y. Versatile supramolecular binding modes of 1,4-diiodotetrafluorobenzene for molecular cocrystal engineering. *J. Mol. Struct.* **2019**, *1187*, 132–137. [CrossRef]

24. Grosu, I.G.; Pop, L.; Miclăuş, M.; Hădade, N.D.; Terec, A.; Bende, A.; Socaci, C.; Barboiu, M.; Grosu, I. Halogen bonds (N···I) at work: Supramolecular catemeric architectures of 2,7-dipyridylfluorene with *ortho*-, *meta*-, or *para*-diiodotetrafluorobenzene isomers. *Cryst. Growth Des.* **2020**, *20*, 3429–3441. [CrossRef]

25. Uran, E.; Fotović, L.; Bedeković, N.; Stilinović, V.; Cinčić, D. The amine group as halogen bond acceptor in cocrystals of aromatic diamines and perfluorinated iodobenzenes. *Crystals* **2021**, *11*, 529. [CrossRef]

26. Nieland, E.; Komisarek, D.; Hohloch, S.; Wurst, K.; Vasylyeva, V.; Weingart, O.; Schmidt, B.M. Supramolecular networks by imine halogen bonding. *Chem. Commun.* **2022**, *58*, 5233–5236. [CrossRef]

27. Hajjar, C.; Nag, T.; Al Sayed, H.; Ovens, J.S.; Bryce, D.L. Stoichiomorphic halogen-bonded cocrystals: A case study of 1,4-diiodotetrafluorobenzene and 3-nitropyridine. *Can. J. Chem.* **2022**, *100*, 245–251. [CrossRef]

28. Yeo, C.I.; Tan, Y.S.; Kwong, H.C.; Lee, V.S.; Tiekink, E.R.T. I···N halogen bonding in 1: 1 co-crystals formed between 1,4-diiodotetrafluorobenzene and the isomeric *n*-pyridinealdazines (*n* = 2, 3 and 4): Assessment of supramolecular association and influence upon solid-state photoluminescence properties. *CrystEngComm* **2022**, *24*, 7579–7591. [CrossRef]

29. Aakeröy, C.B.; Wijethunga, T.K.; Desper, J.; Đaković, M. Crystal engineering with iodoethynylnitrobenzenes: A group of highly effective halogen-bond donors. *Cryst. Growth Des.* **2015**, *15*, 3853–3861. [CrossRef]

30. Aakeröy, C.B.; Welideniya, D.; Desper, J. Ethynyl hydrogen bonds and iodoethynyl halogen bonds: A case of synthon mimicry. *CrystEngComm* **2017**, *19*, 11–13. [CrossRef]

31. Wijethunga, T.K.; Đakovic, M.; Desper, J.; Aakeröy, C.B. A new tecton with parallel halogen-bond donors: A path to supramolecular rectangles. *Acta Cryst. B* **2017**, *73*, 163–167. [CrossRef]

32. Fourmigue, M. Coordination chemistry of anions through halogen-bonding interactions. *Acta Cryst. B* **2017**, *73*, 138–139. [CrossRef] [PubMed]

33. Szell, P.M.J.; Gabidullin, B.; Bryce, D.L. 1,3,5-Tri(iodoethynyl)-2,4,6-trifluorobenzene: Halogen-bonded frameworks and NMR spectroscopic analysis. *Acta Cryst. B* **2017**, *73*, 153–162. [CrossRef] [PubMed]

34. Szell, P.M.J.; Siiskonen, A.; Catalano, L.; Cavallo, G.; Terraneo, G.; Priimagi, A.; Bryce, D.L.; Metrangolo, P. Halogen-bond driven self-assembly of triangular macrocycles. *New J. Chem.* **2018**, *42*, 10467–10471. [CrossRef]

35. Bosch, E.; Kruse, S.J.; Groeneman, R.H. Infinite and discrete halogen bonded assemblies based upon 1,2-bis(iodoethynyl)benzene. *CrystEngComm* **2019**, *21*, 990–993. [CrossRef]

36. Reddy, C.M.; Kirchner, M.T.; Gundakaram, R.C.; Padmanabhan, K.A.; Desiraju, G.R. Isostructurality, polymorphism and mechanical properties of some hexahalogenated benzenes: The nature of halogen···halogen interactions. *Chem. Eur. J.* **2006**, *12*, 2222–2234. [CrossRef]

37. Raffo, P.A.; Suarez, S.; Fantoni, A.C.; Baggio, R.; Cukiernik, F.D. Polymorphism of a widely used building block for halogen-bonded assemblies: 1,3,5-trifluoro-2,4,6-triiodobenzene. *Acta Cryst. C* **2017**, *73*, 667–673. [CrossRef]

38. Bartashevich, E.; Sobalev, S.; Matveychuk, Y.; Tsirelson, V. Variations of quantum electronic pressure under the external compression in crystals with halogen bonds assembled in Cl_3-, Br_3-, I_3-synthons. *Acta Cryst. B* **2020**, *76*, 514–523. [CrossRef]

39. Dominikowska, J.; Rybarczyk-Pirek, A.J.; Guerra, C.F. Lack of cooperativity in the triangular X_3 halogen-bonded synthon? *Cryst. Growth Des.* **2021**, *21*, 597–607. [CrossRef]

40. Saha, A.; Rather, S.A.; Sharada, D.; Saha, B.K. C–X···X–C *vs* C–H···X–C, which one is the more dominant interaction in crystal packing (X = halogen)? *Cryst. Growth Des.* **2018**, *18*, 6084–6090. [CrossRef]

41. Lo, R.; Fanfrlík, J.; Lepšík, M.; Hobza, P. The properties of substituted 3D-aromatic neutral carboranes: The potential for σ-hole bonding. *Phys. Chem. Chem. Phys.* **2015**, *17*, 20814–20821. [CrossRef]

42. Fanfrlík, J.; Holub, J.; Růžičková, Z.; Řezáč, J.; Lane, P.D.; Wann, D.A.; Hnyk, D.; Růžička, A.; Hobza, P. Competition between halogen, hydrogen and dihydrogen bonding in brominated carboranes. *ChemPhysChem* **2016**, *17*, 3373–3376. [CrossRef]

43. Beau, M.; Lee, S.; Kim, S.; Han, W.-S.; Jeannin, O.; Fourmigué, M.; Aubert, E.; Espinosa, E.; Jeon, I.-R. Strong σ-hole activation on icosahedral carborane derivatives for a directional halide recognition. *Angew. Chem. Int. Ed.* **2021**, *60*, 366–370. [CrossRef] [PubMed]

44. Kalinin, V.N.; Ol´shevskaya, V.A. Some aspects of the chemical behavior of icosahedral carboranes. *Russ. Chem. Bull.* **2008**, *57*, 815–836. [CrossRef]

45. Suponitsky, K.Y.; Anisimov, A.A.; Anufriev, S.A.; Sivaev, I.B.; Bregadze, V.I. 1,12-Diiodo-*ortho*-carborane: A classic textbook example of the dihalogen bond. *Crystals* **2021**, *11*, 396. [CrossRef]

46. Vaca, A.; Teixidor, F.; Kivekäs, R.; Sillanpää, R.; Viñas, C. A solvent-free regioselective iodination route of *ortho*-carboranes. *Dalton Trans.* **2006**, 4884–4885. [CrossRef] [PubMed]

47. Puga, A.V.; Teixidor, F.; Sillanpää, R.; Kivekäs, R.; Viñas, C. Iodinated *ortho*-carboranes as versatile building blocks to design intermolecular interactions in crystal lattices. *Chem. Eur. J.* **2009**, *15*, 9764–9772. [CrossRef]

48. Barberà, G.; Viñas, C.; Teixidor, F.; Rosair, G.M.; Welch, A.J. Self-assembly of carborane molecules *via* C–H···I hydrogen bonding: The molecular and crystal structures of 3-I-1,2-*closo*-$C_2B_{10}H_{11}$. *J. Chem. Soc. Dalton Trans.* **2002**, 3647–3648. [CrossRef]

49. Safronov, A.V.; Sevryugina, Y.V.; Jalisatgi, S.S.; Kennedy, R.D.; Barnes, C.L.; Hawthorne, M.F. Unfairly forgotten member of the iodocarborane family: Synthesis and structural characterization of 8-iodo-1,2-dicarba-*closo*-dodecaborane, its precursors, and derivatives. *Inorg. Chem.* **2012**, *51*, 2629–2637. [CrossRef]

50. Barbera, G.; Vaca, A.; Teixidor, F.; Sillanpää, R.; Kivekäs, R.; Viñas, C. Designed synthesis of new *ortho*-carborane derivatives: From mono- to polysubstituted frameworks. *Inorg. Chem.* **2008**, *47*, 7309–7316. [CrossRef]

51. Ramachandran, B.M.; Knobler, C.B.; Hawthorne, M.F. Synthesis and structural characterization of symmetrical *closo*-4,7-I_2-1,2-$C_2B_{10}H_{10}$ and $[(CH_3)_3NH][nido$-2,4-I_2-7,8-$C_2B_9H_{10}]$. *Inorg. Chem.* **2006**, *45*, 336–340. [CrossRef]

52. Batsanov, A.S.; Fox, M.A.; Howard, J.A.K.; Hughes, A.K.; Johnson, A.L.; Martindale, S.J. 9,12-Diiodo-1,2-dicarba-*closo*-dodecaborane(12). *Acta Cryst. A* **2003**, *59*, O74–O76. [CrossRef] [PubMed]

53. Rudakov, D.A.; Kurman, P.V.; Potkin, V.I. Synthesis and deborination of polyhalo-substituted *ortho*-carboranes. *Russ. J. Gen. Chem.* **2011**, *81*, 1137–1142. [CrossRef]

54. Zheng, Z.; Jiang, W.; Zinn, A.A.; Knobler, C.B.; Hawthorne, M.F. Facile electrophilic iodination of icosahedral carboranes. Synthesis of carborane derivatives with boron-carbon bonds via the palladium-catalyzed reaction of diiodocarboranes with Grignard reagents. *Inorg. Chem.* **1995**, *34*, 2095–2100. [CrossRef]

55. Struchkov, Y.T.; Stanko, V.I.; Klimova, A.I.; Kon'kova, G.S. X-ray data on some derivatives of barene and neobarene. *J. Struct. Chem.* **1965**, *8*, 888–890. [CrossRef]

56. Zefirov, Y.V.; Zorky, P.M. New applications of van der Waals radii in chemistry. *Russ. Chem. Rev.* **1995**, *64*, 415–428. [CrossRef]

57. Smith, H.D.; Knowles, T.A.; Schroeder, H. Chemistry of decaborane-phosphorus compounds. V. Bromocarboranes and their phosphination. *Inorg. Chem.* **1965**, *4*, 107–111. [CrossRef]

58. Zhidkova, O.B.; Druzina, A.A.; Anufriev, S.A.; Suponitsky, K.Y.; Sivaev, I.B.; Bregadze, V.I. Synthesis and crystal structure of 9,12-dibromo-*ortho*-carborane. *Molbank* **2022**, *2022*, M1347. [CrossRef]

59. Puga, A.V.; Teixidor, F.; Sillanpää, R.; Kivekäs, R.; Arca, M.; Barbera, G.; Viñas, C. From mono- to poly-substituted frameworks: A way of tuning the acidic character of C_c-H in *o*-carborane derivatives. *Chem. Eur. J.* **2009**, *15*, 9755–9763. [CrossRef]

60. Sivaev, I.B.; Anufriev, S.A.; Shmalko, A.V. How substituents at boron atoms affect the *CH*-acidity and the electron-withdrawing effect of the *ortho*-carborane cage: A close look on the ^{1}H NMR spectra. *Inorg. Chim. Acta* **2023**, *547*, 121339. [CrossRef]

61. Potenza, J.A.; Lipscomb, W.N. Molecular structure of carboranes. Molecular and crystal structure of *o*-$B_{10}Br_3H_7C_2H_2$. *Inorg. Chem.* **1966**, *5*, 1478–1482. [CrossRef]

62. Potenza, J.A.; Lipscomb, W.N. Molecular structure of carboranes. Molecular and crystal structure of *o*-$B_{10}Br_4H_6C_2(CH_3)_2$. *Inorg. Chem.* **1966**, *5*, 1483–1488. [CrossRef]

63. Anufriev, S.A.; Timofeev, S.V.; Zhidkova, O.B.; Suponitsky, K.Y.; Sivaev, I.B. Synthesis, crystal structure, and some transformations of 9,12-dichloro-*ortho*-carborane. *Crystals* **2022**, *12*, 1251. [CrossRef]

64. Turner, M.J.; McKinnon, J.J.; Wolff, S.K.; Grimwood, D.J.; Spackman, P.R.; Jayatilaka, D.; Spackman, M.A. *CrystalExplorer17*; University of Western Australia: Perth, WA, Australia, 2017.

65. Bader, R.F.W. *Atoms in Molecules. A Quantum Theory*; Clarendon Press: Oxford, UK, 1990.

66. Espinosa, E.; Molins, E.; Lecomte, C. Hydrogen bond strengths revealed by topological analyses of experimentally observed electron densities. *Chem. Phys. Lett.* **1998**, *285*, 170–173. [CrossRef]

67. Espinosa, E.; Alkorta, I.; Rozas, I.; Elguero, J.; Molins, E. About the evaluation of the local kinetic, potential and total energy densities in closed-shell interactions. *Chem. Phys. Lett.* **2001**, *336*, 457–461. [CrossRef]

68. Keith, T.A. *AIMAll, Version 15.05.18*; TK Gristmill Software: Overland Park, KS, USA, 2015.

69. Suponitsky, K.Y.; Burakov, N.I.; Kanibolotsky, A.L.; Mikhailov, V.A. Multiple noncovalent bonding in halogen complexes with oxygen organics. I. Tertiary amides. *J. Phys. Chem. A* **2016**, *120*, 4179–4190. [CrossRef]

70. Anufriev, S.A.; Sivaev, I.B.; Suponitsky, K.Y.; Bregadze, V.I. Practical synthesis of 9-methylthio-7,8-*nido*-carborane [9-MeS-7,8-$C_2B_9H_{11}]^-$. Some evidences of BH···X hydride-halogen bonds in 9- $XCH_2(Me)S$-7,8-$C_2B_9H_{11}$ (X = Cl, Br, I). *J. Organomet. Chem.* **2017**, *849–850*, 315–323. [CrossRef]

71. Dmitrienko, A.O.; Karnoukhova, V.A.; Potemkin, A.A.; Struchkova, M.I.; Kryazhevskikh, I.A.; Suponitsky, K.Y. The influence of halogen type on structural features of compounds containing α-halo-α,α-dinitroethyl moieties. *Chem. Heterocycl. Comp.* **2017**, *53*, 532–539. [CrossRef]

72. Gidaspov, A.A.; Zalomlenkov, V.A.; Bakharev, V.V.; Parfenov, V.E.; Yurtaev, E.V.; Struchkova, M.I.; Palysaeva, N.V.; Suponitsky, K.Y.; Lempertd, D.B.; Sheremetev, A.B. Novel trinitroethanol derivatives: High energetic 2-(2,2,2-trinitroethoxy)-1,3,5-triazines. *RSC Adv.* **2016**, *6*, 34921–34934. [CrossRef]

73. Palysaeva, N.V.; Gladyshkin, A.G.; Vatsadze, I.A.; Suponitsky, K.Y.; Dmitriev, D.E.; Sheremetev, A.B. N-(2-Fluoro-2,2-dinitroethyl)azoles: Novel assembly of diverse explosophoric building block-s for energetic compounds design. *Org. Chem. Front.* **2019**, *6*, 249–255. [CrossRef]

74. Zhao, D.; Xie, Z. [3-N$_2$-*o*-C$_2$B$_{10}$H$_{11}$][BF$_4$]: A useful synthon for multiple cage boron functionalizations of *o*-carborane. *Chem. Sci.* **2016**, *7*, 5635–5639. [CrossRef]

75. Armarego, W.L.F.; Chai, C.L.L. *Purification of Laboratory Chemicals*, 6th ed.; Butterworth-Heinemann: Oxford, UK, 2009. [CrossRef]

76. *APEX2 and SAINT*; Bruker AXS Inc.: Madison, WI, USA, 2014.

77. Sheldrick, G.M. Crystal structure refinement with SHELXL. *Acta Cryst. C* **2015**, *71*, 3–8. [CrossRef] [PubMed]

78. Frisch, M.J.; Trucks, G.W.; Schlegel, H.B.; Scuseria, G.E.; Robb, M.A.; Cheeseman, J.R.; Montgomery, J.A.; Kudin, K.N., Jr.; Burant, J.C.; Millam, J.M.; et al. *Gaussian 03, Revision E.01*; Gaussian, Inc.: Wallingford, UK, 2004.

molecules

Dihydrogen Bonding—Seen through the Eyes of Vibrational Spectroscopy

Marek Freindorf, Margaret McCutcheon [†], Nassim Beiranvand [†] and Elfi Kraka *

Computational and Theoretical Chemistry Group (CATCO), Department of Chemistry, Southern Methodist University, 3215 Daniel Ave, Dallas, TX 75275-0314, USA; mfreindorf@smu.edu (M.F.); mmccutcheon@smu.edu (M.M.); nbeiranvand@smu.edu (N.B.)
* Correspondence: ekraka@smu.edu
† These authors contributed equally to this work.

Abstract: In this work, we analyzed five groups of different dihydrogen bonding interactions and hydrogen clusters with an H_3^+ kernel utilizing the local vibrational mode theory, developed by our group, complemented with the Quantum Theory of Atoms–in–Molecules analysis to assess the strength and nature of the dihydrogen bonds in these systems. We could show that the intrinsic strength of the dihydrogen bonds investigated is primarily related to the protonic bond as opposed to the hydridic bond; thus, this should be the region of focus when designing dihydrogen bonded complexes with a particular strength. We could also show that the popular discussion of the blue/red shifts of dihydrogen bonding based on the normal mode frequencies is hampered from mode–mode coupling and that a blue/red shift discussion based on local mode frequencies is more meaningful. Based on the bond analysis of the $H_3^+(H_2)_n$ systems, we conclude that the bond strength in these crystal–like structures makes them interesting for potential hydrogen storage applications.

Keywords: dihydrogen bonding; local vibrational mode analysis; blue/red shifts; hydride complexes; hydrogen storage

Citation: Freindorf, M.; McCutcheon, M.; Beiranvand, N.; Kraka, E. Dihydrogen Bonding—Seen through the Eyes of Vibrational Spectroscopy. *Molecules* **2023**, *28*, 263. https://doi.org/10.3390/molecules28010263

Academic Editors: Qingzhong Li, Steve Scheiner and Zhiwu Yu

Received: 6 December 2022
Revised: 20 December 2022
Accepted: 20 December 2022
Published: 28 December 2022

1. Introduction

Hydrogen bonding (HB) is one of the most prominent non–covalent chemical interactions [1–5], being responsible for the shape and structure of biopolymers found in nature, such as proteins and nucleic acids [6–8], but also being a major player in transition metal catalysis [9], solid–state chemistry and materials science [10]. Generally, HB occurs between a positively charged hydrogen atom of a proton donor, and a lone pair of an electronegative atom, π electron cloud, or a transition metal center, representing the corresponding proton acceptor. However, in recent years, in addition to this *conventional* HB interaction, another intriguing HB type has been discussed, where the two atoms directly involved in the weak interaction are both hydrogen atoms [11–14]. This interaction was coined by Richardson [15] *dihydrogen bonding* (DHB will be used for dihydrogen bonds in this study). This describes the attraction between a protonic ($H^{\delta+}$) and a hydridic ($H^{\delta-}$) hydrogen atom, where the protonic H atom acts as the proton donor as in conventional HBs, and the hydridic H atom takes over the role of the proton acceptor [11].

Interaction energies of DHBs are generally between 1 and 7 kcal/mol, and because of their similarity to HBs, they are often treated as a subclass of HBs [16]. It has been suggested that DHBs are predominantly of electrostatic nature [16–19], accompanied in stronger DHBs with a substantial covalent contribution [16,20–22]; although, this has not been quantified so far. DHBs can influence molecular properties in the gas phase, in solution and in the solid state, yielding broad potential utilities in catalysis and materials sciences. In the solid state, DHBs show the unique ability to lose H_2, which could be utilized for the design of low–cost, high–capacity hydrogen storage materials [23–25]. In particular, a better understanding of the transformation from a weak H—H electrostatic interaction to a strong covalent bond of H_2 type could open new routes for technologies using hydrogen as an environmentally clean and efficient fuel.

The first experimental observation of the attractive interaction between two hydrogen atoms was reported in early 1930 by Zachariasen and Mooney [26] in the crystal structure of ammonium hypophosphite ($NH_4H_2PO_2$), showing a close contact between the H atoms of the hypophosphite group and those of ammonium. More than 30 years later, Burg [27] reported the direct intermolecular interaction between the H atoms of the NH and BH_3 groups in liquid $(CH_3)_2NH-HBH_2$, suggesting the presence of an $H-H$ interaction comparable to that of a HB. This interaction was formally categorized as a true HB by Brown in the late 1960s [28–30] on the basis of the IR spectra of boron coordination compounds.

DHB is particularly attractive for molecular systems involving metal hydrides acting as proton acceptors. The first experimental evidence for the formation of a DBH involving a metal hydride was presented independently by Crabtree [31] and Morris [32]. Later, Crabtree et al. [33,34] investigated the $H-H$ interaction in molecular complexes involving iridium hydrides, exploring the strength of this interaction with experimental and computational means. By using the Cambridge Crystallographic Database [35] and spectroscopic data of aminoboranes, Richardson et al. [15] and later Crabtree et at. [36] showed that the $NH-HB$ distances and heats of formation are in the ranges of 1.7–2.2 Å and 3–7 kcal/mol, respectively, comparable to that of conventional HBs.

Over the years, DHB has been also the subject of many theoretical studies utilizing different levels of theoretical approaches. Comprehensive overviews can be found in Refs. [1–3,14,37,38] and studies cited therein. Furthermore, two new data sets have been recently reported for benchmarking [39]. The focus of these theoretical studies ranges from energy decomposition and bond energies [40–42], geometries [1,11], normal mode frequency shifts [9,43], studies of the topological features of the total electron density $\rho(\mathbf{r})$ [44–47], and electron localization function (ELF) investigations [48,49] to the analysis of electrostatic potential [50–52], which led to a number of open questions. In particular, what is missing so far is a quantitative bond strength measure coupled with the assessment of the covalent character, which allows the classification of DHB as (i) weak $H\cdots H$ interactions of the van der Waals type, (ii) moderate $H\cdots H$ interactions mostly of electrostatic type, and (iii) strong $H-H$ interactions with substantial covalent character.

Therefore, the focus of our study was to introduce a quantitative bond strength measure based on the local vibrational mode analysis (LMA), originally developed by Konkoli and Cremer [53–57], paired with Weinhold's Natural Bond Orbital (NBO) population analysis [58–60], and Bader's Quantum Theory of Atoms–In–Molecules (QTAIM) analysis [61–64], in order to obtain a deeper understanding of DHBs in general, in particular their strength, and to unravel similarities/differences compared to conventional HBs. Figure 1 shows the set of DHB complexes investigated in this study (organized in six specific groups) covering a broad spectrum of DHBs, ranging from weak interactions of electrostatic origin to strong interactions with covalent character. Group I includes hydrides interacting with simple organic neutral and cationic molecules as well intramolecular DHB in boron complexes, and Group II shows DHB in ammonia–borane complexes. DHB in Fe complexes are included in Group III, while DHB in noble gas complexes are involved in Group IV. We also reassessed the $H-H$ interaction in aromatic hydrocarbons such as phenanthrene, dibenz[a]anthracene, and biphenyl [22,65] in Group V, where both hydrogen atoms are equally charged, which has been called the $H-H$ *bond* in the literature, to contrast this situation from DHB [11], leading to some controversy [66]. Group VI of our study is made up of DHB in $H_3^+(H_2)_n$ hydrogen clusters.

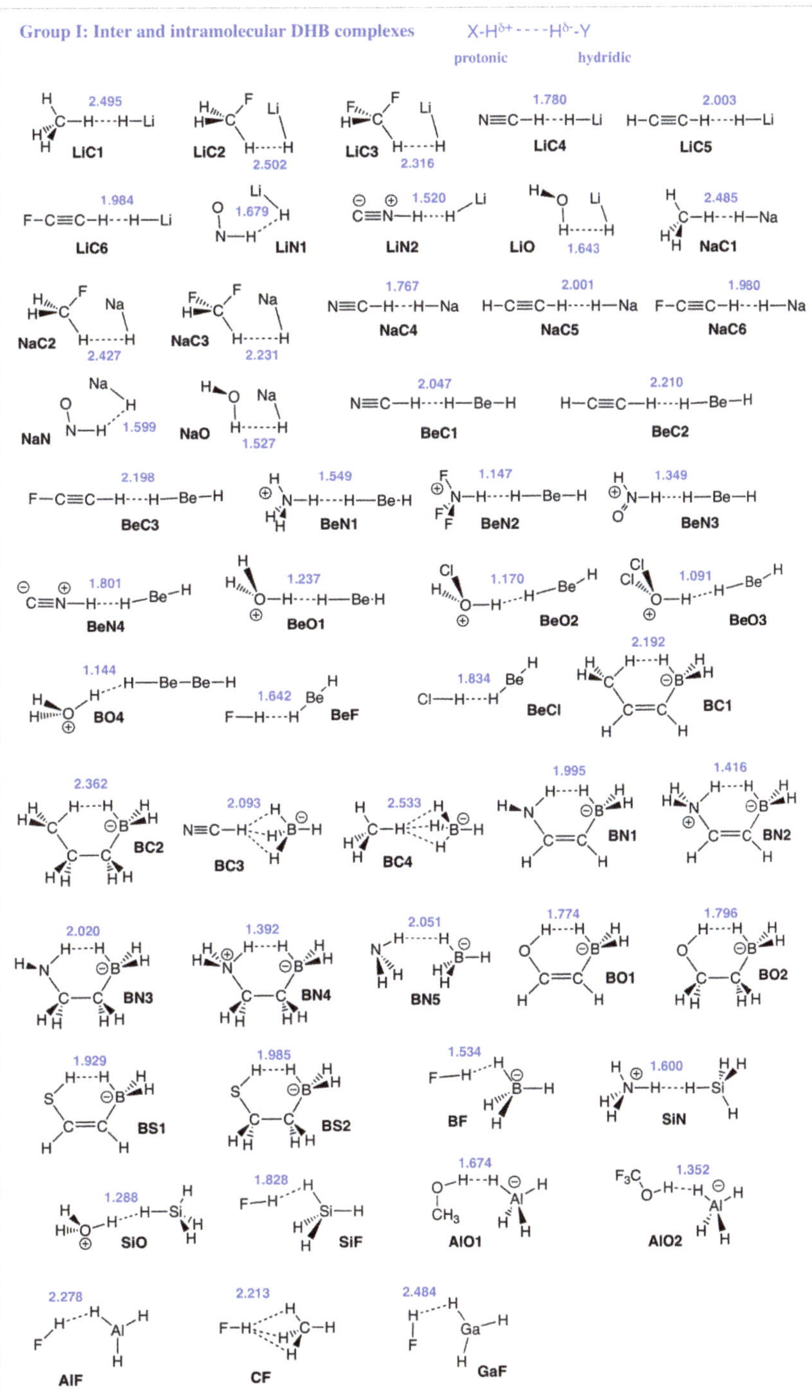

Figure 1. *Cont.*

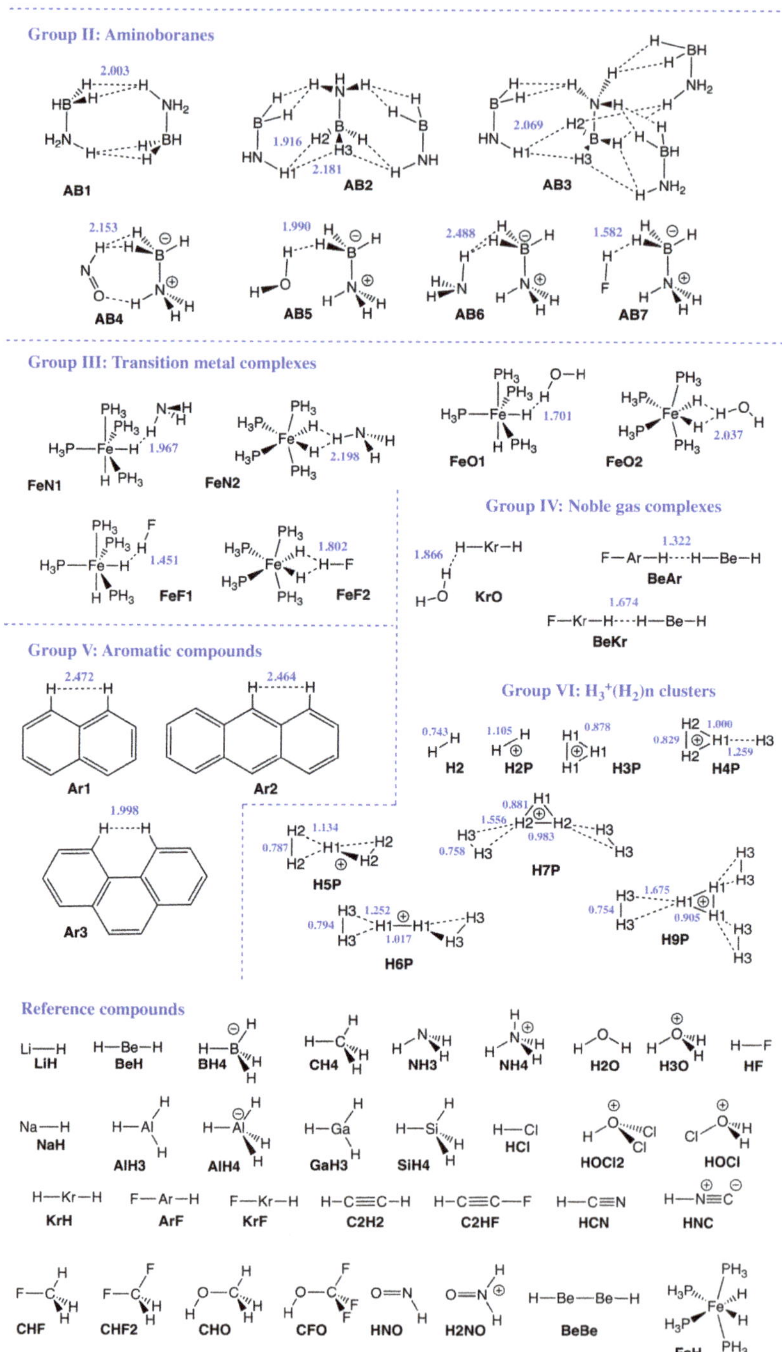

Figure 1. Sketches of Group I–Group VI members and reference compounds investigated in this study. DHB distances (in Å) are given in blue. Short names of the complexes used throughout the manuscript are given in black.

The manuscript is structured in the following way: In the results and discussion part, general findings for Group I–Group V are summarized, followed by a more specific discussion of each individual group and the description of H–H interactions in $H_3^+(H_2)_n$ clusters (Group VI, Figure 1). In the computational methods section, a short summary of LMA is given, and the model chemistry applied and software utilized are described. In the final part, the conclusions and future aspects of this work are presented.

2. Results

2.1. Overall Trends

In the following, general trends found for Group I–Group VI members are discussed referring to the data collected in Figures 2–9 and Tables 1–3. In addition, bond lengths R in Å, local mode force constants k^a in mdyn/Å, local mode vibrational frequencies ω_a in cm^{-1}, binding energies E in kcal/mol, electron densities at the bond critical points ρ_c in e/Å^3 and energy densities at bond critical points H_c in Hartree/Å^3 of protonic and hydridic parts of DHBs and DHBs for all complexes and references compounds are compiled in Table S2 of the Supplementary Materials.

Strength of DHBs: Figure 2a–c shows BSO n values for the DHBs, protonic and hydridic parts of DHBs; in Figure 2d the relationship between protonic and hydridic local mode force constants k^a is shown; and in Figure 2e,f the pairwise correlation between local mode force constants k^a(Protonic), k^a(Hydridic) and k^a(DHB) is shown. BSO n (DHB) values range from 0.14 (**FeF2**) to 0.41 (**BeO4**), which is in a similar range as we previously found for HBs (BSO n = 0.135, N–H$\cdots$F; BSO n = 0.33, N–H$\cdots$N; see Ref. [67]), quantifying that they are both of comparable strength with the former being somewhat stronger. The protonic parts of DHBs ranging from 0.49 for **BeO3** to 0.99 for **CF** (being close to the FH reference value of BSO n = 1) are stronger than their hydridic counter parts with BSO n values of 0.41 for **NaO** and 0.86 for **Ar3**, respectively. One could argue that a weaker protonic part of DHB caused by charge transfer to the hydridic bond should lead to the strengthening of the latter and vice versus, resulting in a correlation between the strengths of protonic and hydridic parts of DHBs. As is obvious from Figure 2d, there is no such correlation between k^a(Protonic) and k^a(Hydridic). On the contrary, according to Figure 2d, there is some trend that weaker protonic parts of DHBs are paired with weaker hydridic bonds, as in the case of **BeO3** and that stronger protonic parts of DHBs are paired with stronger hydridic parts of DHBs, as in the case of **CF**. According to Figure 2e, there is also a tendency that weaker protonic parts of DHBs correspond to stronger DHBs, e.g., as found for **BeO3**, whereas stronger protonic parts of DHBs correspond to weaker DHBs, indicating an larger electron density transfer from a weak protonic part of DHB to DHB, that the donor bond transfers electron density to the DHB. However, as depicted in Figure 2f, this tendency is less pronounced for the hydridic parts of DHB.

Overall, these findings suggest that if one wants to design a DHB with a certain strength, the focus should be on modifying and fine–tuning the protonic rather than the hydridic part of DHB guided by the local mode force constants as a sensitive tool monitoring the changes in the bond strengths.

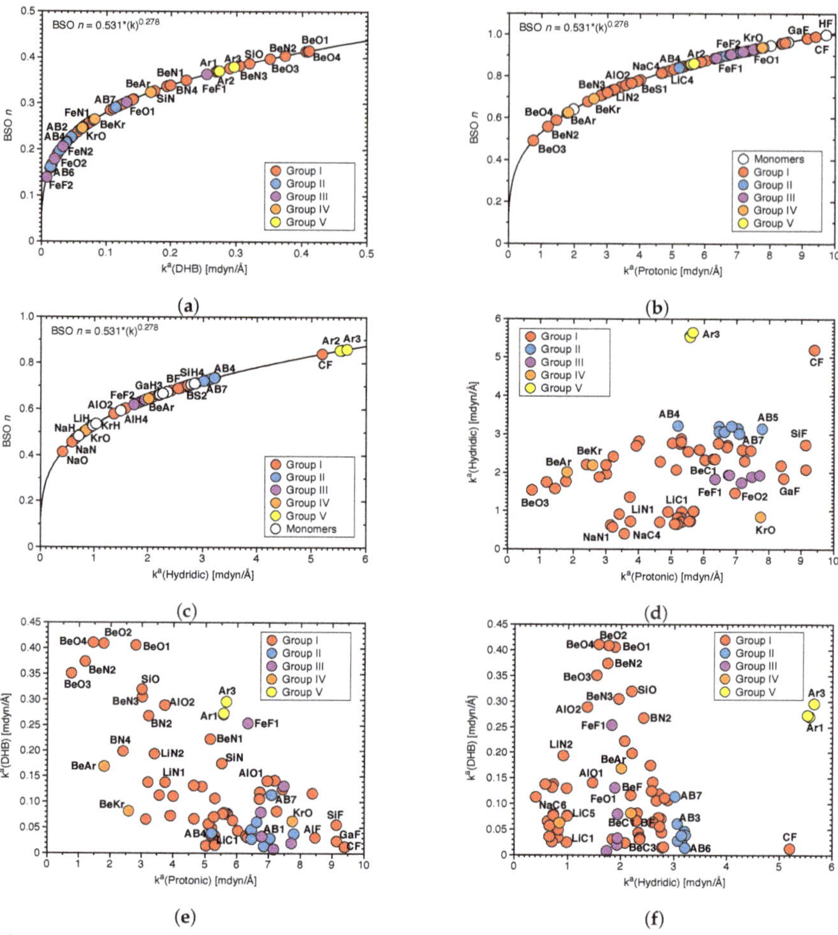

Figure 2. (**a**) BSO n values for DHB, (**b**) BSO n values for protonic parts of DHBs, and (**c**) BSO n values for hydridic parts of DHBs calculated from the corresponding local mode force constants as described in the text. (**d**) Correlation between k^a(Protonic) and k^a(Hydridic), (**e**) between k^a(Protonic) and k^a(DHB), (**f**) between k^a(Hydridic) and k^a(DHB).

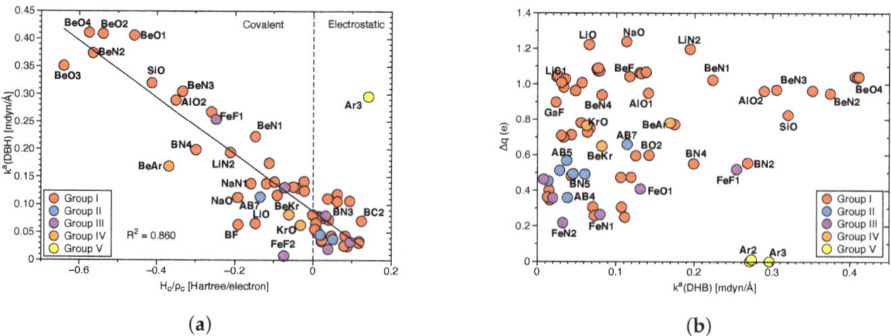

Figure 3. (**a**) Correlation of k^a(DHB) and normalized energy density H_c/ρ_c. (**b**) Correlation of charge differences Δq (see Table 1 for definition) and k^a(DHB).

Covalency, protonic and hydridic charges: The covalency of the DHBs investigated in this work was assessed by the normalized energy density H_c/ρ_c (see Table S2 in the Supplementary Materials). In Figure 3a k^a(DHB) is correlated with H_c/ρ_c which ranges from covalent $H_c/\rho_c = -0.6$ Hartree/electron, to electrostatic $H_c/\rho_c = 0.2$ Hartree/electron in accordance with the range of H_c/ρ_c values -0.6 to 0.3 Hartree/electron for the HBs investigated in our previous work [67], disproving the assumption that DHBs are generally more electrostatic than their HB counterparts. Although there is no strong correlation ($R^2 = 0.860$) between these two quantities, Figure 3a shows that stronger DHBs such as **BeO4** or **BeN2** are of covalent character whereas weak BDHs such as **BN3** or **BC2** are electrostatic in nature. In summary, the Cremer–Kraka criterion of covalent bond offers an efficient tool to assess the covalency of DHB interactions, which as our study shows can be quite different, depending on the electronic structure of the DHB complex and the nature of the protonic and hydridic parts of DHB.

In Tables 1 and 2 protonic and hydridic H–atom NBO charges and their differences for Group I–V complexes are reported, and in Table S1 of Supplementary Materials the NBO hydrogen charges for the reference compounds are shown.

Table 1. Protonic and hydridic NBO charges (in e) for Group I. Protonic hydrogen charge q($H^{\delta+}$); hydridic hydrogen charge q($H^{\delta-}$); charge difference Δq = q($H^{\delta+}$) − q($H^{\delta-}$).

Complex	q($H^{\delta+}$)	q($H^{\delta-}$)	Δq	Complex	q($H^{\delta+}$)	q($H^{\delta-}$)	Δq	Complex	q($H^{\delta+}$)	q($H^{\delta-}$)	Δq
LiC1	0.236	−0.804	1.040	BeN1	0.467	−0.556	1.024	BN5	0.404	−0.078	0.482
LiC2	0.213	−0.769	0.982	BeN2	0.418	−0.530	0.948	BO1	0.494	−0.104	0.598
LiC3	0.189	−0.776	0.964	BeN3	0.417	−0.552	0.969	BO2	0.491	−0.104	0.595
LiC4	0.249	−0.815	1.064	BeN4	0.439	−0.500	0.939	BS1	0.166	−0.083	0.249
LiC5	0.258	−0.816	1.074	BeO1	0.505	−0.537	1.042	BS2	0.166	−0.092	0.258
LiC6	0.269	−0.816	1.086	BeO2	0.521	−0.515	1.036	BF	0.565	−0.162	0.727
LiN1	0.326	−0.740	1.066	BeO3	0.497	−0.467	0.964	SiN	0.462	−0.310	0.772
LiN2	0.428	−0.769	1.197	BeO4	0.558	−0.484	1.042	SiO	0.534	−0.292	0.826
LiO	0.493	−0.729	1.222	BeF	0.558	−0.484	1.042	SiF	0.552	−0.229	0.781
NaC1	0.238	−0.809	1.048	BeCl	0.275	−0.475	0.750	AlO1	0.504	−0.445	0.949
NaC2	0.214	−0.814	1.028	BC1	0.227	−0.079	0.306	AlO2	0.513	−0.449	0.962
NaC3	0.190	−0.818	1.008	BC2	0.218	−0.088	0.306	AlF	0.592	−0.417	1.009
NaC4	0.246	−0.812	1.059	BC3	0.262	−0.140	0.402	GaF	0.575	−0.323	0.898
NaC5	0.258	−0.817	1.075	BC3	0.262	−0.063	0.325	CF	0.548	0.187	0.362
NaC6	0.271	−0.819	1.090	BC4	0.259	−0.087	0.346	CF	0.548	0.210	0.339
NaN1	0.313	−0.756	1.068	BC4	0.259	−0.071	0.330				
NaO	0.480	−0.759	1.239	BN1	0.380	−0.093	0.474				
BeC1	0.229	−0.484	0.713	BN2	0.461	−0.092	0.554				
BeC3	0.244	−0.465	0.709	BN4	0.466	−0.085	0.551				

As obvious from these data, the complexes investigated in this work display a large range of hydrogen charges reflecting the breadth of the different DHB scenarios covered. Protonic H–atom charges range from 0.166 e in **BS1** and **BS2** to 0.592 e in **AlF** (see Table 1) and hydridic H–atom charges from -0.061 e in **AB2** (see Table 2) to -0.819 e in **NaC6** (see Table 1). It is noteworthy that some of the hydridic charges in transition metal and aromatic compounds are positive, which will be further elucidated below. Figure 3b shows the correlation between Δq and the DHB bond strength as reflected by k^a(DHB). As obvious from the large scattering of data points, there is no direct correlation between these two quantities, revealing that differences in the atomic charges is only one of the components determining the strength of these complex interaction. Therefore, Δq, although convenient, cannot serve as a direct DHB strength measure.

Table 2. Protonic and hydridic NBO charges (in e) for Group II–V. Protonic hydrogen charge $q(H^{\delta+})$; hydridic hydrogen charge $q(H^{\delta-})$; charge difference $\Delta q = q(H^{\delta+}) - q(H^{\delta-})$.

Group	Complex	$q(H^{\delta+})$	$q(H^{\delta-})$	Δq	Group	Complex	$q(H^{\delta+})$	$q(H^{\delta-})$	Δq
Aminoboranes	AB1	0.448	−0.065	0.512	Transition	FeN1	0.370	0.106	0.264
		0.448	−0.065	0.512	metals	FeN2	0.350	0.132	0.218
	AB2	0.445	−0.061	0.506		FeO1	0.483	0.074	0.409
		0.445	−0.064	0.509		FeO2	0.490	0.133	0.357
		0.441(H1)	(H2) − 0.052	0.493		FeF1	0.567	0.047	0.519
		0.441(H1)	(H3) − 0.062	0.503		FeF2	0.592	0.130	0.462
	AB3	0.443	−0.058	0.502	Noble gases	KrO	0.464	−0.300	0.765
		0.435(H1)	(H2,H3) − 0.056	0.491		BeAr	0.270	−0.510	0.781
	AB4	0.296	−0.063	0.359		BeKr	0.147	−0.502	0.649
		0.296	−0.063	0.359	Aromatic	Ar1	0.204	0.204	0.000
	AB5	0.494	−0.074	0.568	compounds	Ar2	0.204	0.193	0.011
	AB6	0.388	−0.065	0.453		Ar3	0.202	0.202	0.000
	AB7	0.571	−0.088	0.659					

DHB binding energies, bond lengths and bond angles and local mode force constants: Figure 4 shows how BEs and geometrical features that are frequently used to characterize the strength of a DHB, correlate with the local mode force constants k^a(DHB). Because BE includes geometry relaxation and electron density reorganization effects of the fragments upon DHB dissociation, as discussed above, there is no significant correlation between BE and k^a(DHB) as expected ($R^2 = 0.6186$; see Figure 4a), in particular not for the broad range of complexes investigated in this work, leading to a variety of different fragments formed during the breakage of the DHB and disassociation into fragments. Nevertheless, we observed some trends: e.g., larger BEs are associated with larger k^a(DHB) values, reflecting stronger DHBs for the **BeO** complexes, whereas weak DHBs are characterized by small BE and k^a(DHB) values, such as **LiC1** or **LiC2**.

Figure 4b shows an overall power relationship between DHB lengths R(DHB) and k^a(DHB) following the generalized Badger rule of Cremer and co–workers [68]. Again, no significant correlation ($R^2 = 0.7014$) is observed despite some trends; one of the strongest DHBs found in this study, namely for complexes **BeO1**, **BeO2**, and **BeO3** are characterized by smaller R(DHB) values (1.237, 1.170 and 1.091 Å, respectively), whereas the weakest DHBs found e.g., for **LiC1** and **BC4** have the longer DHBs with values of 2.495 Å and 2.533 Å, respectively. However, there are also outliers, such as **FeF2** which also has one of the weakest DHBs but a DHB length of 1.802 Å. Also obvious is that the Group V members do not follow this generalized Badger type rule caused by the specific topology of the bay H atoms forming these interactions. A more detailed discussion if these interaction can be considered as DHBs, can be found below.

The overall shortest DHB within Group I–IV is R(DHB) 1.091 Å (**BeO3**), which is still 0.348 Å longer than the R(HH) distance of 0.743 Å of the hydrogen molecule **H2**. Therefore, in our set of complexes we did not find any realization of a $X^{\delta+} \cdots H\text{–}H \cdots Y^{\delta-}$ situation, suggested by Bakhmutov [11].

Figure 4c,d check the hypothesis that the angle $X\text{–}H^{\delta+} \cdots H^{\delta-}$ of the protonic fragment is closer to linear, whereas the hydridic fragment $H^{\delta+} \cdots H^{\delta-}\text{–}Y$ is more bent and if there is a correlation between these angles and the DHB bond strength. We do not find any such correlation reflected by the large scattering of data points in Figure 4c,d. Furthermore, we also cannot confirm that the protonic part of the complex is more linear, whereas the hydridic part is more bent.

Local mode versus normal mode frequency red/blue shifts: Frequently normal mode frequency red/blue shifts are utilized to characterize DHB and HB. Although the frequency shifts of the DHB or HB normal mode stretching frequencies can be used to detect these bonds [43,69–71], they are not necessarily suited to assess the bond strength of these interactions as we discussed for HBs [67]. The normal mode frequency shift $\Delta\omega = \omega(complex) - \omega(monomer)$ can only be related to the thermochemical strength of the HB or DHB in question, if the HB or DHB normal stretching modes in both complex and monomer are localized, which is generally not the case [67,72]. Therefore, we draw upon to local mode frequency shifts instead.

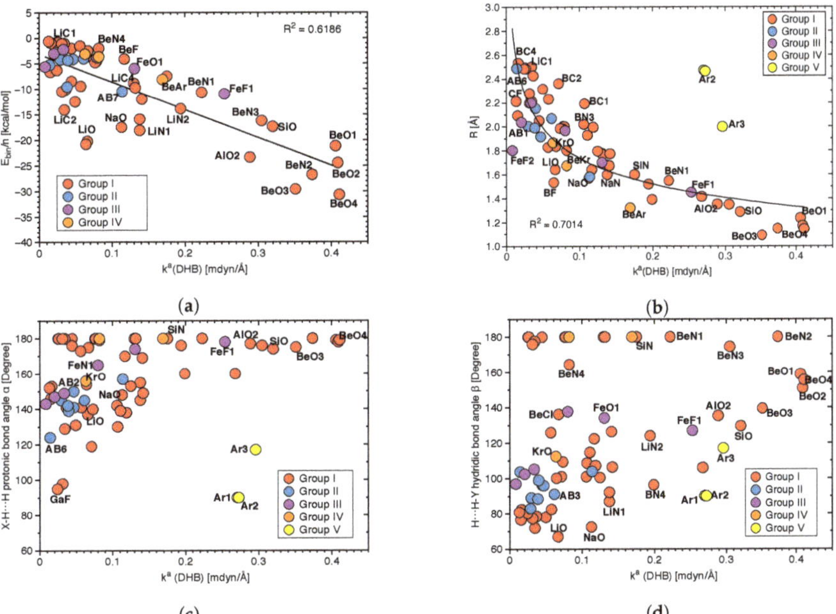

Figure 4. (**a**) Correlation between binding energy E_{bin} scaled by the number of DHBs n and k^a(DHB), (**b**) correlation between distance R(DHB) and k^a(DHB), (**c**) correlation between protonic bond angle (X–H$^{\delta+}\cdots$H$^{\delta-}$) and k^a(DHB), (**d**) correlation between hydridic bond angle (H$^{\delta+}\cdots$H$^{\delta-}$–Y) and k^a(DHB).

Figure 5a depicts the local mode frequency shift $\Delta\omega$ of the protonic parts of DHB and Figure 5b that of the hydridic parts presented as a bar diagram. For both the DHB donor and acceptor, we found almost exclusively red shifts, i.e., the weakening of the bond in question upon complexation; however, the magnitude of the shifts differ significantly; protonic parts of DHBs shifts range from 139 to -2421 cm^{-1}, whereas hydridic parts shifts cover a considerably smaller range, namely from 18 to -406 cm^{-1}. It is interesting to note that largest red shifts reflecting large changes in DHB donor and acceptor upon DHB formation occur for **BeO3** and **BeO4**, i.e., the two complexes with the strongest DHBs.

In the following, we focus on evaluating to what extent the normal vibrational modes are localized in the DHB donor and/or acceptor of these complexes, which forms the necessary prerequisite for a red/blue shift discussion based on normal vibrational modes. As assessment tool we used in this work CMN to probe the local character of both protonic and hydridic normal stretching modes in complex and references molecules. As an example the CMN for the complex **BeO3** and its two references **HOCL2** and **BeH** is shown in Figure 6a,b.

The normal vibrational stretching mode of the DHB protonic part of reference **HOCL2** (O1H4, yellow bar in Figure 6a) with a frequency of 3589 cm^{-1} is 100% localized. In contrast, the hydridic normal vibrational stretching modes of reference **BeH** with frequencies of 2203 and 1995 cm^{-1} are composed of 50% Be2H3 (red color) and 50% Be2H1 (orange color), reflecting the symmetry of the **BeH** reference molecule. Upon complex formation, the protonic part of DHB is considerably weakened, and most importantly, as reflected in Figure 6a, it strongly couples with other modes in the **BeO3** complex. In **BeO3**, the normal mode corresponding to the complex normal vibration of 1386 cm^{-1} has 47% O5H4 (the protonic part of DHB, yellow color) and 39% H3H4 (DHB, purple color); see Figure 6a. The normal vibrational modes corresponding to 2286 cm^{-1} and 2101 cm^{-1} have 14% and 58% B3H3 (the hydridic bond) character. This clearly shows (i) the strong influence of the protonic part of DHB on DHB and (ii) that a red/blue shift discussion based on normal vibrational mode frequencies is rather questionable, and one has to resort to local mode frequencies for a meaningful analysis, which is further elucidated in the following.

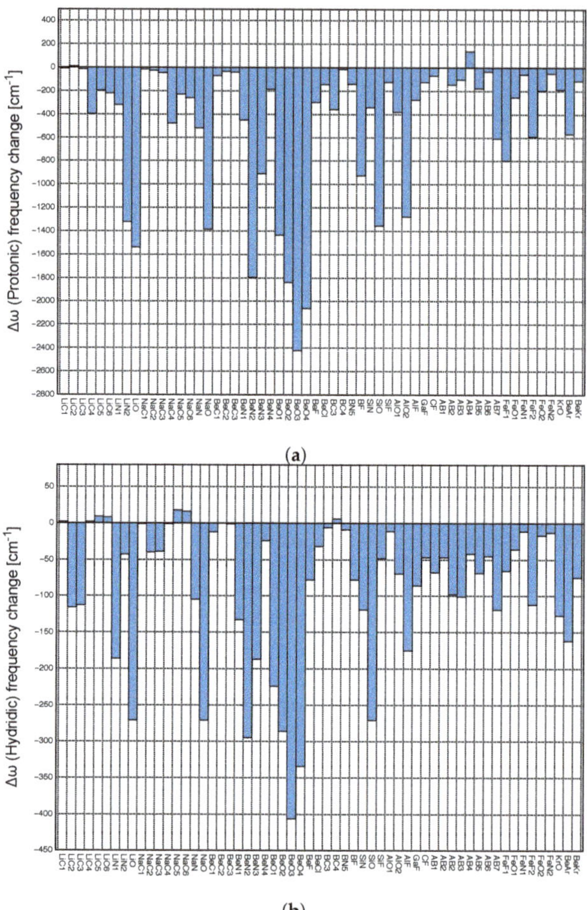

(a)

(b)

Figure 5. (a) Local mode frequency shifts $\Delta\omega$ of the protonic parts of DHBs. (b) Local mode frequency shifts $\Delta\omega$ of the hydridic parts of DHBs, where a positive sign denotes a blue–shift and a negative sign a red–shift.

Figure 7a presents the relationship between protonic and hydridic local mode frequency shifts $\Delta\omega$, and Figure 7b the relationship between the corresponding relative shifts $\Delta\omega/\omega$, involving Group I–Group IV members.

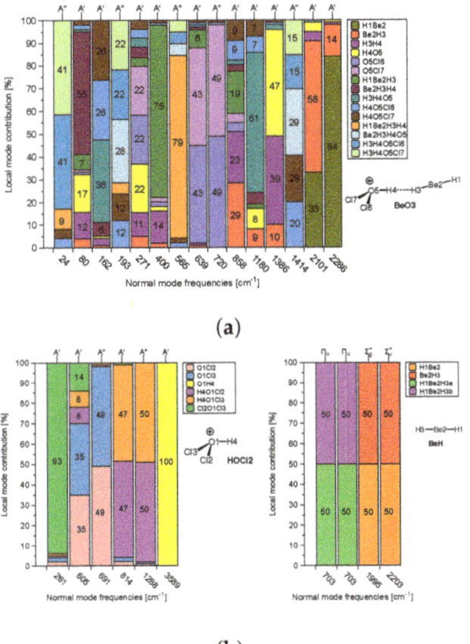

(a)

(b)

Figure 6. (**a**) CNM analysis for the **BeO3** complex. (**b**) CNM analysis for the two reference molecules **HOCL2** (DHB donor) and **BeH** (DHB acceptor) forming the complex.

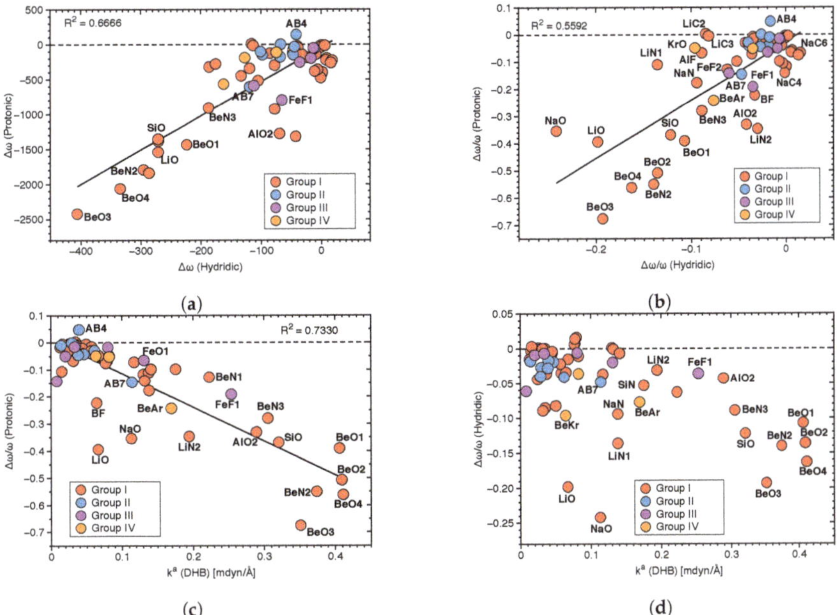

(a)

(b)

(c)

(d)

Figure 7. (**a**) Correlation between local mode frequency shifts $\Delta\omega$(Hydridic) and $\Delta\omega$(Protonic). (**b**) Correlation between the relative frequency shifts $\Delta\omega/\omega$(Hydridic) and $\Delta\omega/\omega$(Protonic). (**c**) Correlation between $\Delta\omega/\omega$(Protonic) and k^a(DHB). (**d**) Correlation between $\Delta\omega/\omega$(Hydridic) and k^a(DHB).

There is no significant correlation between hydridic and protonic frequency shifts (R^2 = 0.6666 and 0.5592, respectively; see Figure 7a,b), although some trends are visible. Overall, larger protonic frequency shifts are in line with larger hydridic frequency shifts, as found for **BeO3**, **BeO4** and on the other end of the spectrum smaller protonic frequency shifts are in line with smaller hydridic frequency shifts, as found for **AB4**. However, there are also a number of outliers, such as **LiO** and **NaO** complexes. Because of the large difference in the frequency shift values for protonic and hydridic parts of DHBs, for the following correlations only the latter were used.

Next, we explored the question if there is a correlation between the frequency shifts of the protonic and hydridic parts of DHB, and the strength of the DHB, as suggested in the literature [43,69,70]. Figure 7c shows the correlation between the relative frequency shifts $\Delta\omega/\omega$(Protonic) and the local mode force constant k^a(DHB), and Figure 7d shows the same situation for $\Delta\omega/\omega$(Hydridic). We found some general trends for the protonic parts of DHBs shifts ($R^2 = 0.7330$). Stronger DHBs are characterized by larger red shifts, as found for **BeO2**, **BeO4** and weaker DBH are characterized by small red shifts or even small blue shifts, such as in the case of **AB4**. Although the same overall trend can be observed for the hydridic frequency shifts scattering of data points, and the outliers **LiO** and **NaO** are more pronounced. This shows that the strength of the DHB cannot be directly deduced from the frequency shifts of the protonic or hydridic parts of DHBs upon complex formation.

Figure 8a shows the correlation between protonic frequency shifts and binding energies and Figure 8b the corresponding correlation for the hydridic frequency shifts. Again, in both cases no significant correlation could be found ($R^2 = 0.7768$ and 0.5933, respectively; see Figure 8a,b), except the general trend that larger binding energies corresponding to stronger DHBs are characterized by larger red shifts, whereas smaller binding energies corresponding to weaker DHBs are characterized by smaller red shifts or even a small blue shift. These findings are consistent with the observed correlation between frequency shifts and local mode DHB force constants. Figure 8c,d show the correlation between frequency shifts and scaled energy density H_c/ρ_c for the protonic and hydridic parts of DHB, respectively. In both cases, there is no correlation between these two quantities in contrast to H_c/ρ_c(DHB), which shows at least a weak correlation with k^a(DHB) (see Figure 3a).

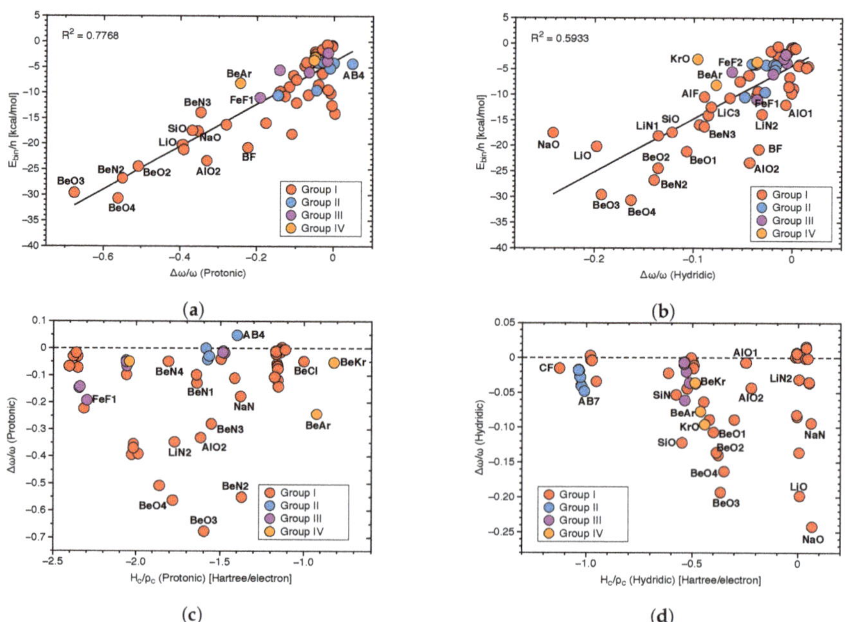

Figure 8. (**a**) Correlation between E_{bin}/n and $\Delta\omega/\omega$(Protonic). (**b**) Correlation between E_{bin}/n and $\Delta\omega/\omega$(Hydridic). (**c**) Correlation between H_c/ρ_c(Protonic) and $\Delta\omega/\omega$(Protonic). (**d**) Correlation of H_c/ρ_c(Hydridic) and $\Delta\omega/\omega$(Hydridic).

2.2. Some Group Specific Highlights

Group I: This group represents DHBs being built from hydrides, such as LiH, NaH, BeH_2, BH_4^-, and AlH_4^- interacting with simple organic molecules—for example, H_2O, HNO, CH_4, CFH_3, and CF_2H_2—and also some cations, such as H_3O^+, NH_4^+, CFH_4^+, taken from the work of Bakhmutov [11]. With 52 members, this is the largest group investigated in this work, with DHB lengths stretching from 1.091 Å to 2.533 Å and BSOs from 0.159 to 0.415. Among them, the strongest H−H interactions were observed for the molecular BeO clusters and the weakest interaction was observed for **CF** sharing three DHBs. It is interesting to note that **BC4** also sharing three DHBs has a longer DHB length of 2.533 Å, compared to 2.484 Å found for the DHBs of **CF**; however, both **CF** and **BC4** DHBs are of comparable strengths (BSO n = 0.159 versus BSO n = 0.168, respectively), another exemption of the generalized Badger rule.

As discussed above, we observed larger changes in the local mode frequencies of the protonic parts of DHB than for the hydridic parts upon DHB complex formation, a trend which also applies to Group I representatives. An interesting example is the **BeO3** complex that has a DHB binding energy of −29.61 kcal/mol. The local mode frequency of the protonic H–O bond has a value of 3584 cm^{-1} in the **HOCL2** monomer, while in the DHB complex this frequency has a value of 1163 cm^{-1}, which corresponds to a large red–shift of −2421 cm^{-1}. On the other hand, in the similar **BeF** DHB complex with a DHB binding energy of −4.08 kcal/mol, the red–shift of the protonic bond H–F is with a value of −298 cm^{-1} almost ten times smaller. This large difference in the red–shifts can be explained by the different electron density transfer from the hydridic Be–H bond into the antibonding σ^* orbital of the protonic H–O and H–F bonds. According to our calculations, the population of the H–O σ^* orbital in the **BeO3** complex has a value of 0.248 e, while the population of the H–F σ^* orbital in the **BeF** has only a value of 0.019 e, which is more than ten times smaller. These results quantify the findings of a previous study, suggesting that the experimentally observed red–shifts in DHBs depend on the amount of electron density transfer into the antibonding σ^* orbitals of the protonic bonds [43]. The larger this transfer is the weaker becomes the protonic part of DHB. This is also reflected in the series of *charge assisted* DHBs [11] **BeO1**, **BeO2**, and **BeO3**. Starting from H_3O^+ successive chlorination, i.e., increasing the electrophilicity of the protonic group leads to a considerable decrease of the DHB length (1.237, 1.170, and 1.091 Å, respectively), however the DHB strength remains fairly unchanged (BSO n values of 0.414, 0.414, and 0.397).

Group II: This group includes dimer (**AB1**), trimer (**AB2**), and tetramer (**AB3**) complexes of ammonia–borane H_3B-NH_3 as well as some reference complexes (**AB4–AB7**). Ammonia borane polymers have been suggested as storage medium for hydrogen [73], therefore we assessed these systems how, starting from **AB1**, the strength of the H−H interaction changes. According to the results of our study the H−H interaction between the monomeric units increases as reflected by the BSO values (0.199, 0.226, and 0.244, for **AB1**, **AB2**, and **AB3**, respectively), whereas the DHB distances remain fairly constant (2.003, 1.916, and 2.069 Å, for **AB1**, **AB2**, and **AB3** respectively). These results suggest an increasing DHB strength trend in larger ammonia borane clusters and ammonia borane polymers. An interesting alternative is **AB7**, namely paring ammonia borane with HF leading to the shortest and strongest DHB in this group (BSO = 0.291 and R(DHB) = 1.582 Å). A follow–up study is planned.

Group III: Group 3 includes transition metal clusters made up of iron–phosphine complexes forming either one or two DHBs with ammonia, water, or HF. DHB length range from 1.451 to 2.198 Å wherein forming two DHB interactions (**FeN2**, **FeO2**, and **FeF2**) leads to longer and weaker DHB interactions than those for the complexes with one DHB (**FeN1**, **FeO1**, and **FeF1**) which parallels our observations for Group I members. Unsurprisingly, HF forms the shortest and strongest DHB interactions in this group (**FeF1**: R(DHB) = 1.451 Å, BSO n = 0.363). As compared to other groups, Group III shows smaller local mode red shifts for both protonic and hydridic parts of DHBs, with the smallest protonic and hydridic part shifts found for **FeN1** (−0.017 and −0.006 cm^{-1}, respectively). Group III members overall show the best correlation between protonic and hydridic frequency shifts in relation to the other groups, likely because the shifts themselves are so small. We also observe a stronger correlation between frequency shifts and BEs than for any of the other groups. However, Group III members are also the most similar compared to the members in other groups, which may cause the better correlation.

The BEs of Group III range from -3.81 to -11.14 kcal/mol for **FeN1** and **FeF2**, respectively, which are among the smallest BEs of the entire set of complexes investigated in this work. Thus, we assert that such transition metal complexes do not tend to form strong DHBs. This difference from other groups can be explained by the presence of d–orbitals, which can withdraw electron density from antibonding σ^* orbitals of the hydridic parts of DHBs, leading to complexes with predominately electrostatic character, as reflected by normalized energy density H_c/ρ_c shown in Figure 3, with the exception of **FeF1** and **FeF2**. This is an interesting aspect to be further explored in the future.

Group IV: This group features four noble gas complexes, whose DHB bond lengths range from 1.322 to 1.866 Å, with BSO n values between 0.246 and 0.324. Although these DHB strengths are on the lower end of the spectrum, they are predominantly covalent in nature, as reflected by the normalized energy density H_c/ρ_c values shown in Figure 3. For Group IV members, we find exclusively local mode red shifts indicating the weakening of both the protonic and hydridic part of DHB upon complexation. **BeAr** shows with -0.243 the most significant relative protonic shift in this group, as well as the largest BSO n value (0.324). This can be explained by the linearity of the DHB as well as the presence of fewer shielding electrons in argon. As with the other groups, there is a weak correlation between the frequency shifts and BE.

Group V: Naphthalene, anthracene and phenanthrene **Ar1**–**Ar3** have been added to the set of DHB complexes in order to compare the H−H interaction in these aromatic systems, where both hydrogen atoms involved are equally charged, with the DHB properties of the other groups investigated in this work. In previous work [66] we could disprove arguments in favor of DHB based on the existence of a bond path and bond critical point between these two H atoms and the stability of anthracene [11,22,65] as dubious [66]. The sole existence of a bond path and a bond critical point does not necessarily imply the existence of a chemical bond or weak chemical interaction, it may be just an artifact of a closer contact between the two atoms in question [66,74–80]. In addition, previous attempts to explain the higher stability of phenanthrene via a maximum electron density path between the bay H atoms are misleading in view of the properties of the electron density distribution in the bay region. The 6.8 kcal/mol larger stability of phenanthrene relative to anthracene predominantly (84%) results from its higher resonance energy, which is a direct consequence of the topology of ring annulation [66]. In summary, these observations clearly disqualify the H−H in these systems as DHBs. With the higher level of theory used in this study, we could no longer find a bond path between the bay H atoms for **Ar1** and **Ar2**. The bond path and bond critical point found for **Ar3** are an artifact of the closed contact (R(HH) = 1.998 Å). The relative large BSO n values of 0.370, 0.370 and 0.379 (see Figure 2a) are also an artifact of the topology of the bay H atoms. In addition, Figure 2d,f as well as Figures 3 and 4b,c clear identify these complexes as outliers.

Group VI: This is a special group devoted to modeling of the strength of DHB in hydrogen clusters, which have been suggested in the interstellar space [81]. Figure 9a–c present results of our investigations for molecular clusters **H4P**, **H5P**, **H6P**, **H7P**, and **H9P** involving diatomic hydrogen **H2**, and triatomic hydrogen cation **H3P**. The BSO data as a function of the local mode force constants k^a of DHBs for these clusters are shown in Figure 9a, whose values are based on the reference molecules presented in the computational details of this study. According to Figure 9a, the strongest DHB (k^a = 5.835 mdyn/Å) is for diatomic hydrogen **H2**, and the weakest (k^a = 0.101 mdyn/Å) is for DHB between the H_3^+ and H_2 units of **H7P**. The strengths of DHB for this Group investigated in our study correlate (R^2 = 0.9218) with the DHB lengths, as shown in Figure 9b. The shortest DHB (R = 0.743 Å) is for **H2**, and the longest (R = 1.675 Å) for DHB between the H_3^+ and H_2 units of **H9P**.

Similarly, according to Figure 9c there is a correlation observed in our study between the strengths of DHB and the covalent characters of DHB expressed by the normalized energy densities H_c/ρ_c, although this correlation is less perfect (R^2 = 0.8981). The strongest DHB of **H2** has the most negative normalized energy density (H_c/ρ_c = −1.125 Hartree/electron), while the weakest DHB between the H_3^+ and H_2 units of **H7P** has one of the less negative normalized energy density (H_c/ρ_c = −0.194 Hartree/electron). Generally, we observed in our investigation of this Group, formation of weak DHBs between the H_3^+ and H_2 units (light blue and orange colors in **H7P** and **H9P** in Figure 9), whose strength is smaller than the strength of DHBs within the H_3^+ unit (dark blue, red, and brown colors in **H7P** and **H9P** in Figure 9), and the strength of DHB in the H_2 unit of these clusters (violet color in **H7P** and **H9P** in Figure 9). The **H6P** cluster represents formation of weak DHBs (yellow color in **H6P** in Figure 9) between the H_2^+ and H_2 units (dark blue color in **H6P** in Figure 9), while the **H5P** cluster shows formation of weak DHBs between the H^+ and H_2 units (red color in **H5P** in Figure 9). However the **H4P** cluster presents formation of weak DHB between the H_3^+ unit and the H atom (orange color in **H4P** in Figure 9). Generally, according to our calculations, the hydrogen clusters are formed between the positively charged unit of the cluster (H^+ in **H5P**, H_2^+ in **H6P**, and H_3^+ in **H7P** and **H9P**) and the H_2 unit. Therefore, we conclude that the results of this study are suggesting possible applications for hydrogen storage devices, where a positively charged central kernel of a molecular complex is surrounded by H_2 molecules, which can easily dissociate.

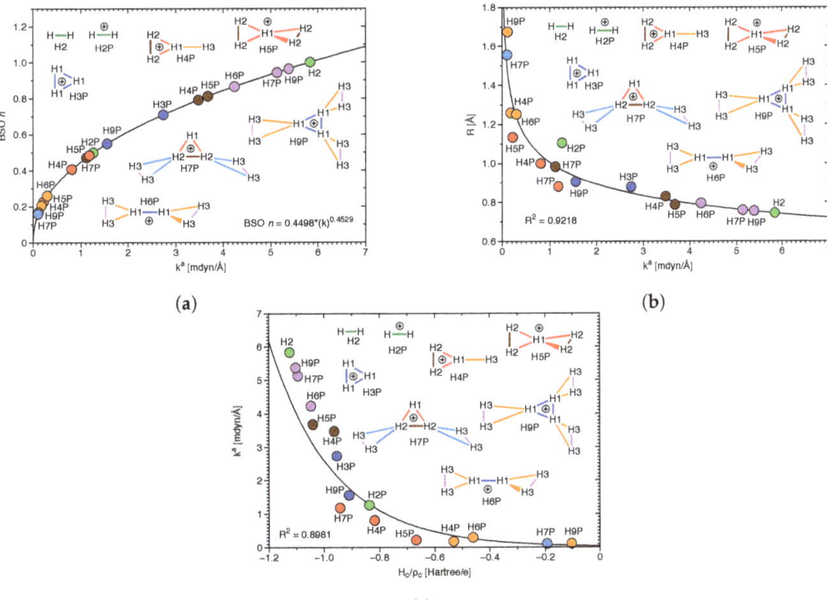

Figure 9. (a) BSO n(HH) as a function of the local mode force constant k^a(HH); power relationship based on H_2 and H_2^+ references see text. (b) Correlation of distance R(HH) and k^a(HH). (c) Correlation of k^a(HH) and normalized energy density H_c/ρ_c.

Table 3. NBO charges (e) for the hydrogen clusters.

	Complex	Atom	q	Bond	Δq^a
Group VI	H2		0.000	H-H	0.000
	H2P		0.500	H-H	0.000
	H3P		0.333	H1-H1	0.000
	H4P	H1	0.161	H1-H2	0.010
		H2	0.171	H2-H2	0.000
		H3	0.496	H3-H1	0.334
	H5P	H1	0.281	H1-H2	0.102
		H2	0.180	H2-H2	0.000
	H6P	H1	0.219	H1-H1	0.000
		H3	0.141	H1-H3	0.078
	H7P	H1	0.209	H1-H2	0.075
		H2	0.284	H2-H2	0.000
		H3	0.056	H2-H3	0.228
				H3-H3	0.000
	H9P	H1	0.262	H1-H3	0.227
		H3	0.035	H3-H3	0.000

3. Methods and Computational Details

As a major assessment tool of H–H bond strength, we utilized LMA in this work. A comprehensive review of the underlying theory and recent applications can be found in Refs. [72,82]. Therefore, in the following, only some essentials are summarized.

Normal vibrational modes are generally delocalized as a result of kinematic and electronic coupling [83–85]. A particular vibrational stretching mode between two atoms of interest can couple with other vibrational modes such as bending or torsion, which inhibits the direct correlation between normal stretching frequency or associated normal mode force constant and bond strength, as well as the comparison between normal stretching vibrational modes in related molecules. As a consequence, the normal stretching force constant cannot be used as a direct bond strength measure. One needs to derive local counterparts that are free from any mode–mode coupling. Konkoli, Cremer and co–workers solved this problem by solving the mass–decoupled analogue of Wilson's equation of vibrational spectroscopy [53–57], leading to local vibrational modes, associated local mode frequencies and local mode force constants.

There is a 1:1 relationship between the normal vibrational modes and a complete non–redundant set of local vibrational modes via an *adiabatic connection scheme* (ACS) [86], which forms the foundation for a new method of analyzing vibrational spectra, the characterization of normal mode (CNM) procedure [55–57,66,72,86–91]. CNM decomposes each normal mode into local mode contributions leading to a wealth of information about structure and bonding. Recently, CNM has been successfully applied in the investigation of pK_a probes [92] and to assess the qualification of *vibrational Stark effect* probes [93].

Zou and Cremer showed that the local stretching force constant k^a reflects the curvature of the PES in the direction of the bond stretching [94]. This important result qualifies the local stretching force constants k^a as unique quantitative measures of the intrinsic strength of chemical bonds and/or weak chemical interactions based on vibrational spectroscopy. Therefore, LMA has been extensively applied for the study of chemical bonding and non–bonded interactions, as documented in Refs. [72,82] and references cited therein, including the characterization of HB ranging from conventional HBs [67], unconventional HBs [95,96], HBs in water and ice [87,97–102], and HBs in biomolecules [103–106] to HBs in catalysis [107,108]—a list which has been extended by DHB in this work. The local vibrational modes theory allows us to analyze a variety of internal coordinates in addition to bond lengths, such as angles between bonds, dihedral bond angles, puckering coordinates, and others. However, this study is focused on the bond lengths leading to the local mode force constants directly related to the strength of DHB.

The widely accepted definition of a DHB includes the following characteristics: DHBs $H^{\delta+} \cdots H^{\delta-}$ are formed from two H–atoms with opposite partial charges and a distance R(DHB) between the two H–atoms that is smaller than the sum of their van der Waals radii (2.4 Å); the angle $X-H^{\delta+} \cdots H^{\delta-}$ of the protonic fragment is closer to linear, whereas the hydridic fragment $H^{\delta+} \cdots H^{\delta-}-Y$ is more bent [11]. For a larger separation, it has been suggested to apply a criterion based on the topological electron density: The density ρ_c at the bond critical point r_c between the two H–atoms should be small, and the Laplacian $\nabla^2(\rho_c)$ should be small and positive. Whereas bond energies are assumed to be similar for both HBs and DHBs, DHBs should have a smaller covalent character, i.e., a more pronounced electrostatic nature [11]. As discussed above, in addition to R(DHB) [1,11], the red/blue shifts of the normal mode frequency associated with the DHB [9,43], as well as components of the binding energies (BE) determined with different energy decomposition methods [40–42] have been applied as a measure of the DHB bond strength.

However a caveat is necessary: both the bond dissociation energies (BDE) and BE [109–111] as well as associated energy decomposition schemes (EDA) [112–114], bond distances [115,116] and normal mode frequency shifts, are not suited to measure direct bond strength. The BDE is a reaction parameter that includes all changes taking place during the dissociation process. Accordingly, it includes any (de)stabilization effects of the fragments to be formed. It reflects the energy needed for bond breaking, but also contains energy contributions due to geometry relaxation and electron density reorganization in the dissociation fragments [67]. Therefore, the BDE is not a suitable measure of the intrinsic strength of a chemical bond and its use may lead to misjudgments, as documented in the literature [117–122]. In addition, EDA schemes are not free from arbitrariness [123]. Additionally, the bond length is not always a qualified bond strength descriptor. Numerous cases have been reported, illustrating that a shorter bond is not always a stronger bond [54,55,57,124,125].

The use of normal mode stretching frequencies and/or frequency shifts as convenient bond strength measures (both are accessible experimentally and computationally) is problematic, because of the coupling of the stretching mode with other vibrational modes, which hampers a direct comparison between bonds in strongly different systems, as pointed out above. The topological analysis of the electron density can be useful to uncover possible attractive contacts between two atoms via the existence of a maximum electron density path (i.e., bond path), with a bond critical point connecting the two nuclei under consideration [126]. However, the sole existence of a bond path and a bond critical point does not necessarily imply the existence of a chemical bond; in particular, the QTAIM description of weak chemical interactions may be problematic [46,66,74–76,78,79]. Using the Laplacian $\nabla^2(\rho_c)$ as a complementary measure is problematic too [127–129], because it does not reflect the complex interplay between kinetic and potential energy, which accompanies the bond forming process [130,131].

Therefore, in this work we utilized LMA to derive a quantitative DHB bond strength measure, in particular local mode force constants and related bond strength orders. It is convenient to base the comparison of the bond strength for a set of molecules on a chemically more prevalent bond strength order (BSO n) rather than on a comparison of local force constant values. Both are connected via a power relationship according to the generalized Badger rule derived by Cremer and co–workers [68]: BSO n = A $(k^a)^B$. Constants A and B can be determined from two reference compounds with known BSO n values and the requirement that, for a zero force constant, the BSO n must be zero. For the protonic parts of DHBs, the hydridic parts of DHBs, and the H–H interaction in DHBs investigated in this study, we use as references the FH bond in the FH molecule with BSO n = 1 and the FH bond in the $[F \cdots H \cdots F]^-$ anion with BSO n = 0.5 [67,99,103,106]. For the ωB97X-D [132,133]/aug-cc-pVTZ [134,135] model chemistry, applied in this study, this led to k^a(FH) = 9.719 mdyn/Å, k^a(F$\cdots$H) = 0.803 mdyn/Å and A = 0.531 and B = 0.278. For the DHBs in hydrogen clusters of Group VI investigated in this work, we used as references

H_2 (BSO n = 1, k^a = 5.840 mdyn/Å) and H_2^+ (BSO n = 0.5, k^a = 1.263 mdyn/Å), leading to $A = 0.450$ and $B = 0.453$.

LMA was complemented with QTAIM [61,62,78,126], where in this work we applied the Cremer–Kraka criterion [127–129] of covalent bonding to assess the covalent/electrostatic character of DHBs. The Cremer–Kraka criterion is composed of two conditions: (i) the necessary condition—existence of a bond path and bond critical point $\mathbf{r_c} = c$ between the two atoms under consideration; (ii) the sufficient condition—the energy density $H(\mathbf{r_c}) = H_c$ is smaller than zero. $H(\mathbf{r})$ is defined as $H(\mathbf{r}) = G(\mathbf{r}) + V(\mathbf{r})$, where $G(\mathbf{r})$ is the kinetic energy density and $V(\mathbf{r})$ is the potential energy density. A negative $V(\mathbf{r})$ corresponds to a stabilizing accumulation of density, whereas the positive $G(\mathbf{r})$ corresponds to the depletion of electron density [128]. As a result, the sign of H_c indicates which term is dominant [129]. If $H_c < 0$, the interaction is considered covalent in nature, whereas $H_c > 0$ is indicative of electrostatic interactions. In addition to the QTAIM analysis, we calculated Natural Bond Orbital (NBO) [58,59] atomic charges and the corresponding charge transfer between the two hydrogen atoms involved in DHB, and we introduced local mode frequency shifts [67] as a more reliable parameter to distinguish different DHB situation than normal mode frequency shifts, which are contaminated by mode–mode coupling.

The geometries of all group members and references compounds (see Figure 1) were optimized using the ωB97X–D density functional [132,133] combined with Dunning's aug-cc–pVTZ basis set [134,135]. The ωB97X–D functional was chosen as it has been proven to reliably describe weak (long–range) intermolecular interactions covering the diverse range of molecules considered in particular in combination with the augmented basis set [133,136]. Using the same model chemistry as applied in our comprehensive study on HBs [137] also allows the direct comparison of HB and DHB features. Furthermore, this model chemistry has been recently assessed and characterized as useful for the description of DHB [39]. Geometry optimizations were followed by a normal mode analysis and LMA. Geometry optimizations, normal mode analyses and NBO analyses were performed with the Gaussian09 program package [138]. For the subsequent LMA the program LModeA [139] was utilized. The QTAIM analysis was performed with the AIMAll package [140].

4. Conclusions

In this computational study, we investigated the strength and nature of dihydrogen bonding (DHB) in a variety of molecular complexes and clusters, categorized into six different groups. As an assessment tool, we used the Local Mode Analysis (LMA) based on DFT calculations involving dispersion corrections and a flexible basis set. LMA, which was used to determine the precise strength of individual DHBs in these complexes and clusters, was supported by the topological analysis of the electron density, quantifying the covalent versus electrostatic character of these DHBs. Along with the analysis of DHB, we also investigated the strength and the nature of the protonic bonds $X–H^{\delta+}$, as well as the hydridic bonds $H^{\delta-}–Y$ forming the DHBs (i.e., $X–H^{\delta+} \cdots H^{\delta-}–Y$) for Group I–V members. According to our results, the strength of DHB is modulated by the strength of the protonic bonds $X–H^{\delta+}$ rather than by the strength of the hydridic bonds $H^{\delta-}–Y$, which was also confirmed for Group VI members involving hydrogen clusters. In accordance with our calculations, the hydrogen clusters were formed by weak DHB between the central positively charged H_3^+ kernel of the cluster and the surrounding H_2 molecules, which can easily dissociate, suggesting a possible application for hydrogen storage.

We hope that our study inspires the community to add the Local Mode Analysis to their repertoire for the investigation of chemical bonds and intermolecular interactions.

Supplementary Materials: The following supporting information can be downloaded at: https://www.mdpi.com/article/10.3390/molecules28010263/s1. Table S1: NBO hydrogen charges for reference compounds; Table S2: With bond lengths R in Å, local mode force constants k^a in mdyn/Å, local mode vibrational frequencies ω_a in cm^{-1}, binding energies E in kcal/mol, electron densities at the bond critical points ρ_c in e/Å^3 and energy densities at critical points H_c in Hartree/Å^3 of protonic and hydridic HBs and DHBs for all complexes and references compounds.

Author Contributions: Conceptualization, E.K. and M.F.; methodology, M.F. and E.K.; validation, E.K. and M.F.; formal analysis, M.F., N.B. and E.K.; investigation, M.F., N.B. and E.K.; resources, E.K.; supervision, E.K.; data curation, M.F. and M.M.; writing—original draft preparation, M.F. and E.K.; writing—review and editing, E.K., M.F. and M.M.; funding acquisition, E.K. All authors have read and agreed to the published version of the manuscript.

Funding: This research was funded by the National Science Foundation NSF, grant CHE 2102461.

Institutional Review Board Statement: Not applicable.

Informed Consent Statement: Not applicable.

Data Availability Statement: All data supporting the results of this work are presented in tables and figure of the manuscript and in the Supplementary Materials.

Acknowledgments: This work was supported by the National Science Foundation, Grant CHE 2102461. We thank the Center for Research Computation at SMU for providing generous high-performance computational resources.

Conflicts of Interest: The authors declare no conflict of interest.

Sample Availability: Not applicable.

References

1. Karas, L.J.; Wu, C.H.; Das, R.; Wu, J.I.C. Hydrogen bond design principles. *Wiley Interdiscip. Rev. Comput. Mol. Sci.* **2020**, *10*, e1477. [CrossRef] [PubMed]
2. Grabowski, S.J. *Understanding Hydrogen Bonds*; Theoretical and Computational Chemistry Series; The Royal Society of Chemistry: London, UK, 2021.
3. Scheiner, S. Understanding noncovalent bonds and their controlling forces. *Int. J. Chem. Phys.* **2020**, *153*, 140901. [CrossRef] [PubMed]
4. Arunan, E. One Hundred Years After The Latimer and Rodebush Paper, Hydrogen Bonding Remains an Elephant! *Indian J. Sci.* **2020**, *100*, 249–255. [CrossRef]
5. Gibb, B.C. The Centenary (maybe) of The Hydrogen Bond. *Nat. Chem.* **2020**, *12*, 665–667. [CrossRef] [PubMed]
6. Mallamace, D.; Fazio, E.; Mallamace, F.; Corsaro, C. The role of hydrogen bonding in the folding/unfolding process of hydrated lysozyme: A review of recent NMR and FTIR results. *Int. J. Mol. Sci.* **2018**, *19*, 3825. [CrossRef]
7. Bulusu, G.; Desiraju, G.R. Strong and Weak Hydrogen Bonds in Protein–Ligand Recognition. *J. Indian Inst. Sci.* **2019**, *100*, 31–41. [CrossRef]
8. Mobika, J.; Rajkumar, M.; Sibi, S.L.; Priya, V.N. Investigation on hydrogen bonds and conformational changes in protein/polysaccharide/ceramic based tri-component system. *Spectrochim. Acta A Mol. Biomol. Spectrosc.* **2021**, *244*, 118836. [CrossRef]
9. Belkova, N.V.; Epstein, L.M.; Filippov, O.A.; Shubina, E.S. Hydrogen and dihydrogen bonds in the reactions of metal hydrides. *Chem. Rev.* **2016**, *116*, 8545–8587. [CrossRef]
10. Gadwal, I. A Brief Overview on Preparation of Self-Healing Polymers and Coatings via Hydrogen Bonding Interactions. *Macromol* **2021**, *1*, 18–36. [CrossRef]
11. Bakhmutov, V.I. *Dihydrogen Bonds*; John Wiley & Sons: Hobroken, NJ, USA, 2008.
12. Crabtree, R.H. Dihydrogen complexation. *Chem. Rev.* **2016**, *116*, 8750–8769. [CrossRef]
13. Kubas, G.J. Dihydrogen complexes as prototypes for the coordination chemistry of saturated molecules. *Proc. Natl. Acad. Sci. USA* **2007**, *104*, 6901–6907. [CrossRef] [PubMed]
14. Grabowski, S.J. Molecular Hydrogen as a Lewis Base in Hydrogen Bonds and Other Interactions. *Molecules* **2021**, *25*, 3294. [CrossRef] [PubMed]
15. Richardson, T.; de Gala, S.; Crabtree, R.H.; Siegbahn, P.E.M. Unconventional hydrogen bonds: Intermolecular B-H$\cdots$H-N interactions. *J. Am. Chem. Soc.* **1995**, *117*, 12875–12876. [CrossRef]
16. Custelcean, R.; Jackson, J.E. Dihydrogen bonding: Structures, energetics, and dynamics. *Chem. Rev.* **2001**, *101*, 1963–1980. [CrossRef]
17. Alkorta, I.; Elguero, J. Non-conventional hydrogen bonds. *Chem. Soc. Rev.* **1998**, *27*, 163–170. [CrossRef]
18. Epstein, L.M.; Shubina, E.S. New types of hydrogen bonding in organometallic chemistry. *Coord. Chem. Rev.* **2002**, *231*, 165–181. [CrossRef]
19. Solimannejad, M.; Amlashi, L.M.; Alkorta, I.; Elguero, J. XeH$_2$ as a proton-accepting molecule for dihydrogen bonded systems: A theoretical study. *Chem. Phys. Lett.* **2006**, *422*, 226–229. [CrossRef]
20. Alkorta, I.; Elguero, J.; Mo, O.; Yanez, M.; Del Bene, J.E. Ab Initio study of the structural, energetic, bonding, and IR spectroscopic properties of complexes with dihydrogen bonds. *J. Phys. Chem. A* **2002**, *106*, 9325–9330. [CrossRef]

21. Alkorta, I.; Zborowski, K.; Elguero, J.; Solimannejad, M. Theoretical study of dihydrogen bonds between $(XH)_2$, X = Li, Na, BeH, and MgH, and weak hydrogen bond donors (HCN, HNC, and HCCH). *J. Phys. Chem. A* **2006**, *110*, 10279–10286. [CrossRef]

22. Grabowski, S.J.; Sokalski, W.A.; Leszczynski, J. Wide spectrum of H$\cdots$H interactions: Van der Waals contacts, dihydrogen bonds and covalency. *Chem. Phys.* **2007**, *337*, 68–76. [CrossRef]

23. Castilla-Martinez, C.A.; Moury, R.; Demirci, U.B. Amidoboranes and hydrazinidoboranes: State of the art, potential for hydrogen storage, and other prospects. *Int. J. Hydrog. Energy* **2020**, *45*, 30731–30755. [CrossRef]

24. Magos-Palasyuk, E.; Litwiniuk, A.; Palasyuk, T. Experimental and theoretical evidence of dihydrogen bonds in lithium amidoborane. *Sci. Rep.* **2020**, *10*, 17431. [CrossRef] [PubMed]

25. Zhou, W.; Jin, S.; Dai, W.; Lyon, J.T.; Lu, C. Theoretical study on the structural evolution and hydrogen storage in NbH_n (n = 2–15) clusters. *Int. J. Hydrog. Energy* **2021**, *46*, 17246–17252. [CrossRef]

26. Zachariasen, W.H.; Mooney, R.C.L. The structure of the hypophosphite group as determined from the crystal lattice of ammonium hypophosphite. *J. Chem. Phys.* **1934**, *2*, 34–38. [CrossRef]

27. Burg, A.B. Enhancement of P-H bonding in a phosphine monoborane. *Inorg. Chem.* **1964**, *3*, 1325–1327.

28. Brown, M.P.; Heseltine, R.W. Co-ordinated BH_3 as a proton acceptor group in hydrogen bonding. *Chem. Comm.* **1968**, 1551–1552. [CrossRef]

29. Brown, M.P.; Heseltine, R.W.; Sutcliffe, L.H. Studies of aminoboranes. Part II. Pyrolysis of monoalkylamine-boranes: Some aminoborane trimers (1,3,5-trialkylcyclotriborazanes) and dimers. *J. Chem. Soc.* **1968**, 612–616. [CrossRef]

30. Brown, M.P.; Heseltine, R.W.; Smith, P.A.; Walker, P.J. An infrared study of co-ordinated BH_3 and BH_2 groups as proton acceptors in hydrogen bonding. *J. Chem. Soc.* **1970**, 410–414. [CrossRef]

31. Lee, J.C., Jr.; Rheingold, A.L.; Muller, B.; Pregosin, P.S.; Crabtree, R.H. Complexation of an amide to iridium via an iminol tautomer and evidence Ir-H$\cdots$H-O hydrogen bond. *J. Chem. Soc.* **1994**, 1021–1022. [CrossRef]

32. Lough, A.J.; Park, S.; Ramachandran, R.; Morris, R.H. Switching on and off a new intramolecular hydrogen-hydrogen interaction and the heterolytic splitting of dihydrogen. Crystal and molecular structure. *J. Am. Chem. Soc.* **1994**, *116*, 8356–8357. [CrossRef]

33. Peris, E.; Lee, J.C.; Rambo, J.R.; Eisenstein, O.; Crabtree, R.H. Factors affecting the strength of N-H$\cdots$H-Ir hydrogen bonds. *J. Am. Chem. Soc.* **1995**, *117*, 3485–3491. [CrossRef]

34. Clot, E.; Eisenstein, O.; Crabtree, R.H. How hydrogen bonding affects ligand binding and fluxionality in transition metal complexes: A DFT study on interligand hydrogen bonds involving HF and H_2O. *New J. Chem.* **2001**, *25*, 66–72. [CrossRef]

35. Groom, C.R.; Bruno, I.J.; Lightfoot, M.P.; Ward, S.C. The Cambridge Structural Database. *Acta Cryst. B* **2016**, *72*, 171–179. [CrossRef] [PubMed]

36. Crabtree, R.H.; Siegbahn, P.E.M.; Eisenstein, O.; Rheingold, A.L.; Koetzle, T.F. A new intermolecular interaction: Unconventional hydrogen bonds with element hydride bonds as proton acceptor. *Acc. Chem. Res.* **1996**, *29*, 348–354. [CrossRef] [PubMed]

37. Jayaraman, A. 100th Anniversary of Macromolecular Science Viewpoint: Modeling and Simulation of Macromolecules with Hydrogen Bonds: Challenges, Successes, and Opportunities. *ACS Macro Lett.* **2020**, *2*, 656–665. [CrossRef] [PubMed]

38. Bauer, C.A. How to Model Inter- and Intramolecular Hydrogen Bond Strengths with Quantum Chemistry. *J. Chem. Inf. Model.* **2019**, *59*, 3735–3743. [CrossRef]

39. Fanfrlík, J.; Pecina, A.; Řezáč, J.; Lepšík, M.; Sárosi, M.B.; Hnyk, D.; Hobza, P. Benchmark data sets of boron cluster dihydrogen bonding for the validation of approximate computational methods. *ChemPhysChem* **2020**, *21*, 2599–2604. [CrossRef]

40. Van der Lubbe, S.C.C.; Guerra, C.F. The Nature of Hydrogen Bonds: A Delineation of the Role of Different Energy Components on Hydrogen Bond Strengths and Lengths. *Chem. Asian J.* **2019**, *14*, 2760–2769.

41. Karimi, S.; Sanchooli, M.; Shoja-Hormozzah, F. Estimation of resonance assisted hydrogen bond (RAHB) energies using properties of ring critical points in some dihydrogen-bonded complexes. *J. Mol. Struct.* **2021**, *1242*, 130710. [CrossRef]

42. Deshmukh, M.M.; Gadre, S.R. Molecular Tailoring Approach for the Estimation of Intramolecular Hydrogen Bond Energy. *Molecules* **2021**, *26*, 2928. [CrossRef]

43. Uchida, M.; Shimizu, T.; Shibutani, R.; Matsumoto, Y.; Ishikawa, H. A comprehensive infrared spectroscopic and theoretical study on phenol- ethyldimethylsilane dihydrogen-bonded clusters in the S_0 and S_1 states. *J. Chem. Phys.* **2020**, *153*, 104305. [CrossRef] [PubMed]

44. Matta, C.F.; Huang, L.; Massa, L. Characterization of a trihydrogen bond on the basis of the topology of the electron density. *J. Phys. Chem. A* **2011**, *115*, 12451–12458. [CrossRef] [PubMed]

45. Nakanishi, W.; Hayashi, S.; Narahara, K. Atoms–in–Molecules Dual Parameter Analysis of Weak to Strong Interactions: Behaviors of Electronic Energy Densities versus Laplacian of Electron Densities at Bond Critical Points. *J. Phys. Chem. A* **2008**, *112*, 13593–13599. [CrossRef] [PubMed]

46. Shahbazian, S. Why Bond Critical Points Are Not 'Bond' Critical Points. *Chem. Eur. J.* **2018**, *24*, 5401–5405. [CrossRef] [PubMed]

47. Mo, Y. Can QTAIM Topological Parameters Be a Measure of Hydrogen Bonding Strength. *J. Phys. Chem. A* **2012**, *116*, 5240–5246. [CrossRef] [PubMed]

48. Fuster, F.; Grabowski, S.J. Intramolecular Hydrogen Bonds: The QTAIM and ELF Characteristics. *J. Phys. Chem. A* **2011**, *115*, 10078–10086. [CrossRef]

49. Alikhani, M.E.; Fuster, F.; Silvi, B. What Can Tell the Topological Analysis of ELF on Hydrogen Bonding? *Struct. Chem.* **2005**, *16*, 204–210. [CrossRef]

50. Shahbazian, S. A Molecular Electrostatic Potential Analysis of Hydrogen, Halogen and Dihydrogen Bonds. *J. Phys. Chem. A* **2014**, *118*, 1697–1705.
51. Murray, J.S.; Politzer, P. Hydrogen Bonding: A Coulombic σ? Hole Interaction. *J. Indian Inst. Sci.* **2020**, *100*, 21–30. [CrossRef]
52. Murray, J.S.; Politzer, P. The Electrostatic Potential: An Overview. *WIREs Comput. Mol. Sci.* **2011**, *1*, 153–163. [CrossRef]
53. Konkoli, Z.; Cremer, D. A New Way of Analyzing Vibrational Spectra. I. Derivation of Adiabatic Internal Modes. *Int. J. Quantum Chem.* **1998**, *67*, 1–9. [CrossRef]
54. Konkoli, Z.; Larsson, J.A.; Cremer, D. A New Way of Analyzing Vibrational Spectra. II. Comparison of Internal Mode Frequencies. *Int. J. Quantum Chem.* **1998**, *67*, 11–27. [CrossRef]
55. Konkoli, Z.; Cremer, D. A New Way of Analyzing Vibrational Spectra. III. Characterization of Normal Vibrational Modes in terms of Internal Vibrational Modes. *Int. J. Quantum Chem.* **1998**, *67*, 29–40. [CrossRef]
56. Konkoli, Z.; Larsson, J.A.; Cremer, D. A New Way of Analyzing Vibrational Spectra. IV. Application and Testing of Adiabatic Modes within the Concept of the Characterization of Normal Modes. *Int. J. Quantum Chem.* **1998**, *67*, 41–55. [CrossRef]
57. Cremer, D.; Larsson, J.A.; Kraka, E. New Developments in the Analysis of Vibrational Spectra on the Use of Adiabatic Internal Vibrational Modes. In *Theoretical and Computational Chemistry*; Parkanyi, C., Ed.; Elsevier: Amsterdam, The Netherlands, 1998; pp. 259–327.
58. Reed, A.; Curtiss, L.; Weinhold, F. Intermolecular Interactions from A Natural Bond Orbital, Donor-Acceptor Viewpoint. *Chem. Rev.* **1988**, *88*, 899–926. [CrossRef]
59. Weinhold, F.; Landis, C.R. *Valency and Bonding: A Natural Bond Orbital Donor-Acceptor Perspective*; Cambridge University Press: Cambridge, UK, 2005.
60. Glendening, E.D.; Badenhoop, J.K.; Reed, A.E.; Carpenter, J.E.; Bohmann, J.A.; Morales, C.M.; Landis, C.R.; Weinhold, F. *NBO6*; Theoretical Chemistry Institute, University of Wisconsin: Madison, WI, USA, 2013.
61. Bader, R. *Atoms in Molecules: A Quantum Theory*; Clarendron Press: Oxford, UK, 1995.
62. Bader, R.F.W. The Quantum Mechanical Basis of Conceptual Chemistry. *Monatsh. Chem.* **2005**, *136*, 819–854. [CrossRef]
63. Popelier, P.L.A. On Quantum Chemical Topology. In *Applications of Topological Methods in Molecular Chemistry*; Springer International Publishing: Berlin/Heidelberg, Germany, 2016; pp. 23–52.
64. Matta, C.F.; Boyd, R.J. An Introduction to the Quantum Theory of Atoms in Molecules. In *The Quantum Theory of Atoms in Molecules*; Wiley-VCH Verlag GmbH & Co. KGaA: Weinheim, Germany, 2007; pp. 1–34.
65. Matta, C.F.; Hernndez-Trujillo, J.; Tang, T.H.; Bader, R.F.W. Hydrogen hydrogen bonding: A stabilizing interaction in molecules and crystals. *Chem. Europ. J.* **2003**, *9*, 1940–1951. [CrossRef] [PubMed]
66. Kalescky, R.; Kraka, E.; Cremer, D. Description of Aromaticity with the Help of Vibrational Spectroscopy: Anthracene and Phenanthrene. *J. Phys. Chem. A* **2013**, *118*, 223–237. [CrossRef]
67. Freindorf, M.; Kraka, E.; Cremer, D. A Comprehensive Analysis of Hydrogen Bond Interactions Based on Local Vibrational Modes. *Int. J. Quantum Chem.* **2012**, *112*, 3174–3187. [CrossRef]
68. Kalescky, R.; Kraka, E.; Cremer, D. Identification of the Strongest Bonds in Chemistry. *J. Phys. Chem. A* **2013**, *117*, 8981–8995. [CrossRef]
69. Grabowski, S. Halogen bond and its counterparts: Bent's rule explains the formation of nonbonding interactions. *J. Phys. Chem. A* **2011**, *115*, 12340. [CrossRef] [PubMed]
70. Zhang, X.; Zeng, Y.; Li, X.; Meng, L.; Zheng, S. Comparison in the complexes of oxygen-containing σ-electron donor with hydrogen halide and dihalogen molecules. *J. Mol. Struct. THEOCHEM* **2010**, *950*, 27–35. [CrossRef]
71. De Oliveira, B.G.; Ramos, M.N. Dihydrogen Bonds and Blue–Shifting Hydrogen Bonds. AH···CF$_3$ and TH$_2$···HCF$_3$ Model Systems with A = Li or Na and T = Be or Mg. *Int. J. Quant. Chem.* **2010**, *110*, 307–316. [CrossRef]
72. Kraka, E.; Zou, W.; Tao, Y. Decoding Chemical Information from Vibrational Spectroscopy Data: Local Vibrational Mode Theory. *WIREs: Comput. Mol. Sci.* **2020**, *10*, 1480. [CrossRef]
73. Li, S.F.; Tang, Z.W.; Tan, Y.B.; Yu, X.B. Polyacrylamide Blending with Ammonia Borane: A Polymer Supported Hydrogen Storage Composite. *J. Phys. Chem. C* **2012**, *116*, 1544–1549. [CrossRef]
74. Jabłoński, M. On the Uselessness of Bond Paths Linking Distant Atoms and on the Violation of the Concept of Privileged Exchange Channels. *ChemistryOpen* **2019**, *8*, 497–507. [CrossRef]
75. Jabłoński, M. Bond paths between distant atoms do not necessarily indicate dominant interactions. *J. Comput. Chem.* **2018**, *39*, 2183–2195. [CrossRef]
76. Wick, C.R.; Clark, T. On bond-critical points in QTAIM and weak interactions. *J. Mol. Model.* **2018**, *24*, 142. [CrossRef]
77. Jabłoński, M. A Critical Overview of Current Theoretical Methods of Estimating the Energy of Intramolecular Interactions. *Molecules* **2020**, *25*, 5512. [CrossRef]
78. Bader, R.F.W. Bond Paths Are Not Chemical Bonds. *J. Phys. Chem. A* **2009**, *113*, 10391–10396. [CrossRef]
79. Stöhr, M.; Voorhis, T.V.; Tkatchenko, A. Theory and practice of modeling van der Waals interactions in electronic–structure calculations. *Chem. Soc. Rev.* **2019**, *48*, 4118–4154. [CrossRef] [PubMed]
80. Hazrah, A.S.; Nanayakkara, S.; Seifert, N.A.; Kraka, E.; Jäger, W. Structural study of 1- and 2-naphthol: New insights into the non-covalent H–H interaction in cis-1-naphthol. *Phys. Chem. Chem. Phys.* **2022**, *24*, 3722–3732. [CrossRef] [PubMed]
81. Barbatti, M.; Nascimento, M.A.C. Does the H$_5^+$ hydrogen cluster exist in dense interstellar clouds? *Int. J. Quantum Chem.* **2012**, *112*, 3169–3173. [CrossRef]

82. Kraka, E.; Quintano, M.; Force, H.W.L.; Antonio, J.J.; Freindorf, M. The Local Vibrational Mode Theory and Its Place in the Vibrational Spectroscopy Arena. *J. Phys. Chem. A* **2022**, *126*, 8781–8798. [CrossRef] [PubMed]

83. Wilson, E.B.; Decius, J.C.; Cross, P.C.M. *Molecular Vibrations. The Theory of Infrared and Raman Vibrational Spectra*; McGraw-Hill: New York, NY, USA, 1955; pp. 59–136.

84. Califano, S. *Vibrational States*; Wiley: London, UK, 1976.

85. Kelley, J.D.; Leventhal, J.J. *Problems in Classical and Quantum Mechanics: Normal Modes and Coordinates*; Springer: Berlin/Heidelberg, Germany, 2017; pp. 95–117.

86. Zou, W.; Kalescky, R.; Kraka, E.; Cremer, D. Relating Normal Vibrational Modes to Local Vibrational Modes with the Help of an Adiabatic Connection Scheme. *J. Chem. Phys.* **2012**, *137*, 084114. [CrossRef]

87. Kalescky, R.; Zou, W.; Kraka, E.; Cremer, D. Local Vibrational Modes of the Water Dimer—Comparison of Theory and Experiment. *Chem. Phys. Lett.* **2012**, *554*, 243–247. [CrossRef]

88. Zou, W.; Kalescky, R.; Kraka, E.; Cremer, D. Relating Normal Vibrational Modes to Local Vibrational Modes: Benzene and Naphthalene. *J. Mol. Model.* **2012**, *19*, 2865–2877. [CrossRef]

89. Freindorf, M.; Kraka, E. Critical Assessment of the FeC and CO Bond strength in Carboxymyoglobin—A QM/MM Local Vibrational Mode Study. *J. Mol. Model.* **2020**, *26*, 281. [CrossRef]

90. Yannacone, S.; Freindorf, M.; Tao, Y.; Zou, W.; Kraka, E. Local Vibrational Mode Analysis of π-Hole Interactions between Aryl Donors and Small Molecule Acceptors. *Crystals* **2020**, *10*, 556. [CrossRef]

91. Yannacone, S.; Sayala, K.D.; Freindorf, M.; Tsarevsky, N.V.; Kraka, E. Vibrational Analysis of Benziodoxoles and Benziodazolotetrazoles. *PhysChem* **2021**, *1*, 45–68. [CrossRef]

92. Quintano, M.; Kraka, E. Theoretical insights into the linear relationship between pK_a values and vibrational frequencies. *Chem. Phys. Lett.* **2022**, *803*, 139746. [CrossRef]

93. Verma, N.; Tao, Y.; Zou, W.; Chen, X.; Chen, X.; Freindorf, M.; Kraka, E. A Critical Evaluation of Vibrational Stark Effect (VSE) Probes with the Local Vibrational Mode Theory. *Sensors* **2020**, *20*, 2358. [CrossRef] [PubMed]

94. Zou, W.; Cremer, D. C$_2$ in a Box: Determining its Intrinsic Bond Strength for the X$^1\Sigma^+_g$ Ground State. *Chem. Eur. J.* **2016**, *22*, 4087–4097. [CrossRef] [PubMed]

95. Cremer, D. Stereochemistry of the Ozonolysis of Alkenes: Ozonide-versus Carbonyl Oxide-Control. *Angew. Chem. Int. Ed.* **1981**, *20*, 888–889. [CrossRef]

96. Zou, W.; Zhang, X.; Dai, H.; Yan, H.; Cremer, D.; Kraka, E. Description of an Unusual Hydrogen Bond between Carborane and a Phenyl Group. *J. Organomet. Chem.* **2018**, *856*, 114–127. [CrossRef]

97. Cremer, D. From Configuration Interaction to Coupled Cluster Theory: The Quadratic Configuration Interaction Approach. *WIREs Comput. Mol. Sci.* **2013**, *3*, 482–503. [CrossRef]

98. Kalescky, R.; Zou, W.; Kraka, E.; Cremer, D. Vibrational Properties of the Isotopomers of the Water Dimer Derived from Experiment and Computations. *Aust. J. Chem.* **2014**, *67*, 426. [CrossRef]

99. Tao, Y.; Zou, W.; Jia, J.; Li, W.; Cremer, D. Different Ways of Hydrogen Bonding in Water - Why Does Warm Water Freeze Faster than Cold Water? *J. Chem. Theory Comput.* **2017**, *13*, 55–76. [CrossRef]

100. Tao, Y.; Zou, W.; Kraka, E. Strengthening of Hydrogen Bonding With the Push-Pull Effect. *Chem. Phys. Lett.* **2017**, *685*, 251–258. [CrossRef]

101. Delgado, A.A.A.; Sethio, D.; Kraka, E. Assessing the Intrinsic Strengths of Ion-Solvent and Solvent-Solvent Interactions for Hydrated Mg^{2+} Clusters. *Inorganics* **2021**, *9*, 31. [CrossRef]

102. Nanayakkara, S.; Tao, Y.; Kraka, E. Capturing Individual Hydrogen Bond Strengths in Ices via Periodic Local Vibrational Mode Theory: Beyond the Lattice Energy Picture. *J. Chem. Theory Comput.* **2022**, *18*, 562–579. [CrossRef] [PubMed]

103. Lyu, S.; Beiranvand, N.; Freindorf, M.; Kraka, E. Interplay of Ring Puckering and Hydrogen Bonding in Deoxyribonucleosides. *J. Phys. Chem. A* **2019**, *123*, 7087–7103. [CrossRef]

104. Yannacone, S.; Sethio, D.; Kraka, E. Quantitative Assessment of Intramolecular Hydrogen Bonds in Neutral Histidine. *Theor. Chem. Acc.* **2020**, *139*, 125. [CrossRef]

105. Verma, N.; Tao, Y.; Kraka, E. Systematic Detection and Characterization of Hydrogen Bonding in Proteins via Local Vibrational Modes. *J. Phys. Chem. B* **2021**, *125*, 2551–2565. [CrossRef]

106. Beiranvand, N.; Freindorf, M.; Kraka, E. Hydrogen Bonding in Natural and Unnatural Base Pairs - Explored with Vibrational Spectroscopy. *Molecules* **2021**, *26*, 2268. [CrossRef]

107. Makoś, M.Z.; Freindorf, M.; Sethio, D.; Kraka, E. New Insights into Fe–H$_2$ and Fe–H$^-$ Bonding of a [NiFe] Hydrogenase Mimic—A Local Vibrational Mode Study. *Theor. Chem. Acc.* **2019**, *138*, 76. [CrossRef]

108. Freindorf, M.; Yannacone, S.; Oliveira, V.; Verma, N.; Kraka, E. Halogen Bonding Involving I$_2$ and d^8 Transition-Metal Pincer Complexes. *Crystals* **2021**, *11*, 373. [CrossRef]

109. Luo, Y.R. *Comprehensive Handbook of Chemical Bond Energies*; Taylor and Francis: Boca Raton, FL, USA, 2007.

110. Moltved, K.A.; Kepp, K.P. Chemical Bond Energies of 3d Transition Metals Studied by Density Functional Theory. *J. Chem. Theory Comput.* **2018**, *14*, 3479–3492. [CrossRef]

111. Morse, M.D. Predissociation measurements of bond dissociation energies. *Acc. Chem. Res.* **2018**, *52*, 119–126. [CrossRef]

112. Stasyuk, O.A.; Sedlak, R.; Guerra, C.F.; Hobza, P. Comparison of the DFT-SAPT and canonical EDA Schemes for the energy decomposition of various types of noncovalent interactions. *J. Chem. Theory Comput.* **2018**, *14*, 3440–3450. [CrossRef]

113. Levine, D.S.; Head-Gordon, M. Energy decomposition analysis of single bonds within Kohn-Sham density functional theory. *Proc. Natl. Acad. Sci. USA* **2017**, *114*, 12649–12656. [CrossRef]
114. Zhao, L.; Hermann, M.; Schwarz, W.H.E.; Frenking, G. The Lewis electron-pair bonding model: Modern energy decomposition analysis. *Nat. Rev. Chem.* **2019**, *3*, 49–62. [CrossRef]
115. Zhao, L.; Pan, S.; Holzmann, N.; Schwerdtfeger, P.; Frenking, G. Chemical Bonding and Bonding Models of Main-Group Compounds. *Chem. Rev.* **2019**, *119*, 8781–8845. [CrossRef] [PubMed]
116. Gavezzotti, A. Comparing the strength of covalent bonds, intermolecular hydrogen bonds and other intermolecular interactions for organic molecules: X-ray diffraction data and quantum chemical calculations. *New J. Chem.* **2016**, *40*, 6848–6853. [CrossRef]
117. Cremer, D.; Kraka, E. From Molecular Vibrations to Bonding, Chemical Reactions, and Reaction Mechanism. *Curr. Org. Chem.* **2010**, *14*, 1524–1560. [CrossRef]
118. Kalescky, R.; Kraka, E.; Cremer, D. Are Carbon-Halogen Double and Triple Bonds Possible? *Int. J. Quantum Chem.* **2014**, *114*, 1060–1072. [CrossRef]
119. Kalescky, R.; Zou, W.; Kraka, E.; Cremer, D. Quantitative Assessment of the Multiplicity of Carbon-Halogen Bonds: Carbenium and Halonium Ions with F, Cl, Br, and I. *J. Phys. Chem. A* **2014**, *118*, 1948–1963. [CrossRef]
120. Oliveira, V.; Kraka, E.; Cremer, D. Quantitative Assessment of Halogen Bonding Utilizing Vibrational Spectroscopy. *Inorg. Chem.* **2016**, *56*, 488–502. [CrossRef]
121. Setiawan, D.; Sethio, D.; Cremer, D.; Kraka, E. From Strong to Weak NF Bonds: On the Design of a New Class of Fluorinating Agents. *Phys. Chem. Chem. Phys.* **2018**, *20*, 23913–23927. [CrossRef]
122. Sethio, D.; Oliveira, V.; Kraka, E. Quantitative Assessment of Tetrel Bonding Utilizing Vibrational Spectroscopy. *Molecules* **2018**, *23*, 2763. [CrossRef]
123. Andrés, J.; Ayers, P.W.; Boto, R.A.; Carbó-Dorca, R.; Chermette, H.; Cioslowski, J.; Contreras-García, J.; Cooper, D.L.; Frenking, G.; Gatti, C.; et al. Nine questions on energy decomposition analysis. *J. Comput. Chem.* **2019**, *40*, 2248–2283. [CrossRef] [PubMed]
124. Kraka, E.; Cremer, D. Characterization of CF Bonds with Multiple-Bond Character: Bond Lengths, Stretching Force Constants, and Bond Dissociation Energies. *ChemPhysChem* **2009**, *10*, 686–698. [CrossRef] [PubMed]
125. Kaupp, M.; Danovich, D.; Shaik, S. Chemistry is about energy and its changes: A critique of bond-length/bond-strength correlations. *Coord. Chem. Rev.* **2017**, *344*, 355–362. [CrossRef]
126. Bader, R.F.W. A Bond Path: A Universal Indicator of Bonded Interactions. *J. Phys. Chem. A* **1998**, *102*, 7314–7323. [CrossRef]
127. Cremer, D.; Kraka, E. Chemical Bonds without Bonding Electron Density? Does the Difference Electron-Density Analysis Suffice for a Description of the Chemical Bond? *Angew. Chem. Int. Ed.* **1984**, *23*, 627–628. [CrossRef]
128. Cremer, D.; Kraka, E. A Description of the Chemical Bond in Terms of Local Properties of Electron Density and Energy. *Croat. Chem. Acta* **1984**, *57*, 1259–1281.
129. Kraka, E.; Cremer, D. Chemical Implication of Local Features of the Electron Density Distribution. In *Theoretical Models of Chemical Bonding. The Concept of the Chemical Bond*; Maksic, Z.B., Ed.; Springer Verlag: Berlin/Heidelberg, Germany, 1990; Volume 2, pp. 453–542.
130. Ruedenberg, K. The physical nature of the chemical bond. *Rev. Mod. Phys.* **1962**, *44*, 326–376. [CrossRef]
131. Levine, D.; Head-Gordon, M. Clarifying the quantum mechanical origin of the covalent chemical bond. *Nat. Commun.* **2020**, *11*, 4893. [CrossRef]
132. Chai, J.D.; Head-Gordon, M. Long-Range Corrected Hybrid Density Functionals with Damped Atom–Atom Dispersion Corrections. *Phys. Chem. Chem. Phys.* **2008**, *10*, 6615–6620. [CrossRef]
133. Chai, J.D.; Head-Gordon, M. Systematic Optimization of Long-Range Corrected Hybrid Density Functionals. *J. Chem. Phys.* **2008**, *128*, 084106. [CrossRef]
134. Kendall, R.A.; Dunning, T.H., Jr.; Harrison, R.J. Electron Affinities of the First-Row Atoms Revisited. Systematic Basis Sets and Wave Functions. *J. Chem. Phys.* **1992**, *96*, 6796–6806. [CrossRef]
135. Kendall, R.A.; Dunning, T.H., Jr.; Harrison, R.J. Gaussian basis sets for use in correlated molecular calculations. IX. The atoms gallium through krypton. *J. Chem. Phys.* **1999**, *110*, 7667–7676.
136. Sato, T.; Nakai, H. Density Functional Method Including Weak Interactions: Dispersion Coefficients Based On the Local Response Approximation. *J. Chem. Phys.* **2009**, *131*, 224104. [CrossRef] [PubMed]
137. López, C.S.; Faza, O.N.; Freindorf, M.; Kraka, E.; Cremer, D. Solving the Pericyclic-Pseudo pericyclic Puzzle in the Ring-Closure Reactions of 1,2,4,6-Heptatetraene Derivatives. *J. Org. Chem.* **2015**, *81*, 404–414. [CrossRef]
138. Frisch, M.J.; Trucks, G.W.; Schlegel, H.B.; Scuseria, G.E.; Robb, M.A.; Cheeseman, J.R.; Scalmani, G.; Barone, V.; Petersson, G.A.; Nakatsuji, H.; et al. *Gaussian 16*; Gaussian Inc.: Wallingford, CT, USA, 2016.
139. Zou, W.; Tao, Y.; Freindorf, M.; Makoś, M.Z.; Verma, N.; Cremer, D.; Kraka, E. *Local Vibrational Mode Analysis (LModeA*; Computational and Theoretical Chemistry Group (CATCO), Southern Methodist University: Dallas, TX, USA, 2022.
140. Keith, T.A. *AIMAll Version 17.11.14*; K Gristmill Software: Overland Park, KS, USA, 2017.

Article

Thermodynamics and Spectroscopy of Halogen- and Hydrogen-Bonded Complexes of Haloforms with Aromatic and Aliphatic Amines

Emmanuel Adeniyi [1], Olivia Grounds [1], Zachary Stephens [1], Matthias Zeller [2] and Sergiy V. Rosokha [1,*]

[1] Department of Chemistry, Ball State University, Muncie, IN 47306, USA
[2] Department of Chemistry, Purdue University, West Lafayette, IN 47907, USA
*   Correspondence: svrosokha@bsu.edu

Abstract: Similarities and differences of halogen and hydrogen bonding were explored via UV–Vis and ^{1}H NMR measurements, X-ray crystallography and computational analysis of the associations of CHX_3 (X=I, Br, Cl) with aromatic (tetramethyl-p-phenylenediamine) and aliphatic (4-diazabicyclo[2,2,2]octane) amines. When the polarization of haloforms was taken into account, the strengths of these complexes followed the same correlation with the electrostatic potentials on the surfaces of the interacting atoms. However, their spectral properties were quite distinct. While the halogen-bonded complexes showed new intense absorption bands in the UV–Vis spectra, the absorptions of their hydrogen-bonded analogues were close to the superposition of the absorption of reactants. Additionally, halogen bonding led to a shift in the NMR signal of haloform protons to lower ppm values compared with the individual haloforms, whereas hydrogen bonding of CHX_3 with aliphatic amines resulted in a shift in the opposite direction. The effects of hydrogen bonding with aromatic amines on the NMR spectra of haloforms were ambivalent. Titration of all CHX_3 with these nucleophiles produced consistent shifts in their protons' signals to lower ppm values, whereas calculations of these pairs produced multiple hydrogen-bonded minima with similar structures and energies, but opposite directions of the NMR signals' shifts. Experimental and computational data were used for the evaluation of formation constants of some halogen- and hydrogen-bonded complexes between haloforms and amines co-existing in solutions.

Keywords: halogen bonding; hydrogen bonding; haloforms; Amines; NMR spectroscopy; UV–Vis spectroscopy; DFT calculations; X-ray crystallography

Citation: Adeniyi, E.; Grounds, O.; Stephens, Z.; Zeller, M.; Rosokha, S.V. Thermodynamics and Spectroscopy of Halogen- and Hydrogen-Bonded Complexes of Haloforms with Aromatic and Aliphatic Amines. *Molecules* **2022**, *27*, 6124. https://doi.org/10.3390/molecules27186124

Academic Editors: Qingzhong Li, Steve Scheiner and Zhiwu Yu

Received: 27 August 2022
Accepted: 14 September 2022
Published: 19 September 2022

Publisher's Note: MDPI stays neutral with regard to jurisdictional claims in published maps and institutional affiliations.

1. Introduction

While hydrogen bonding (HB) has been extensively studied for more than a century [1,2], similar interactions involving other atoms have captivated the attention of the chemical community only during the last two decades [3–5]. Halogen bonding (XB) is a prominent example of these newly recognized interactions. It has already become a powerful tool for molecular recognition, crystal engineering, catalysis and many other applications [6–8]. The ubiquity of molecules containing both hydrogen and halogen substituents and the similarity in the factors determining XB and HB strength implies that many of them might be involved in both types of interactions. In fact, many X-ray structural studies revealed a co-existence of XB and HB bonds, or a dominance of one of them in co-crystals of such molecules with various nucleophiles [9–13]. However, the experimental characterization of these competing or complementary intermolecular interactions in solutions represents a challenging task which requires an accurate knowledge of the distinctions between XB and HB complexes formed by the same molecule with the same nucleophile.

The haloforms, CHX_3 (X=I, Br or Cl) are characterized by the areas of positive potentials (σ-holes) along the extensions of the C–H and C–X bonds (Figure 1) [14]. As such,

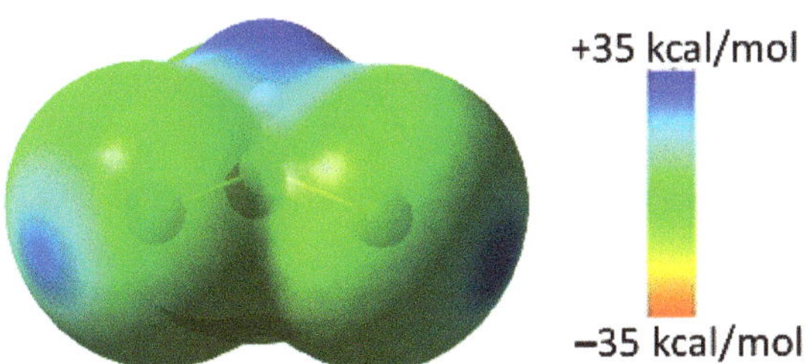

Figure 1. Surface electrostatic potential of iodoform (at 0.001 a.u. electron density) showing areas of positive potentials (σ-holes) along the extensions of C–H and C–I bonds.

These simple molecules can form both HB and XB complexes via the attraction of nucleophiles (Lewis base) B to their halogen or hydrogen substituents (Equations (1) and (2)):

$$CHX_3 + B \overset{K_{XB}}{\rightleftharpoons} [X_2HC\text{-}X \cdots B] \tag{1}$$

$$CHX_3 + B \overset{K_{HB}}{\rightleftharpoons} [X_3C\text{-}H \cdots B] \tag{2}$$

where K_{XB} and K_{HB} are formation constants of the corresponding complex. Most frequently, such complexes were studied using NMR spectroscopy [15–18]. The earlier NMR studies of the complexes of halides with haloforms by Green and Martin showed that CHI_3 forms predominantly XB complexes in which the signal of the proton is shifted to lower ppm values [17]. The HB associations (in which NMR signals of the haloforms' protons are shifted to higher values) prevail in the solutions of halides with $CHBr_3$ or $CHCl_3$. Similar conclusions were obtained by Bertrán and Rodrígues based on NMR studies of the interaction of haloforms with aza-containing solvents [18]. In the recent study by Schulz et al., the relative strengths of the XB/HB interactions of haloimidazolium derivatives were measured experimentally, and the quantitative comparison of the interaction energies and free energies of different association modes were derived from quantum mechanical calculations and molecular dynamics simulations [19].

UV–Vis spectroscopy represents another method with a high potential for the differentiation of HB and XB interactions [14]. Our recent study showed that a combination of NMR and UV–Vis measurements together with computational analysis allows quantitative characterization of the concurrent XB and HB complexes between haloforms and (pseudo-)halide anions in solutions [14]. We have also demonstrated that the common anesthetic, halothane, acts as a XB and HB donor in solutions, which gives an atomic rationale for its eudismic ratio [20]. To establish generalities and limitations of the application of the NMR and UV–Vis spectroscopies for the identification and characterization of the co-existing XB and HB complexes, we turned in the current work to complex formation of haloforms with aromatic and aliphatic amines (which represent an important class of XB and HB acceptors). While the crystallographic literature contains a number of X-ray structures of either XB and HB complexes with these neutral nucleophiles [21–25], and these interactions were compared via computational analysis [26–29], the efforts of differentiating and characterizing the concurrent formation of the corresponding XB and HB complexes in solutions are lacking. As such, we carried out liquid-phase measurements and computational analysis of the interaction of haloforms with N,N,N',N'-tetramethyl-p-phenylenediamine (TMPD) and 4-diazabicyclo[2,2,2]octane (DABCO) (Scheme 1).

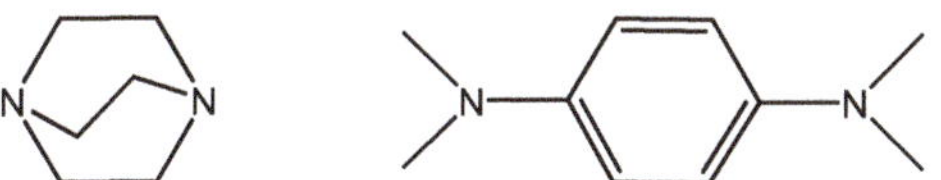

Scheme 1. Structures of DABCO (**left**) and TMPD (**right**).

The general character of the spectral features observed with these nucleophiles were verified via measurements of interactions of haloforms with other aliphatic and aromatic amines (trimethylamine and various *N,N*-dimethylanilines). Such a study facilitates the identification and quantitative characterization of the co-existing XB and HB complexes involving the same pair of reactants in chemical and biochemical systems, and clarification of the factors which determine the strengths and preferences of one or another mode of interaction.

2. Results and Discussion

2.1. UV–Vis Study of Interaction of CHX_3 with Amines

Although the UV–Vis spectrum of DABCO is transparent at $\lambda > 300$ nm, an addition of this amine to an acetonitrile solution containing iodoform resulted in an increase in absorption in the 300–400 nm range. The subtraction of the absorption of each component showed that this increase is related to the appearance of a pair of close absorption bands (Figure 2). In solutions with a constant concentration of iodoform, the intensity of these bands increased with the increase in concentration of DABCO. The addition of DABCO to bromoform also resulted in the appearance of a new absorption band in the UV–Vis spectra, as described earlier [24]. This new band is substantially blue-shifted compared with that observed with CHI_3, and it was partially overshadowed by the absorption of DABCO itself. In comparison, solutions of DABCO and $CHCl_3$ did not show any new absorption beyond 280 nm (spectra of the solutions containing $CHCl_3$ and DABCO are the same as the spectra of individual DABCO), and the strong absorption of DABCO hinders measurements below this wavelength.

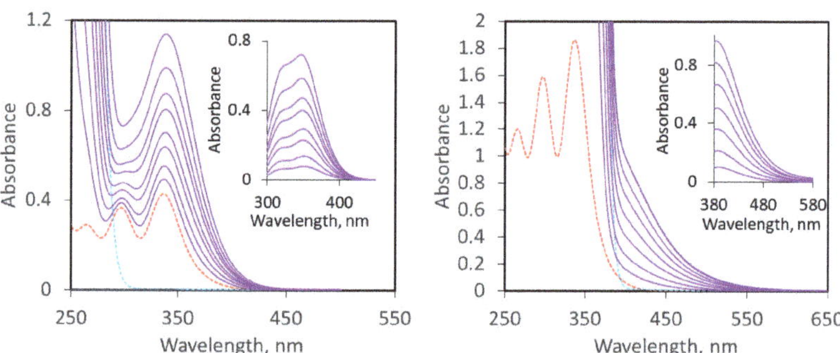

Figure 2. Spectra of solutions with a constant concentration of CHI_3 and various concentrations of DABCO (**left**) and TMPD (**right**). Spectrum of the solutions of individual reactants are shown as dashed red (CHI_3) or blue (DABCO and TMPD) lines. Inserts: Spectra of the complexes obtained by subtraction of the absorption of components from the spectra of their mixtures.

Similar UV–Vis measurements of interactions of haloforms with TMPD were hindered by the strong absorption of TMPD in the 200–380 nm range. As such, only part of the new absorption of the complex of TMPD with CHI_3 could be observed at $\lambda > 380$ nm, and the corresponding bands of the complexes with $CHBr_3$ or with $CHCl_3$ are apparently overshadowed by the absorption of TMPD. As such, the spectra of the solutions containing either $CHBr_3$ or $CHCl_3$ and TMPD are the same as the spectra of individual TMPD. Appearances of new absorptions were also observed upon the addition of the other *N,N*-dimethylanilines or trimethylamine to CHI_3 (Figures S1–S4 in the Supplementary Materials). A Benesi–Hildebrand treatment [30] of the variations of the absorption intensities with concentrations of amines produced straight lines with $R^2 > 0.99$, and the data were also well fit by 1:1 binding isotherms (Figures S5 and S6 in the Supplementary Materials). However, such treatments did not take into account the presence of the two equilibria (Equations (1) and (2)). To further elucidate competitions of XB and HB, we carried out NMR measurements of the analogous solutions.

2.2. ^1H NMR Study of the Interaction of CHX_3 with Amines

The addition of DABCO to the solution of CHI_3 in deuterated acetonitrile resulted in the shift in the signal of the haloform's proton to lower ppm values (Figure 3), indicating an increased shielding of this proton. A similar addition of DABCO to bromoform or chloroform produced a shift in its proton signal to higher ppm values. These results are consistent with earlier studies of the interaction of haloform with halide anions [17], and they suggest a prevalence of XB complexes in solutions of iodoforms with aliphatic amines, and a domination of HB in similar solutions with bromoform or chloroform. The opposite shifts in the proton signals in the XB and HB complexes with DABCO are apparently related to the polarization of haloform by this electron-rich nucleophile. In XB complexes, it results in the shift in electron density from the bonded halogen to the unbonded halogen and hydrogen atoms, increasing the shielding of the latter. In contrast, the polarization of HB complex results in a shift in the electron density from the bonded proton, decreasing its shielding. However, an addition of TMPD to any of the haloforms under study produced a shift to lower ppm values, indicating an increased shielding of this proton (Figure 3). NMR measurements of interactions of haloforms with the other aliphatic and aromatic amines confirm the trends observed with DABCO and TMPD. Specifically, the addition of trimethylamine to CHI_3 resulted in the shift in the proton signal lower ppm and similar experiments with $CHBr_3$ or $CHCl_3$ resulted in a shift in the opposite direction. On the other hand, the titrations of any of the haloforms with aromatic amines led to shifts in the proton signals to lower ppm values (Figures S7–S9 in the Supplementary Materials). To clarify the results of the UV–Vis and NMR measurements, we turned to the X-ray structural analysis of the associations and computational analysis of the XB and HB complexes of haloforms with amines.

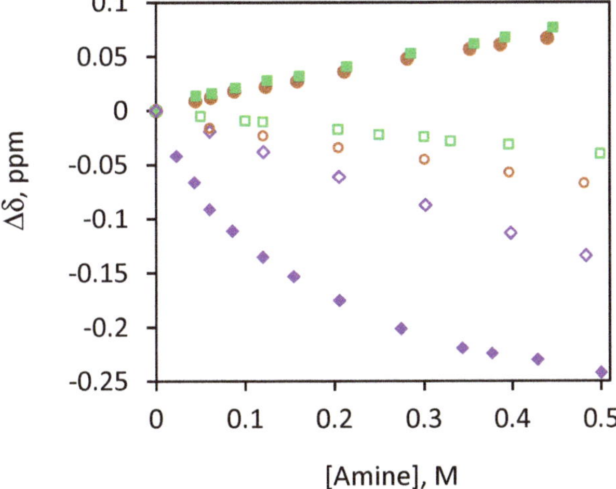

Figure 3. Dependencies of the chemical shifts in the proton of CHI_3 (◇), $CHBr_3$ (o) and $CHCl_3$ (□) (as compared to that in the corresponding isolated molecules) on the concentration of DABCO (filled symbols) or TMPD (open symbols) (in CD_3CN, 22 °C).

2.3. X-ray Structural Analysis of Co-crystals of Iodoform with TMPD or DABCO

Cooling down acetonitrile solutions containing equimolar quantities of iodoform and either TMPD or DABCO led to formation of co-crystals suitable for X-ray structural measurements. X-ray analysis showed that these co-crystals comprise zigzag chains consisting of alternating iodoform and either TMPD or DABCO molecules (Figure 4A,C). Co-crystallization of CHI_3 with TMPD also produced discrete 2:1 complexes (Figure 4B).

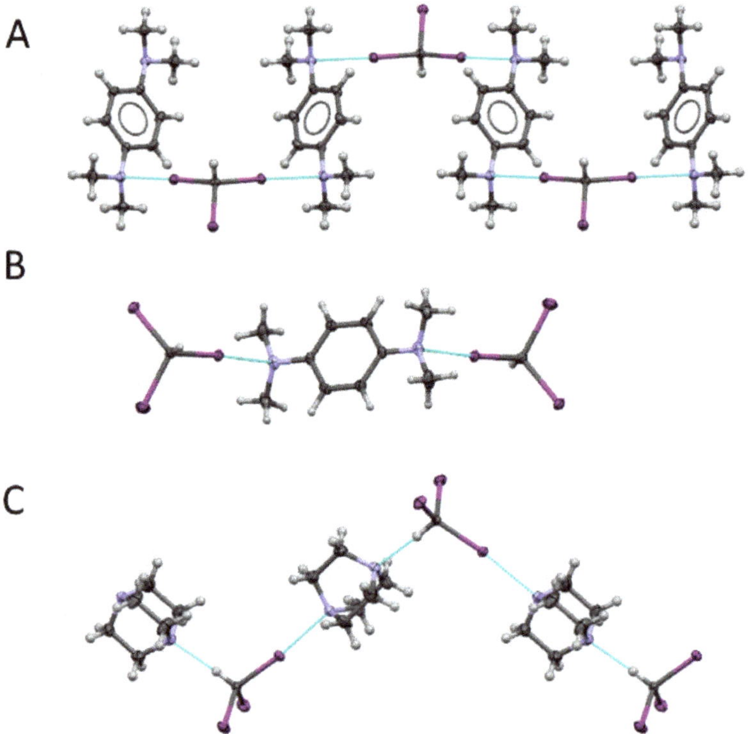

Figure 4. X-ray structures of co-crystals of iodoform with TMPD ((**A**): 1:1, (**B**): 2:1) and DABCO (**C**) showing alternating halogen (and hydrogen)-bonded zigzag chains or discrete 1:2 complexes.

Chains of TMPD with iodoform in their 1:1 co-crystals were formed by I–N halogen bonding between these molecules involving two iodine substituents of each CHI_3 and two amino groups of each TMPD. The I–N distances of 2.902 Å were about 22% shorter than the sum of the van der Waals radii of these atoms, and the C–I–N angles were close to linear (177.9 deg), as is typical for halogen bonding. The (centrosymmetric) discrete 2:1 CHI_3:TMPD complexes also show a pair of halogen bonds with a slightly shorter bond length of 2.842 Å and the C–I–N angles of 172.4 deg. In comparison, DABCO molecules were linked with iodoform by I–N halogen and H–N hydrogen bonding. Both these bonds were close to linear (177.1 deg and 174.2 deg for HB and XB, respectively) and quite short (HB and XB bond length of 2.152 Å and 2.756 Å, respectively). Interestingly, the I–N distances in associations of DABCO with iodoform were shorter than the Br–N distances of 2.877 Å reported earlier [24] in the similar zigzag chains formed by both halogen and hydrogen bonding of this nucleophile with bromoform. This indicates substantially stronger XB involving iodine atoms. Overall, similar to the co-crystals with halide anions, the interaction of haloforms with aromatic or aliphatic amines shows both modes (X–N and H–N) of interactions.

2.4. Computational Analysis of XB and HB Complexes

Surface electrostatic potentials of TMPD and DABCO are illustrated in Figure 5. Both these molecules show areas of negative potentials corresponding to the location of lone pairs on the surface of the nitrogen atoms. The magnitude of the minimum (most negative) potential, $V_{S,min}$, of -36.4 kcal/mol on the surface of DABCO is somewhat higher than that on the surface of the nitrogen atom of TMPD (-33.2 kcal/mol), apparently due to the partial delocalization of nitrogen's lone pairs to the aromatic ring in the latter.

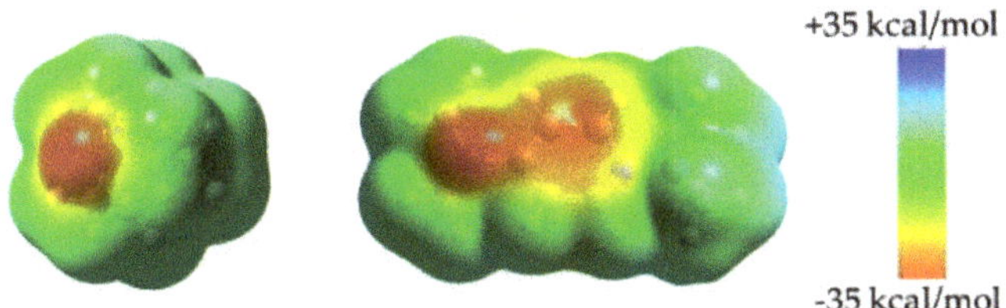

Figure 5. Electrostatic potential (calculated at 0.001 electron bohr^{-3} electronic density) on the molecular surfaces of DABCO (**left**) and TMPD (**right**).

For the TMPD molecule, the negative potential is extended from the surface of the nitrogen atom to the aromatic ring. The center of the aromatic ring shows another minimum potential of about -30 kcal/mol. The locations of $V_{S,min}$ on the surfaces of nitrogen atoms in TMPD and DABCO suggest they would be attracted to σ-holes on the surface of either hydrogen or halogen atoms in haloforms. Indeed, DFT M062X/def2tzvpp calculations (see Experimental for details) produced energy minima corresponding to XB and HB complexes between all three haloforms and nitrogen atoms of aromatic or aliphatic amines. The structural features of halogen (and, for the complexes with DABCO, hydrogen bonds found for these minima (illustrated in Figure 6) were consistent with the geometries obtained via X-ray crystallographic analysis of the solid-state associations. (In addition, calculations of complexes with TMPD produced minima in which hydrogen or halogen substituents of haloforms were directed toward carbon atoms in an aromatic ring or the middle of C–N bonds; vide infra.

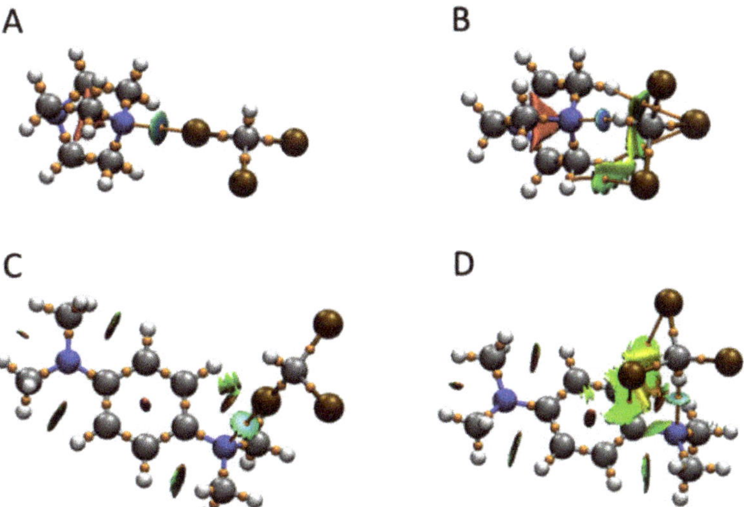

Figure 6. Superposition of the results QTAIM and NCI analyses onto the structures of the XB complexes of CHBr$_3$ with DABCO (**A**) and TMPD (**C**) and HB complexes with DABCO (**B**) and TMPD (**D**). The bond paths and critical $(3,-1)$ points (from QTAIM) are shown as orange lines and spheres, respectively, and blue-green discs indicate areas of bonding interactions (from NCI).

Quantum Theory of Atoms in Molecules (QTAIM) analysis [31] of the optimized structures showed bond paths (orange lines in Figure 6) from nitrogen atoms to halogen or hydrogen substituents of haloforms in XB and HB complexes, respectively. It also revealed $(3,-1)$ bond-critical points (BCPs) along these bond paths (small orange spheres). Bonding interactions between nucleophilic nitrogen atoms and halogen or hydrogen of haloform were further confirmed by the non-covalent index (NCI) analysis [32]. The NCI treatment showed the presence of the blue-colored discs located at the BCPs between nitrogen atoms of amines and halogen or hydrogen atoms of haloforms, which indicates moderately strong intermolecular attraction between these atoms. It should be mentioned that besides the bond paths and BCPs between nitrogen atoms and hydrogens, HB complexes showed bond paths and BCPs between halogen atoms of haloforms and hydrogen substituents or aromatic carbons of amines. The NCI analysis showed green surfaces corresponding to non-bonding or very weak bonding interaction along these bond paths. This indicates that they represent secondary interactions most likely supported by the close approach of haloform to amines. Interaction energies, XB and HB lengths and C–I–N or C–H–N angles for the representative complexes are listed in Table 1.

Table 1. Interaction energies and interatomic distances in the XB and HB complexes.

CHX_3	B	XB Complexes			HB Complexes		
		ΔE, kcal/mol	d_{X-N}, Å	R_{XN} [a]	ΔE, kcal/mol	d_{H-N}, Å	R_{HN} [a]
CHI_3	DABCO	−7.0	2.694	0.72	−4.7	2.051	0.75
	TMPD	−5.5	2.845	0.76	−5.5	2.163	0.79
$CHBr_3$	DABCO	−3.5	2.802	0.79	−4.4	2.043	0.78
	TMPD	−3.3	2.872	0.81	−4.7	2.095	0.76
$CHCl_3$	DABCO	−1.8	2.931	0.89	−4.0	2.095	0.76
	TMPD	−1.9	2.892	0.88	−4.2	2.183	0.79

[a] Normalized interatomic separations $R_{XN} = d_{X\cdots N}/(r_X + r_N)$, where r_X and r_N are van der Waals radii [33].

The characteristics of the BCPs on the XB and HB bond paths obtained from the QTAIM analysis corroborate the similarities of these associations between amines and haloforms. The electron densities and energies at these BCPs are listed in Table 2.

Table 2. Electron densities and energies ($\rho(r)$ and $H(r)$, in a.u.) at BCPs along XB and HB bond paths.

CHX_3	B	XB Complexes		HB Complexes	
		$\rho(r) \times 10^2$	$H(r) \times 10^3$	$\rho(r) \times 10^2$	$H(r) \times 10^3$
CHI_3	DABCO	3.65	−3.70	2.76	−1.11
	TMPD	2.70	−1.11	2.15	0.44
$CHBr_3$	DABCO	2.39	0.51	2.81	−1.31
	TMPD	2.07	0.76	2.16	0.34
$CHCl_3$	DABCO	1.52	1.61	2.53	−0.37
	TMPD	1.66	1.51	2.05	0.57

TD DFT calculations showed that UV–Vis spectra of all XB complexes contain absorption bands (Table 3) which are red-shifted and substantially more intense than the absorption bands in the individual compounds. These bands are related to the transition involving orbitals localized on both haloform and amines. On the other hand, UV–Vis spectra of the optimized XB complexes are very close to that of the superposition of individual components. These results are consistent with the reported data, and they indicate that the appearance of new absorption bands in the UV–Vis range is related to the formation of XB complexes.

Table 3. Calculated UV–Vis and NMR characteristics of the XB and HB complexes [a].

CHX$_3$	B	XB Complexes			HB Complexes		
		λ_{max}, nm	$\varepsilon \times 10^{-3}$, M^{-1} cm^{-1}	$\Delta\delta$, ppm [b]	λ_{max}, nm	$\varepsilon \times 10^{-3}$, M^{-1}cm^{-1}	$\Delta\delta$, ppm [b]
CHI$_3$	DABCO	311	9.0	−1.539	302	1.4	1.847
	TMPD	335	7.2	−0.669	295	4.5	1.506
CHBr$_3$	DABCO	259	7.5	−0.452	212	2.8	2.502
	TMPD	285	8.0	−0.437	245	20.0	1.503
CHCl$_3$	DABCO	216	3.3	−0.230	220	250	2.119
	TMPD	290	3.0	−0.344	247	20.1	1.353

[a] λ, in nm, (ε) for individual compounds are: CHI$_3$: 304 (3050), CHBr$_3$: 223 (2500), CHCl$_3$: 175 (1600), DABCO-TMPD 245 (24,400), δ (in ppm) for CHI$_3$: 7.0287, CHBr$_3$: 7.6825, CHCl$_3$: 7.582. [b] Relative to the protons' chemical shifts in the individual haloforms.

The proton signals of haloforms in the NMR spectra of the optimized XB complexes were shifted to lower ppm values indicating increased shielding of these protons. In the case of all HB associations, the signals were shifted to higher ppm values. These results agree with earlier observations of opposite shifts in the haloforms' proton signals related to halogen and hydrogen bonding [14–17]. However, despite the fact that calculations (and the data on similar associations) suggest that CHBr$_3$ and especially CHCl$_3$ form stronger HB complexes with TMPD, experimental measurements showed a uniform shift in the haloforms' proton signal to lower ppm values upon addition of this amine to any of the haloforms (Figure 3). It should be noted, however, that in contrast to the singular solid-state donor/acceptor arrangement, solution-phase complexes are subject to fluctuations around the optimized minimum (or several local minima) which might affect spectral characteristics. Indeed, an analysis of the potential energy landscape shows that the XB and HB complexes between CHX$_3$ and TMPD are characterized by a shallow minimum. The variations of the X–N or H–N separations by about 0.5 Å are accompanied by energy changes of less than 1 kcal/mol (Figure 7, note that complexes of CHBr$_3$ and CHCl$_3$ with TMPD, and associations of haloforms with DABCO showed similar shallow minima; see Figure S10 in the Supplementary Materials).

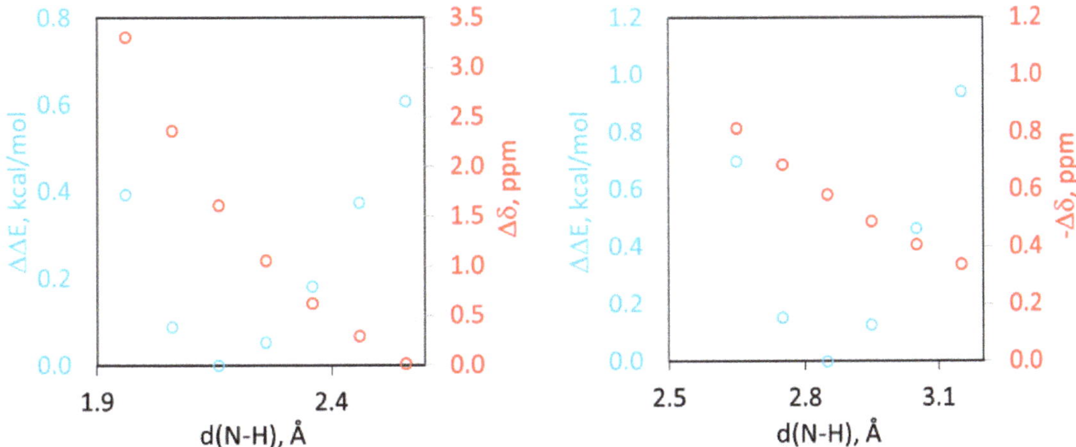

Figure 7. Dependencies of the energies (blue circles) of HB (**left**) and XB (**right**) complexes relative to that of the energy minimum and the chemical shifts in the proton (red circles) relative to that in the individual CHI$_3$ on interatomic separations.

These structural variations are accompanied by changes in NMR spectra, i.e., the increase in the separation is accompanied by a decrease in the difference in the position of the signal in the complex compared with that of the individual haloform. While shallow minima imply the co-existence of assemblies of associations with varying separations, the average distances and NMR shifts for these assemblies seem to be close to those found for the minimum. As such, they would not substantially affect general trends of the NMR shifts. However, the analysis of the potential energy landscape also revealed the presence of the additional minima for the complexes of haloforms with TMPD. Such minima are apparently related to the presence of additional binding sites on the surface of TMPD due increased electron density on the aromatic ring, as shown in Figure 5). The structural overlap (Figure 8) demonstrates that the alternative HB structures were quite similar. The main structural difference was the shift in the position of the protons in the alternative structures from the nitrogen atom toward the aromatic ring, so it was directed toward the middle of the C–N bond or toward the aromatic carbon in ortho-position with respect to the amino group. The differences in energies of these alternative structures and the corresponding minima showing hydrogen bonding with nitrogen atoms were 0.9 kcal/mol, 1.1 kcal/mol and -0.1 kcal/mol for associations with CHI_3, $CHBr_3$ and $CHCl_3$, respectively. Despite such seemingly minor structural and thermodynamic differences, the proton signal in the NMR spectra of the haloform in the alternative complexes were shifted to lower ppm values by 1.41 ppm, 1.21 ppm and 0.72 ppm in complexes of TMPD with CHI_3, $CHBr_3$ and $CHCl_3$, respectively, compared with the signal of the individual haloform molecules.

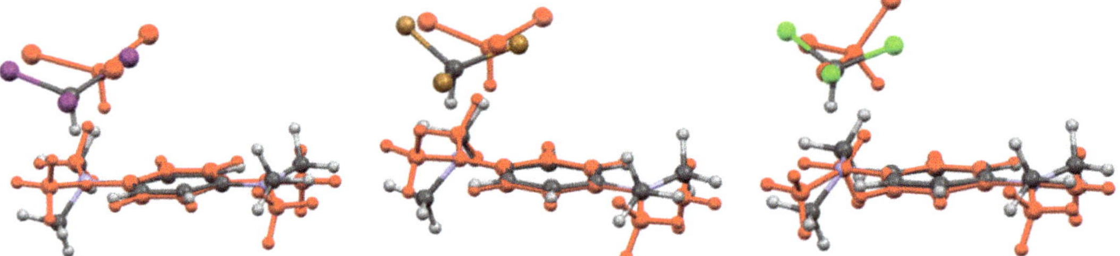

Figure 8. Structural overlap of the associations between TMPD and CHI_3 (**left**), $CHBr_3$ (**middle**) and $CHCl_3$ (**right**) formed via H–N bonding and alternative complexes (shown as red structures).

Computational analysis also revealed the presence of the XB complexes with TMPD in which the halogen atom is directed toward aromatic carbons. Specifically, in the alternative structure of CHI_3 with this molecule, the iodine substituent of iodoform is directed toward a nitrogen-bonded carbon atom (Figure S11 in the Supplementary Materials). The energy of this structure was about 2.0 kcal/mol higher than that of the complex with the I–N bond. Most notably, similarly to the structure with the I–N interaction, the iodoform proton in the alternative XB structure was shifted by about 0.55 ppm to lower ppm values, and its UV–Vis spectrum contained a new absorption band with $\lambda_{max} = 392$ nm, $\varepsilon = 2900$ M^{-1} cm^{-1}.

2.5. Unified Correlation of Strength of the XB and HB Complexes with the Surface Electrostatic Potentials in the Polarized Molecules

The differentiation and simultaneous measurements of XB and HB complexes which are formed by the same pairs of molecules in solutions are challenging tasks that require an accurate knowledge of the distinctions between these two interactions.

In accordance with the results of the X-ray crystallographic analysis, computations of complexes of haloforms with DABCO or TMPD produced energy minima showing I–N or H–N bonding. The data in Table 1 show that XB strength between CHX_3 and amines decreased in the expected order for this interaction with X as I > Br > Cl. For iodoform, this interaction was somewhat stronger in XB complexes with DABCO than that with TMPD. These computational results agree with the experimental X-ray structural data, i.e., shorter I–N separations in the solid-state complexes of CHI_3 with DABCO than that in associations with TMPD, and shorter I–N distances in complexes of iodoform with DABCO than Br–N distances reported in the similar associations with $CHBr_3$. The differences in the ΔE values for complexes of these aromatic and aliphatic amines with either bromoform or chloroform were small, if any. The HB complexes of iodoform with amines were also slightly stronger than with those with bromoform and chloroform, and all energies were within a -5 ± 1 kcal/mol range. To clarify the reasons for the variations of the interaction energies, we compared their values with the changes in the maximum ($V_{S,max}$) and minimum ($V_{S,min}$) electrostatic potentials on the surfaces of haloforms and amines, respectively [34]. The dependence of the ΔE values on the difference $V_{S,max} - V_{S,min}$ (found for the individual haloforms and amines) is shown as open circles in Figure 9, left (red and blue colors denote HB and XB complexes, respectively).

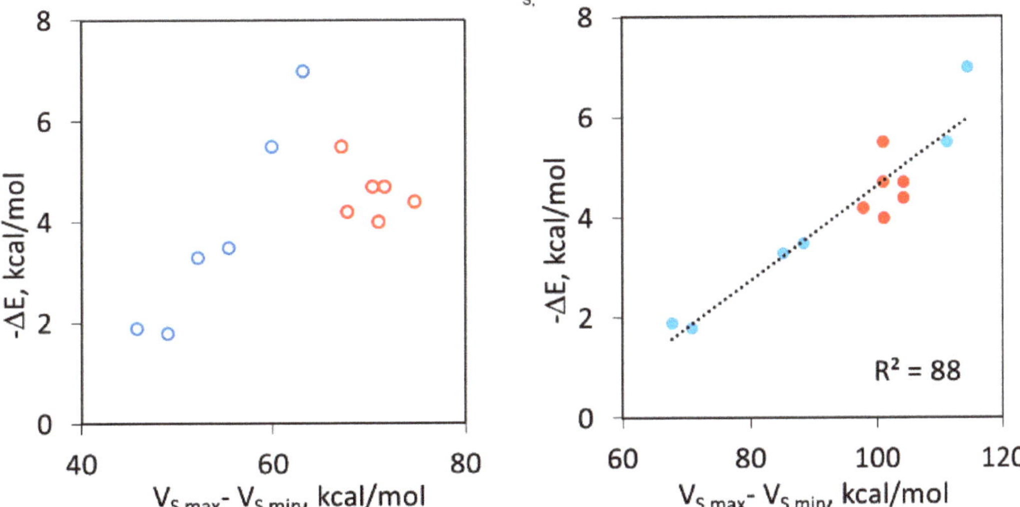

Figure 9. Correlations between the interaction energies in the XB and HB complexes (blue and red circles, respectively) with DABCO and TMPD and the difference of the electrostatic potentials on the surfaces of interacting atoms in the individual molecules (**left**) and in polarized molecules (**right**).

For the XB complexes, the increase in the difference of potentials is accompanied by an increase in the magnitude of (negative) ΔE values. For the HB complexes, however, no such correlation was observed. Furthermore, while the $V_{S,max}$ values on the surfaces of hydrogens atoms are higher than those on the surfaces of halogen in all individual haloforms (which is reflected in the $V_{S,max} - V_{S,min}$ differences), halogen bonding is the dominant mode of interaction of iodoform. As such, the overall R^2 value for the whole set of XB and HB complexes is just 0.33.

It should be noted, however, that the presence of the electron-rich species near the haloforms may substantially affect electron distributions in these species (and the same is true for the amines). Such polarization represents an important factor in the strength of intermolecular complexes [35]. Thus, we evaluated electrostatic potentials on the surfaces of haloforms in the presence of the partial charge located at the positions of the bonded nitrogen atoms in the HB and XB complexes (see Experimental for details). The values of $V_{S,max}$ on the surface of halogen and hydrogen atoms in the presence of the charges are substantially higher than those of the individual molecules (see Table S1 in the Supplementary Materials). The corresponding correlation between ΔE and $V_{S,max} - V_{S,min}$ values calculated using $V_{S,max}$ in the polarized CHX_3 molecules is shown in Figure 9 (right) as the filled circles. While this approach takes into account only the polarization of haloforms, it considerably improves the correlation. The points corresponding to the HB complexes follow the same trend line as the XB associations (with $R^2 = 0.88$ for the whole series). This indicates that once polarization is taken into account, the strengths of the HB and XB complexes between haloforms and aromatic and aliphatic amines can be uniformly related to the electrostatic potentials on the surfaces of HB/XB donors and acceptors.

The variations in the electron densities and energies at BCPs (which are most commonly used for characterization of bonding strength and nature [36]) for different complexes in Table 2 follow the trends observed in energies and bond lengths listed in Table 1. In particular, electron densities at the BCPs for XB complexes of CHI_3 are higher than that of the corresponding HB associations, and the values in the range 0.02–0.04 a.u. are consistent with the strong intermolecular bonding in these complexes [36,37]. For the XB complexes, $\rho(r)$ values decrease from complexes of CHI_3 to those of $CHBr_3$ and $CHCl_3$; however, the values for the XB complexes are rather uniform. As such, the relative values of electron density for the HB complexes of chloroform and bromoform are higher than those for the XB associations. Very small negative or positive values of the energy density $H(r)$ for the complexes in Table 2 are also consistent with the strong intermolecular interactions [36,37], and their variations are consistent with the changes in ΔE.

2.6. Differentiation of XB and HB Complexes Based on Their UV–Vis and NMR Characteristics

While bonding characteristics of the optimized HB and XB complexes were quite similar, the UV–Vis and NMR spectral characteristics of these associations were different. Most notably, formation of the XB complexes is accompanied by the appearance of new intense absorption bands in the UV–Vis spectra of the solutions containing haloforms and amines (although they were overshadowed by the absorption of their components for some systems). In comparison, the electronic spectra of HB complexes are close to the superposition of the spectra of the individual reactants. Additionally, halogen and hydrogen bonding of CHX_3 with aliphatic amines led to the shift in the NMR signals of haloforms in opposite directions, i.e., halogen bonding resulted in the signal shift to the lower ppm values, and hydrogen bonding led to the shift to the higher values. These data were consistent with earlier studies of the interactions of haloforms with the other nucleophiles [14–18]. Thus, the multivariable analysis of the data obtained from the UV–Vis and NMR measurements (as described in our previous work [14]) of the solutions containing constant concentrations of haloforms and variable concentrations DABCO (Figure 10) allowed us to evaluate equilibria constant for the XB and HB complexes co-existing in solutions. For the DABCO complexes with CHI_3, this treatment produced values of $K_X = 3.7 \pm 0.3\ M^{-1}$ and $K_H = 2.0 \pm 0.2\ M^{-1}$ and the formation constants for associations with $CHBr_3$ were $K_X = 0.27 \pm 0.03\ M^{-1}$ and $K_H = 0.12 \pm 0.01\ M^{-1}$.

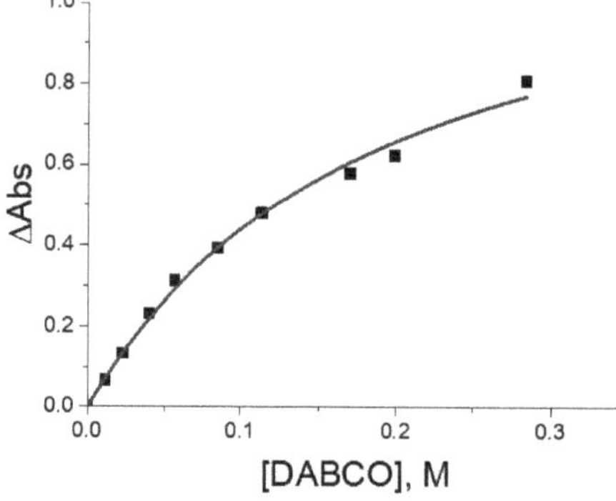

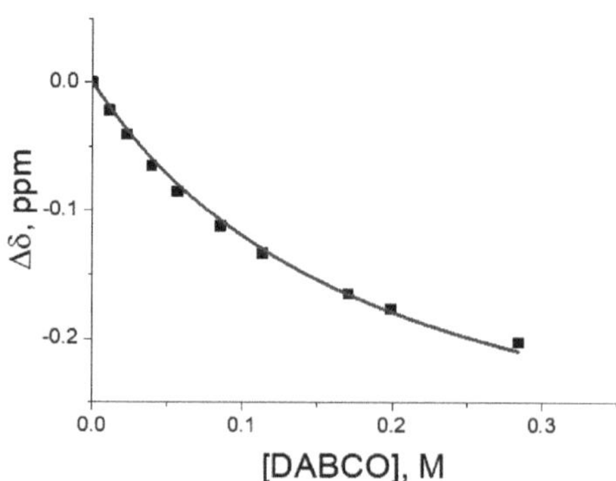

Figure 10. Dependencies of the ΔAbs and Δδ values in the solutions with constant concentration of CHI_3 (0.01 M) and variable concentrations of DABCO. Solid lines show the simultaneous multivariable fitting of the UV–Vis and NMR titrations data.

The differentiation of the effects of HB and XB with aromatic amines using NMR data is complicated by the possible presence of the XB complexes with the rather small differences in energies, but opposite directions of shifts in proton signals (compared with those in the individual molecules). As such, the values of these constants were estimated using UV–Vis spectral data which reflected formation of XB associations Assuming that the ratio $K_H/K_X = \exp(\Delta\Delta E/RT) \approx 1$ (where $\Delta\Delta E = 0$ is a difference of interaction energy of XB and the most stable HB complex for CHI_3/TMPD pair, see Table 1), the values of both formation constants are roughly 0.3 M^{-1}.

3. Materials and Methods

Commercially available haloforms, TMPD and DABCO, were purified by distillation or sublimations.

The UV–Vis measurements were carried out on a Cary 5000 spectrophotometer (Agilent, Santa Clara, CA, USA) in dry (HPLC grade) acetonitrile. NMR measurements were performed on a 400 MHz spectrometer Jeol 400 (Jeol USA Inc., Peabody, MA, USA) in deuterated acetonitrile with internal TMS standard. The intensities of the absorption of [CHX_3, D] complexes, ΔAbs, were obtained by the subtraction of the absorption of the components from the spectrum of the mixtures of CHX_3 and amine. The K_{XB} and K_{HB} values were obtained by the simultaneous nonlinear fitting (using the multiple variable option with Levenberg–Marquardt iteration algorithm in OriginPro 2016) of the dependencies of ΔAbs and $\Delta\delta$ on the concentrations of amines measured at the same concentrations of CHX_3 using Equations (3) and (4), as described in detail earlier [14]:

$$\Delta\text{Abs} = \varepsilon l \times C_{XB} = \varepsilon l \times \{(C^{\circ}_A + C^{\circ}_D + 1/(K_{XB} + K_{HB})) - ((C^{\circ}_A + C^{\circ}_D + 1/(K_{XB} + K_{HB}))^2 - 4C^{\circ}_A C^{\circ}_D)^{0.5}\}/(2\,(1 + K_{HB}/K_{XB})) \quad (3)$$

$$\Delta\delta = \Delta\delta_{XB}/C^{\circ}_D \times C_{XB} + \Delta\delta_{HB}/C^{\circ}_D \times C_{HB} = [\Delta\delta_{XB}/C^{\circ}_D \times \{(C^{\circ}_A + C^{\circ}_D + 1/(K_{XB} + K_{HB})) - ((C^{\circ}_A + C^{\circ}_D + 1/(K_{XB} + K_{HB}))^2 - 4C^{\circ}_A C^{\circ}_D)^{0.5}\}$$
$$+ \Delta\delta_{HB}/C^{\circ}_D \times \{(C^{\circ}_A + C^{\circ}_D + 1/(K_{XB} + K_{HB})) - ((C^{\circ}_A + C^{\circ}_D + 1/(K_{XB} + K_{HB}))^2 - 4C^{\circ}_A C^{\circ}_D)^{0.5}\}]/(2\,(1 + K_{XB}/K_{HB})) \quad (4)$$

where C_{com} is the concentration of the complex, and C°_D and C°_A are initial concentrations of CHX_3 and amine, ε and l are extinction coefficient of the complex and the length of the cell which was used in the UV–Vis measurements, and $\Delta\delta_{\infty} = \delta_{\infty} - \delta_0$ is the difference between the ppm of the CHX_3 proton in the presence of an infinite concentration of amine, δ_{∞} (obtained from the calculations of these complexes) and that of the separate CHX_3, δ_0 and K_{XB} and K_{HB} are formation constants of the XB and HB complexes.

Geometries of the XB and HB complexes and their components were optimized without constraints in acetonitrile via M06-2X/def2tzvpp calculations (with a polarizable continuum model) using the Gaussian 09 suite of programs [38–40]. The interaction energies were determined as: $\Delta E = E_{comp} - (E_{CHX3} + E_A) + BSSE$, where E_{comp}, E_{CHX3} and E_A are sums of the electronic and ZPE of the complex, CHX_3 and amine (DABCO or TMPD) and BSSE is a basis set superposition error [41]. UV–Vis spectra of complexes and trihalomethanes were calculated via TD-DFT calculations, proton NMR shifts were obtained via GIAO calculations using geometries of the complexes optimized in acetonitrile. Molecular electrostatic potentials of the polarized molecules were calculated by placing point charges corresponding to the charge of nitrogen in DABCO (calculated as ESP charges) at the position where such atoms are located in the optimized complexes. Such approximation allowed to take into account various polarizabilities of interacting atoms in XB and HB complexes and led to reasonable correlation between electrostatic potentials and interaction energies. QTAIM and NCI analyses were performed with Multiwfn [42] using wfn files generated by Gaussian 09. The results were visualized using the molecular graphics program VMD [43]. Details of the calculations, energies, geometric and spectral characteristics of HB and XB complexes as well as atomic coordinates of the calculated complexes are listed in the ESI.

The single crystals were measured on a Bruker Quest diffractometer (Bruker AXS, LLC, Madison, WI, USA) with a fixed chi angle, a sealed tube fine focus X-ray tube, single crystal curved graphite incident beam monochromator (Bruker AXS, LLC, Madison, WI, USA), a Photon100 area detector (Bruker AXS, LLC, Madison, WI, USA) and an Oxford Cryosystems low-temperature device (Hanborough House, Oxford, United Kingdom). Examination and data collection were performed with Mo Kα radiation (λ = 0.71073 Å). Reflections were indexed and processed, and the files were scaled and corrected for absorption using APEX3 [44]. The space groups were assigned, and the structures were solved by direct methods using XPREP within the SHELXTL suite of programs [45] and refined by full matrix least squares against F^2 with all reflections using Shelxl2018 [46,47] using the graphical interface Shelxle [46]. If not specified otherwise, H atoms attached to carbon and nitrogen atoms were positioned geometrically and constrained to ride on their parent atoms, with C–H bond distances of 1.00, 0.99 and 0.98 Å for aliphatic C–H, CH_2 and CH_3 moieties, respectively. Methyl H atoms were allowed to rotate but not to tip to best fit the experimental electron density. $U_{iso}(H)$ values were set to a multiple of $U_{eq}(C)$ with 1.5 for CH_3, and 1.2 for C–H units, respectively. Crystallographic, data collection and refinement details are listed in Table S2 in the Supplementary Materials. Complete crystallographic data, in CIF format, have been deposited with the Cambridge Crystallographic Data Centre. CCDC 2202934, 2202935 and 2206546 contain the supplementary crystallographic data for this paper. These data can be obtained free of charge via www.ccdc.cam.ac.uk/data_request/cif.

4. Conclusions

Experimental and computational analysis of interactions of haloforms with aromatic and aliphatic amines highlighted the similarities and distinctions of HB and XB complexes, which will be helpful for the identification and quantitative characterization of these competing interactions in chemical and biochemical systems. We demonstrated that when polarization of haloforms is taken into account, the interaction energies within the HB and XB complexes of CHX_3 molecules with TMPD and DABCO follow the same correlation with the difference electrostatic potentials on the surfaces of the interacting atoms. The electron densities and energies at BCPs along the H–N and N–X bond paths also follow the same trends. These data confirm that the thermodynamics of these moderately strong associations is dominated by electrostatic interactions. However, spectral properties of the XB and HB complexes were quite different. The most consistent distinction is observed in the UV–Vis spectra of the complexes. Halogen bonding is accompanied by an appearance of strong absorption bands related to the formation of the XB associations (which suggests substantial molecular-orbital interactions between haloform and amine within these complexes). In contrast, the spectral of the HB associations were close to the superposition of the spectra of the individual reactants. The effects of intermolecular interactions on the NMR spectra were dependent on the nature of the amine. In particular, the HB and XB associations of haloforms with aliphatic amines led to the opposite shifts in their protons' signals in the NMR spectra. Thus, combination of the UV–Vis and NMR data allows to differentiate XB and HB complexes of haloforms with these amines in solutions. XB with aromatic amines led to the shift in the same direction as the aliphatic ones; however, the corresponding effects of HB of CHX_3 with aromatic amines were complicated by the presence of multiple HB minima in which hydrogens were directed either toward the nitrogen atom or C–N bond or aromatic carbon. The haloforms' protons signals in the NMR spectra of these complexes were shifted in the opposite direction, which hinder the application of this method for quantitative analysis of XB and HB complexes.

Supplementary Materials: The following supporting information can be downloaded at: https://www.mdpi.com/article/10.3390/molecules27186124/s1. Figure S1: UV–Vis spectra of the solutions of CHI_3 and 4-methoxy-N,N-dimethylaniline in CH_3CN. Figure S2: UV–Vis spectra of the solutions of CHI_3 and 3-methoxy-N,N-dimethylaniline in CH_3CN. Figure S3: UV–Vis spectra of the solutions of CHI_3 and 4-(dimethylamino)benzonitrile in CH_3CN. Figure S4: UV–Vis spectra of the solutions of CHI_3 and 3-(dimethylamino)benzonitrile in CH_3CN. Figure S5: Benesi–Hildebrand plots based on the UV–Vis spectra of solutions of CHI_3 with TMPD and DABCO. Figure S6: Fit of spectral changes in solutions of CHI_3 with TMPD and DABCO to 1:1 binding isotherm. Figure S7: Dependencies of the chemical shifts in the protons of CHI_3 on the concentration of added trimethylamine, *N,N*-dimethylaniline, 4-(dimethylamino)benzonitrile, *p*-bromo-*N,N*-dimethylaniline or 3-(dimethylamino)benzonitrile. Figure S8: Dependencies of the chemical shifts in the protons of $CHBr_3$ on the concentration of added trimethylamine, *N,N*-dimethylaniline, 4-(dimethylamino)benzonitrile, *p*-bromo-*N,N*-dimethylaniline or 3-(dimethylamino)benzonitrile. Figure S9: Dependencies of the chemical shifts in the protons of $CHBr_3$ on the concentration of added trimethylamine, *N,N*-dimethylaniline, 4-(dimethylamino)benzonitrile, *p*-bromo-*N,N*-dimethylaniline or 3-(dimethylamino)benzonitrile. Figure S10. Effect of variation of interatomic H–N and X–N separations on the energy of HB and XB complexes and chemical shifts in the haloforms' protons. Figure S11. Alternative structure of XB complex between CHI_3 and TMPD. Table S1: Values of the maximum electrostatic potentials on the surfaces of halogen and hydrogen atoms in individual and polarized haloforms. Table S2. Crystallographic, data collection and refinement details. Table S3: Energies of the HB and XB complexes and their components. Table S4: Atomic coordinate of the HB and XB complexes of haloforms with DABCO and TMPD.

Author Contributions: Conceptualization and methodology, S.V.R., UV–Vis and NMR measurements, E.A., O.G., Z.S., X-ray structural analysis, M.Z., computations, E.A., O.G. and S.V.R.; data curation, E.A., O.G., Z.S., S.V.R.; writing—original draft preparation, S.V.R.; writing—review and editing, E.A., O.G., Z.S., M.Z., S.V.R.; visualization, supervision, project administration, funding acquisition, S.V.R. All authors have read and agreed to the published version of the manuscript.

Funding: This research was funded by the National Science Foundation, grant number CHE-2003603. Calculations were carried out on Ball State University's beowulf cluster, which is supported by The National Science Foundation (MRI-1726017) and Ball State University. X-ray structural measurements were supported by the National Science Foundation through the Major Research Instrumentation Program under Grant No. CHE 1625543 (funding for the single crystal X-ray diffractometer).

Institutional Review Board Statement: Not applicable.

Informed Consent Statement: Not applicable.

Data Availability Statement: Crystallographic data have been deposited with the Cambridge Crystallographic Data Centre and can be obtained free of charge (see above). Atomic coordinates and energies of the calculated complexes are available in the Supplementary Information.

Conflicts of Interest: The authors declare no conflict of interest.

Sample Availability: Substances used in the study are commercially available and they are not available from the authors.

References

1. Vinogradov, S.N.; Linnell, R.H. *Hydrogen Bonding*; Van Nostrand Reinhold: New York, NY, USA, 1971.
2. Scheiner, S. *Hydrogen Bonding: A Theoretical Perspective*; Oxford University Press: New York, NY, USA, 1997.
3. Metrangolo, P.; Neukirch, H.; Pilati, T.; Resnati, G. Halogen bonding based recognition processes: A world parallel to hydrogen bonding. *Acc. Chem. Res.* **2005**, *38*, 386–395. [CrossRef] [PubMed]
4. Politzer, P.; Murray, J.S.; Clark, T. Halogen bonding and other σ-hole interactions: A perspective. *Phys. Chem. Chem. Phys.* **2013**, *15*, 11178–11189. [CrossRef] [PubMed]
5. Scheiner, S. The pnicogen bond: Its relation to hydrogen, halogen, and other noncovalent bonds. *Acc. Chem. Res.* **2013**, *46*, 280–288. [CrossRef]
6. Cavallo, G.; Metrangolo, P.; Milani, R.; Pilati, T.; Priimagi, A.; Resnati, G.; Terraneo, G. The halogen bond. *Chem. Rev.* **2016**, *116*, 2478–2601. [CrossRef] [PubMed]
7. Gilday, L.C.; Robinson, S.W.; Barendt, T.A.; Langton, M.J.; Mullaney, B.R.; Beer, P.D. Halogen bonding in supramolecular chemistry. *Chem. Rev.* **2015**, *115*, 7118–7195. [CrossRef] [PubMed]
8. Pennington, W.T.; Resnati, G.; Taylor, M.S. Halogen bonding: From self-assembly to materials and biomolecules. *CrystEngComm* **2013**, *15*, 3057. [CrossRef]
9. Awwadi, F.F.; Taher, D.; Haddad, S.F.; Turnbull, M.M. Competition between hydrogen and halogen bonding interactions: Theoretical and crystallographic studies. *Cryst. Growth Des.* **2014**, *14*, 1961–1971. [CrossRef]

10. Aakeröy, C.B.; Panikkattu, S.; Chopade, P.D.; Desper, J. Competing hydrogen-bond and halogen-bond donors in crystal engineering. *CrystEngComm* **2013**, *15*, 3125–3136. [CrossRef]

11. Posavec, L.; Nemec, V.; Stilinović, V.; Cinčić, D. Halogen and hydrogen bond motifs in ionic cocrystals derived from 3-halopyridinium halogenides and perfluorinated iodobenzenes. *Cryst. Growth Des.* **2021**, *21*, 6044–6050. [CrossRef]

12. Zapata, F.; Caballero, A.; Molina, P.; Alkorta, I.; Elguero, J. Open bis(triazolium) structural motifs as a benchmark to study combined hydrogen- and halogen-bonding interactions in oxoanion recognition processes. *J. Org. Chem.* **2014**, *79*, 6959–6969. [CrossRef]

13. Aakeröy, C.B.; Fasulo, M.; Schultheiss, N.; Desper, J.; Moore, C. Structural competition between hydrogen bonds and halogen bonds. *J. Am. Chem. Soc.* **2007**, *129*, 13772–13773. [CrossRef] [PubMed]

14. Watson, B.; Grounds, O.; Borley, W.; Rosokha, S.V.; Resolving halogen, vs. hydrogen bonding dichotomy in solutions: Intermolecular complexes of trihalomethanes with halide and pseudohalide anions. *Phys. Chem. Chem. Phys.* **2018**, *2*, 21999–22007. [CrossRef] [PubMed]

15. von der Heiden, D.; Vanderkooy, A.; Erdélyi, M. Halogen bonding in solution: NMR spectroscopic approaches. *Coord. Chem. Rev.* **2020**, *407*, 213147. [CrossRef]

16. Lu, J.; Scheiner, S. Effects of halogen, chalcogen, pnicogen, and tetrel bonds on IR and NMR Spectra. *Molecules* **2019**, *24*, 2822. [CrossRef]

17. Green, R.D.; Martin, J.S. Anion-molecule complexes in solution. I. Nuclear magnetic resonance and infrared studies of halide ion-trihalomethane association. *J. Am. Chem. Soc.* **1968**, *90*, 3659–3668. [CrossRef]

18. Bertrán, J.F.; Rodríguez, M. Detection of halogen bond formation by correlation of proton solvent shifts. 1. Haloforms in *n*-electron donor solvents. *Org. Magn. Reson.* **1979**, *12*, 92–94. [CrossRef]

19. Schulz, N.; Sokkar, P.; Engelage, E.; Schindler, S.; Erdélyi, M.; Sanchez-Garcia, E.; Huber, S.M. The interaction modes of haloimidazolium salts in solution. *Chem. Eur. J.* **2018**, *24*, 3464–3473. [CrossRef]

20. Nayak, S.K.; Terraneo, G.; Piacevoli, Q.; Bertolotti, F.; Scilabra, P.; Brown, J.T.; Rosokha, S.V.; Resnati, G. Molecular bases for anesthesia: Halothane as a halogen and hydrogen bonds donor. *Angew. Chem. Int. Ed.* **2019**, *58*, 12456–12459. [CrossRef]

21. Corradi, E.; Meille, S.V.; Messina, M.T.; Metrangolo, P.; Resnati, G. Halogen bonding versus hydrogen bonding in driving self-assembly processes. *Angew. Chem. Int. Ed.* **2000**, *39*, 1782–1786. [CrossRef]

22. Liantonio, R.; Luzzati, S.; Metrangolo, P.; Pilati, T.; Resnati, G. Perfluorocarbon-hydrocarbon self-assembly. Part 16: Anilines as new electron donor modules for halogen bonded infinite chain formation. *Tetrahedron* **2002**, *58*, 4023–4029. [CrossRef]

23. Raatikainen, K.; Rissanen, K. Interaction between amines and *N*-haloimides: A new motif for unprecedentedly short Br· · ·N and I· · ·N halogen bonds. *CrystEngComm* **2011**, *13*, 6972–6977. [CrossRef]

24. Weinberger, C.; Hines, R.; Zeller, M.; Rosokha, S.V. Continuum of covalent to intermolecular bonding in the halogen-bonded complexes of 1,4-diazabicyclo[2.2.2]octane with bromine-containing electrophiles. *Chem. Commun.* **2018**, *54*, 8060–8063. [CrossRef] [PubMed]

25. Riel, A.M.S.; Rowe, R.K.; Ho, E.N.; Carlsson, A.-C.C.; Rappé, A.K.; Berryman, O.B.; Ho, P.S. Hydrogen bond enhanced halogen bonds: A synergistic interaction in chemistry and biochemistry. *Acc. Chem. Res.* **2019**, *52*, 2870–2880. [CrossRef] [PubMed]

26. Grabowski, S.J. Hydrogen and halogen bonds are ruled by the same mechanisms. *Phys. Chem. Chem. Phys.* **2013**, *15*, 7249–7259. [CrossRef] [PubMed]

27. Wolters, L.P.; Bickelhaupt, F.M. Halogen bonding versus hydrogen bonding: A molecular orbital perspective. *ChemistryOpen* **2012**, *1*, 96–105. [CrossRef] [PubMed]

28. Li, S.; Xu, T.; van Mourik, T.; Früchtl, H.; Kirk, S.R.; Jenkins, S. Halogen and hydrogen bonding in halogenabenzene/NH_3 complexes compared using next-generation QTAIM. *Molecules* **2019**, *24*, 2875. [CrossRef]

29. Nepal, B.; Scheiner, S. Competitive halide binding by halogen versus hydrogen bonding: Bis-triazole pyridinium. *Chem. Eur. J.* **2015**, *21*, 13330–13335. [CrossRef]

30. Benesi, H.A.; Hildebrand, J.H. A spectrophotometric investigation of the interaction of iodine with aromatic hydrocarbons. *J. Am. Chem. Soc.* **1949**, *71*, 2703–2707. [CrossRef]

31. Bader, R.F.W. A quantum theory of molecular structure and its applications. *Chem. Rev.* **1991**, *91*, 893–928. [CrossRef]

32. Johnson, E.R.; Keinan, S.; Mori-Sánchez, P.; Contreras-García, J.; Cohen, A.J.; Yang, W. Revealing noncovalent interactions. *J. Am. Chem. Soc.* **2010**, *132*, 6498–6506. [CrossRef]

33. Bondi, A. Van der Waals volumes and Radii. *J. Phys. Chem.* **1964**, *68*, 441–451. [CrossRef]

34. Politzer, P.; Murray, J.S.; Clark, T. Halogen bonding: An electrostatically-driven highly directional noncovalent interaction. *Phys. Chem. Chem. Phys.* **2010**, *12*, 7748–7757. [CrossRef] [PubMed]

35. Clark, T. Halogen bonds and σ-holes. *Faraday Discuss.* **2017**, *203*, 9–27. [CrossRef] [PubMed]

36. Popelier, P.L.A. The QTAIM Perspective of Chemical Bonding. In *The Chemical Bond: Fundamental Aspects of Chemical Bonding*; John Wiley & Sons: Hoboken, NJ, USA, 2014; pp. 271–308.

37. Miller, D.K.; Loy, C.; Rosokha, S.V. Examining a transition from supramolecular halogen bonding to covalent bonds: Topological analysis of electron densities and energies in the complexes of bromosubstituted electrophiles. *ACS Omega* **2021**, *6*, 23588–23597. [CrossRef]

38. Frisch, M.J.; Trucks, G.W.; Schlegel, H.B.; Scuseria, G.E.; Robb, M.A.; Cheeseman, J.R.; Scalmani, G.; Barone, V.; Mennucci, B.; Petersson, G.A.; et al. *Gaussian 09, Rev. C.01*; Gaussian, Inc.: Wallingford, CT, USA, 2009.

39. Zhao, Y.; Truhlar, D.G. The M06 suite of density functionals for main group thermochemistry, thermochemical kinetics, noncovalent interactions, excited states, and transition elements: Two new functionals and systematic testing of four M06-class functionals and 12 other functionals. *Theor. Chem. Acc.* **2008**, *120*, 215–241.
40. Tomasi, J.; Mennucci, B.; Cammi, R. Quantum mechanical continuum solvation models. *Chem. Rev.* **2005**, *105*, 2999–3093. [CrossRef]
41. Boys, S.F.; Bernardi, F. The calculation of small molecular interactions by the differences of separate total energies. Some procedures with reduced errors. *Mol. Phys.* **1970**, *19*, 553–566. [CrossRef]
42. Lu, T.; Chen, F. Multiwfn: A multifunctional wavefunction analyzer. *J. Comput. Chem.* **2012**, *33*, 580–592. [CrossRef]
43. Humphrey, W.; Dalke, A.; Schulten, K. VMD: Visual molecular dynamics. *J. Mol. Graphics* **1996**, *14*, 33–38. [CrossRef]
44. *Bruker Apex3 v2016.9-0, SAINT V8.37A*. Bruker AXS Inc.: Madison, WI, USA, 2016.
45. *SHELXTL Suite of Programs*, version 6.14; Bruker AXS Inc.: Madison, WI, USA, 2003.
46. Sheldrick, G. Crystal Structure Refinement with SHELXL. *Acta Cryst. C* **2015**, *71*, 3–8. [CrossRef]
47. Hübschle, C.; Sheldrick, G.; Dittrich, B. ShelXle: A Qt Graphical User Interface for SHELXL. *J. Appl. Crystallogr.* **2011**, *44*, 1281. [CrossRef] [PubMed]

MDPI
St. Alban-Anlage 66
4052 Basel
Switzerland
www.mdpi.com

Molecules Editorial Office
E-mail: molecules@mdpi.com
www.mdpi.com/journal/molecules

Disclaimer/Publisher's Note: The statements, opinions and data contained in all publications are solely those of the individual author(s) and contributor(s) and not of MDPI and/or the editor(s). MDPI and/or the editor(s) disclaim responsibility for any injury to people or property resulting from any ideas, methods, instructions or products referred to in the content.